Synthese Library

Studies in Epistemology, Logic, Methodology, and Philosophy of Science

Volume 38

The aim of *Synthese Library* is to provide a forum for the best current work in the methodology and philosophy of science and in epistemology. A wide variety of different approaches have traditionally been represented in the Library, and every effort is made to maintain this variety, not for its own sake, but because we believe that there are many fruitful and illuminating approaches to the philosophy of science and related disciplines.

Special attention is paid to methodological studies which illustrate the interplay of empirical and philosophical viewpoints and to contributions to the formal (logical, set-theoretical, mathematical, information-theoretical, decision-theoretical, etc.) methodology of empirical sciences. Likewise, the applications of logical methods to epistemology as well as philosophically and methodologically relevant studies in logic are strongly encouraged. The emphasis on logic will be tempered by interest in the psychological, historical, and sociological aspects of science.

Besides monographs *Synthese Library* publishes thematically unified anthologies and edited volumes with a well-defined topical focus inside the aim and scope of the book series. The contributions in the volumes are expected to be focused and structurally organized in accordance with the central theme(s), and should be tied together by an extensive editorial introduction or set of introductions if the volume is divided into parts. An extensive bibliography and index are mandatory.

More information about this series at http://www.springer.com/series/6607

Norwood Russell Hanson

Matthew D. Lund
Editor

What I Do Not Believe, and Other Essays

Second Edition

First Edition Edited by
Stephen Toulmin
Michigan State University

And

Harry Woolf
Johns Hopkins University

 Springer

Norwood Russell Hanson (deceased)

Editor
Matthew D. Lund
Department of Philosophy and Religion Studies
Rowan University
Glassboro, NJ, USA

The first edition of this book was published by D. Reidel Publishing Company, Dordrecht, Holland in 1971 publishing with ISBN-13:978-94-010-3110-3. The editors of the first edition were Stephen Toulmin and Harry Woolf.

Synthese Library
ISBN 978-94-024-1741-8   ISBN 978-94-024-1739-5   (eBook)
https://doi.org/10.1007/978-94-024-1739-5

This Springer imprint is published by the registered company Springer Nature B.V.
The registered company address is: Van Godewijckstraat 30, 3311 GX Dordrecht, The Netherlands

Acknowledgments

First Edition

We thank the publishers of *Philosophy of Science* for permission to reprint the papers 'On Elementary Particle Theory' and 'Stability Proofs and Consistency Proofs: A Loose Analogy'; of *Scientia* for 'The Contributions of Other Disciplines to 19th Century Physics' and 'Mental Events Yet Again'; of *Philosophical Studies* for 'It's Actual, so it's Possible' and 'On the Impossibility of any Future Metaphysics'; of *The Journal of Philosophy* for 'A Budget of Cross-Type Inferences, or Invention is the Mother of Necessity' and 'The Irrelevance of History of Science to Philosophy of Science'. Further, we thank the publishers of the *Delaware Symposium on Philosophy of Science* for permission to reprint the paper 'Some Philosophical Aspects of Contemporary Cosmologies'; of *Isis* for 'Leverrier: The Zenith and Nadir of Newtonian Mechanics'; of *The Review of Metaphysics* for 'On Being in two Places at Once'; of the *Journal of the History of Ideas* for 'Copernicus' Role in Kant's Revolution'; of *Mind* for 'On Having the Same Visual Experiences'; of *Analysis* for 'Imagining the Impossible'; of the *Philosophical Quarterly* for 'Good Inductive Reasons'; of *Dialogue* for 'The Idea of a Logic of Discovery'; of *The American Rationalist* for 'The Agnostic's Dilemma'; and of *Continuum* for 'What I Don't Believe'. We gratefully acknowledge that, for the benefit of the late Professor Hanson's children, all publishers concerned had the kindness to waive the reprint fee.

Stephen Toulmin
Harry Woolf

Second Edition

I thank Harper-Collins for the permission to reprint the short book *Observation and Explanation: A Guide to Philosophy of Science* here as a long chapter. I also thank Amy Jaffe Barzach and the University of Hartford Women's Education and

Leadership Fund for the permission to reprint 'The Trial of Galileo'. The process of updating the scholarship and style of the original edition of *What I Do Not Believe, and Other Essays* could not have been accomplished without the generous help of William Sheehan (concerning nineteenth-century astronomy) and Bruce Paternoster (concerning ancient philosophy). I would also like to thank the following individuals at Springer who helped shape this project through many unexpected turns and delays: Natalie Rieborn, Ingrid van Laarhoven, Christi Lue, Ties Nijssen, and Stephen O'Reilly. Finally, I wish to thank the entire Hanson family for all their interest and encouragement throughout this process.

Matthew D. Lund

Introductory Note for the First Edition

This collection of essays by Norwood Russell Hanson is one of a pair of volumes being prepared for the Synthese Library from his posthumous papers. The present book comprises two major items which have not previously been published – the opening essay, entitled 'A Picture Theory of Theory-Meaning', and the set of three Harris Lectures on The Theory of Flight originally delivered at Northwestern University, and edited here by the Rev. Prof. Edward MacKinnon S.J., from a verbatim transcript – together with some of Hanson's less readily accessible, or less well-known published papers and articles. The other, companion book will contain a single connected analysis of the historical development of ideas about scientific explanation, as exemplified in theories about planetary motion from the Greeks up to the seventeenth century. (We have provisionally entitled this companion volume *Constellations and Conjectures*: at the time of Hanson's death it had been almost completely re-edited by the author from an earlier manuscript, and it is being prepared for publication by Professor Willard C. Humphreys jr., who is familiar with Hanson's work in this area.)

In making the selection of essays for this present book, we have been guided by two main considerations. In the first place, it is even truer of Russ Hanson than of most other men that *le style, c'était l'homme même*; and we have tried to choose items which are capable of conveying, to people who never knew Hanson the man, something of the individual flavour of his mind and personality. Robust, pugnacious, intolerant of humbug and self-deceit, he was quick to master any of the techniques (or games) of the scholarly and scientific life, but would never allow them to master him in turn. Thus, a few introductory undergraduate courses aside, Hanson's knowledge of theoretical physics was largely self-taught; yet he was soon capable of discussing the philosophical significance and epistemological status of quantum physics or cosmology with a P.A.M. Dirac or a Fred Hoyle – both of them colleagues of his at St. John's College Cambridge, during the 1950s – on a basis of mutual respect. And he could do so, not just in general or abstract terms, but from a familiarity with specific details of the scientific ideas and arguments involved as extensive and penetrating as that possessed by many university professors of physics.

Similarly elsewhere: while quickly making himself at home in the mysteries of symbolic logic or rational mechanics, Hanson was not a man to lose sight of the deeper intellectual issues underlying those formal systems, or to be stampeded into accepting mere techniques as philosophical or scientific panaceas.

The other aspect of Hanson's work which we have tried to illustrate here is his versatility. Unlike those scholars who build a whole career around a single idea, Russ Hanson was an intellectual prodigal, who turned from field to field with a quite uncommon ease and insouciance. Yet he was not just given to piecemeal polemics against targets of opportunity. Re-reading these essays all together, one comes to see how far his excursions into different academic disciplines were made from a consistent standpoint and in the service of a unified philosophical point of view. Whether he is discussing arguments from logic or theology, psychology or astronomy, aerodynamics or philosophy of language, his attitude is the same: *See it like it is* – or, as Bishop Butler put it, "Things are what they are, and their consequences will be what they will be: why then should we seek to be deceived?" And a few key concepts, notably those of *necessity*, *good reasons* and *understanding*, gave a direction to Hanson's arguments in all these different fields. So, in the last resort, we see him rebutting attacks on the Copenhagen Interpretation of quantum mechanics in just the same terms as he uses to explain his own preference for a frank atheism over a tepid agnosticism: *We must stand openly by what there is reason to believe, until there is sufficiently good reason to believe otherwise.*

Finally: we are glad to have the opportunity of including in this collection the only available record of the work that Russ Hanson was doing in his early forties, on the development of ideas in the theory of flight. He himself was, of course, passionately devoted to this own spare-time occupation as a flyer. He never lost his taste for that combination of physical exhilaration and intellectual mastery which is required of a naval pilot and which he first learned when flying from carriers in the Pacific during the second World War; and in his last years his energies were equally divided between his scholarly work, his family and the Grumman Bearcat which he was grooming for an attack on the speed record for piston-engined airplanes. Still, flying was always as much an intellectual as a physical challenge to him, and in this last phase of his work we can see him attempting to build the results of all his hard work on aerodynamics and airfoil design into the same conceptual framework that he had constructed for the rest of his ideas. We must all be grateful to Ed MacKinnon for the great effort and intelligence he has given to the task of preparing a publishable version of the Harris Lectures for inclusion here. At any rate, the ill-adjusted altimeter which (it seems) was responsible for the crash in April 1967 in which Hanson was killed and his Bearcat destroyed did not rob us entirely of the thoughts on which so much of his final years' work was concentrated.

East Lansing, MI, USA   Stephen Toulmin
Baltimore, MD, USA   Harry Woolf
May 1971

Introduction

In 1967, Norwood Russell Hanson was killed in a plane crash. Despite having passed away at the relatively young age of 42, Hanson had already made indelible contributions to the philosophy of science with his work on observation, the interpretation of quantum theory, and the logic of discovery. He had also created the first History and Philosophy of Science (HPS) Department in the United States and made history of science an earnest concern for philosophers. While Hanson is still recognized for his critical work in the philosophy of science, his scholarly versatility is no longer given its proper due. This expanded edition of *What I Do Not Believe, and Other Essays* presents today's reader with Hanson's best work, some of which was much discussed in his lifetime and shortly after and some of which might have been quite influential had Hanson lived long enough to develop it fully.

Hanson's extraordinary range in intellectual matters is sometimes overlooked because of his prodigious talents in nonacademic domains. Hanson was skilled at nearly everything – boxing, playing the trumpet, drawing, shot-putting, and flying airplanes. He also had the forceful personality and daring to pull off things that others would never have thought possible. Hanson's intellectual versatility was, thus, not achieved through an abridgment or stunting of his other interests and capacities. Yet, as is so admirably expressed by Toulmin and Woolf in the Introductory Note, for all Hanson's wide-ranging inquiry, one detects a singularity of perspective and purpose within the vastness of the subject matters he surveys. Since there is so much of Hanson's thought contained in this volume, an introduction that narrowly summarizes each article would be tedious, if not unreadable. Therefore, I will instead discuss most of the main parts by focusing on one or two of the most significant articles in each part.[1]

Hanson's inventive application of the concepts and history of science to general philosophical problems is striking. For instance, in "A Picture Theory of Theory-Meaning," Hanson brought his considerable knowledge of the history and practice of

---

[1] Since Part VI, The Theory of Flight, is nicely introduced by Edward MacKinnon on pages 331–332, it is not discussed here.

aerodynamics to the questions of how theories originate and how they represent the world. Philosophers of science both before and after Hanson have felt most comfortable in the realm of linguistic representation, and their theories of science have, unsurprisingly, struggled to illuminate other forms of representation. By contrast, Hanson did not confine his analysis to linguistic structures only and instead addresses the epistemic roles played by charts, maps, and curves. In his discussion, Hanson anticipates many of the philosophical points emphasized by Ronald Giere, Hanson's successor at the Indiana University's History and Philosophy of Science Department. It is in this essay where Hanson most explicitly presents his denouement of the thesis of theory-laden observation. Hanson argued that theoretical representation can take many distinct forms (visual and functional analogy, mapping, curves, models, algebraic formulae, diagrams, etc.) and that each of these representations succeeds by sharing the structure of the phenomena. For Hanson, theories and forms of representation are neither wholly "out there" nor "in here." They are "Janus-faced" entities, which point both outward toward the external world and inward to the realm of the mind. Hanson demonstrates not only the richness of different forms of representation but also underscores that our theories cannot be separated from the phenomena. Once a theory promotes our capacity to select out orderly subsets from the overwhelming phenomenological chaos, and these subsets are thereby rendered intelligible, we can no longer experience them at all in any useful sense without the theories.

"A Picture Theory of Theory-Meaning" represents the most successful fusion Hanson ever effected between his professional interests in philosophy of science and his avocational passion for aeronautics. While Hanson indicates that his motivations for the piece come from Wittgenstein's *Tractatus Logico-Philosophicus* and Wisdom's "Logical Constructions," it is clear that his own reflections drawn from aeronautic maps, airfoil designs, and engine parametric and lift charts propel his thought far beyond that of his mentors. Hanson notes repeatedly that picturing is only one way in which theories advance our understanding. At the close of the article, he even explicitly acknowledges that the article might have been better titled the "Structural Representation Theory of What Theories Do." Thus, in this article, one of the last Hanson was to write, he offers a very complete and probing account of how it is that theories represent, and he shows how theory-laden observation is the means through which the new and foreign are made intelligible.

Hanson is best remembered for his thesis that observation is theory-laden, but that specific thesis figures surprisingly little in the pages of this volume. Hanson came increasingly to emphasize that the theory-laden character of observation is merely one consequence of the conceptual and logical layout of science. Since our theories have to make contact with the empirical world somewhere, observations are necessitated – by the conceptual and logical rules of the game of empirical knowledge – to be imbued with theory. Since Hanson saw the theory-ladenness of observation as a consequence of the overarching conceptual structure of science, he was not interested in using the theory-laden observation thesis as a ground for arguing against scientific objectivity or scientific realism, as many other philosophers were inclined to do. Instead, Hanson's primary interest lay in discerning the types of good reasoning that take place within the realm of empirical fact. Philosophers, he thought, had been so enamored of deductive logic that they either translated all empirical reasoning into a deductive mold or ignored empirical reasoning altogether. Either way, traditional philosophy not only leaves empirical reasoning unilluminated but obscured. In showing how different forms of representation facilitate the creation of new theories, Hanson revisits his earlier analysis of those cognitive processes that allow us to see anomalous gestalts suddenly in terms of familiar conceptual arrangements. Such cognitive processes are very significant contributors to our empirical reasoning, and Hanson makes a strong case that it is not only impossible to cleanse their influence from our account of nature but that the attempt to do so robs us of the capacity to understand how science advances. Once we learn how to "read" all these representations – and this is a matter that is far from trivial – we are able to understand parts of the world that were previously slippery and amorphous. The chaotic and strange has coalesced into something stable, predictable, and well-formed. Again, Hanson's distaste for the idea of clamping an interpretation onto raw data emerges forcefully: what the very data are, and which other items of would-be data fade impotently into the irrelevant background, is mediated and settled by the structures through which the data are made intelligible.

Hanson's account has two advantages over the main treatments of theories in philosophy of science. First, Hanson does not suppose that theories are just intellectual posits spawned by scientific whimsy; if that were so, we would be set upon by such a multitude of theories, each with an equal claim to our consideration, that serious testing and development would never be able to get underway. Hanson points out that the creation of a structure of representation capable of rendering the phenomena intelligible is a difficult undertaking, one requiring knowledge, patience, luck, and creativity. He urged that philosophers pay more attention to how such creations were produced and that their concern should be not with "theory-using, but with theory-finding" ([1958] 2010, 3). Second, on Hanson's account, structures of representation are the product of a temporally extended process – there is a beginning, middle, and end to it. Because of this, it is possible to analyze the creation of a theory in stepwise fashion. Since the creation of these structures is therefore an object of study (in principle at least), study of historical processes of theory construction may provide lessons for how theory creation should be pursued. Hanson addressed this theme in his many articles on the logic of discovery.

Hanson's are the eyes of the perennial outsider. Once he masters the new conceptual terrain of cosmology, logic, religion, or the theory of flight, Hanson then extracts the logical commitments implicit in that domain and subjects them to rigorous scrutiny. Hanson's reflections and arguments never exactly emulate those of the true expert in a field, though they often stimulate new paths of thought and speculation that had eluded even the brightest minds. A nice case in point of this thesis is Hanson's discussion of rival cosmological theories in "Some Philosophical Aspects of Contemporary Cosmologies." Though Hanson was a colleague of Fred Hoyle at St. John's College at Cambridge, and the two had many friendly exchanges, Hanson does not provide an advertisement for Hoyle's cosmology. What he does do is show that many of Hoyle's assumptions and motivations are not only naïve but perhaps absurd; however, Hanson then shows the same to be true of the orientation of the

Big Bang theorists, and he makes a convincing case that the rhetoric of each theory renders the other implausible from the get go. Hanson's discussion, however, is not intended to disparage cosmology as a nonscientific pursuit, a mere philosophical idle. Instead, he wants to clarify the far-reaching consequences of what are, initially at least, some rather innocent-looking conceptual assumptions. For instance, are we to interpret the principle of the conservation of energy as applying only to the part of the universe that is observable to us (as Hoyle does) or as applying to the whole of the universe, including areas forever outside of our observational reach (as the Big Bangers do)? In the end, neither theory seems capable of being wholly right or wholly wrong. The lesson to be drawn is that the empirical future of cosmology must ever remain in close contact with its deeply philosophical past.

Throughout his unfortunately short career, Hanson found himself pulled in multiple directions. Often his thought started with a defense of a specific position, to be followed by a slightly modified defense, and another still, until his last position was reached, which could not in truth be said to have been an extension of the original one. While this observation applies to Hanson's work generally, his writings on logic and levels of discourse fit this pattern most dramatically. Early on, Hanson argued that necessary and contingent truths occupy separate logical spaces and that inferential commerce between the two is always fallacious and therefore to be avoided. However, as Hanson proceeded through his typology of logical types, he came to recognize that just as there are different grades of possibility – a position philosophers have long held – so too are there different grades of necessity.[2] According to Hanson, while logical necessities have inconsistent negations, conceptual necessities have unintelligible negations. Hanson argued that many apparent paradoxes in philosophy were due to superimposing the terms and methods of deductive logic onto conceptual problems, the problem of induction being the most illustrious of these manufactured difficulties. On Hanson's account, inductive inference surely cannot be deductively justified (as Hume implied was necessary), but it cannot be dispensed with without making the empirical world unintelligible. Hanson was intent on defining enough of the critical concepts within these non-deductive logical realms to produce the appropriate logics, but he was never able to create a variant system – due either to the lack of time or (more likely) to the difficulty if not impossibility of the task. Instead, Hanson started with a bold, interesting position and then slowly modified it until he was left with an inventory, budget, or anatomy of the new field along with a few directives regarding how to approach it.

Hanson had decried the neglect of Leverrier's story, with all of its importance for both the history of celestial mechanics and scientific methodology, in his first book, *Patterns of Discovery: An Inquiry into the Conceptual Foundations of Science*. He himself finally wrote the history of Leverrier ("Leverrier: The Zenith and Nadir of Newtonian Mechanics"), and it is one of the finest exemplars of Hanson's work as a historian. This is not to say that the essay lacks philosophical interest. Hanson is as

---

[2] For logical reasons, distinct grades of necessity must accompany distinct types of possibility. Necessities are just alternative ways of stating impossibilities.

keen as ever to trace out the patterns of hypothesis generation, but his careful engage-
ment with the fascinating details of the discovery of Neptune (complete with an
explanation for why Leverrier's work was more significant than that of John Couch
Adams, who is ordinarily considered to have been a co-discoverer) display his skill
as a historian. Moreover, the even more fascinating "observations" of Vulcan, and
other supposed intra-Mercurial planetary objects, make for very entertaining reading. While the contemporary reader may feel the need to suppress a scornful smile at
reading of the "planet" Vulcan, Hanson shows that some very rigorous and diverse
theoretical approaches were directed at the problem caused by Mercury's classically
recalcitrant orbit. For the most part, the Vulcan hunters were not quacks but talented
scientists, attempting to work out a thorny problem as best their theoretical and
observational resources would allow. Finally, this article bears some striking the-
matic affinities with another work that spun off the press in 1962: Thomas Kuhn's
The Structure of Scientific Revolutions. Hanson points out that Newtonianism entered
a state of crisis, to use Kuhn's term, after the failure of the Vulcan conjecture: "Even
confidence in the lawgiver Sir Isaac, declined somewhat" (140, this volume); also,
when the Vulcan hypothesis first came into vogue, scores of observers began noticing
intra-Mercurial bodies where they hadn't seen them before.

I will now turn to two of the most influential areas of Hanson's thought – the
relation between philosophy of science and history of science and his advocacy for
a logic of discovery. Definitive essays on these topics ("The Irrelevance of History
of Science to Philosophy of Science" and "The Idea of a Logic of Discovery")
appear in Part IV: Logic of the present volume.[3]

The first essay is largely the fruit of Hanson's institution building. In creating a
space for history and philosophy of science, it was necessary to define the regions
of overlap and difference for the two disciplines. Hanson had always believed that
history and philosophy of science mutually enrich one another. His title for the
article made it sound as though history is not important for philosophy of science;
after all, to say it is "irrelevant" entails that it is unimportant. Or so it would seem.
Let me explain Hanson's specific conception of relevance. Hanson identified philosophy of science with the logic of science; thus, the job of the logician of science
is to appraise the logical character of arguments that crop up in the history of science. Logicians, of course, assess the validity of arguments – whether such arguments have true premises is an extralogical matter. Hanson's piece presents his
interpretation of the maxim that "philosophy of science without history of science
is empty; history of science without philosophy of science is blind."[4] According to

---

[3] Hanson certainly would have placed these articles in the Logic section – during his years at
Indiana University (1958–1963), Hanson preferred the expression logic of science to philosophy
of science. However, the contemporary reader would likely place them in Part I: Philosophy of
Science.

[4] This maxim is usually attributed to Imre Lakatos. Though Hanson was the first to discuss the
maxim in print, he gave credit to Lakatos as the original source of it. Herbert Feigl also discussed
the maxim in isolation from both Hanson and Lakatos. For more details on the history and use of
this maxim, see Lund (2010, 136–137).

Hanson, even though history and philosophy of science are not *logically* related, they are very tightly interconnected. Philosophy of science needs to select its content from the history of science to ensure that it is actually about real, and not pretended or fantastical, science. Historians of science must consult logical canons to appraise the arguments used by historical scientists. Historians don't just chronicle facts; instead, they order their narratives around the significant concepts and arguments that led to critical changes in the history of science. Without an appreciation for what makes the history go, history of science is blind.

While this famous article certainly formulated the outlook of Hanson's History and Logic of Science Department at Indiana University, it did not represent Hanson's last word on the issue. This article was published in 1962, near the midpoint of Hanson's abbreviated career, and Hanson came later to question his earlier position on the genetic fallacy. One commits the genetic fallacy when one assumes that the provenance of a proposition or argument is relevant to its truth or validity. The early Hanson argued that anyone who supposed that the goodness of an argument was a function of the argument's conditions of origin committed the genetic fallacy. Later, Hanson came to believe that, in special circumstances, the structure of an object is explainable, at least in part, by its history; in other words, he came to believe that some genetic arguments may not be fallacious. Such an insight would have led to a less differentiated view of HPS, but Hanson's efforts toward such a synthesis were mere gropings.

The logic of discovery was a central concern for Hanson throughout his career. Initially, he had argued that there can be good reasons for suggesting a hypothesis in the first place and these reasons need not be identical to the reasons for acceptance. As time went by, Hanson drifted away from the strong thesis that there can be a logical method for conjuring up worthwhile hypotheses toward the position that there exists a logical analysis of hypothesis plausibility. This is another area where Hanson seemed to have had strong intuitions running counter to those of mainstream philosophy of science, but where he ended up sticking, to a surprising degree, with the orthodox position. One reason for Hanson's adherence to orthodoxy was an apparent confusion about the relation between the psychological and logical – Hanson assumed that the two forms of analysis were entirely distinct, though he elsewhere argued (especially concerning the relation between history and philosophy of science) that different forms of analysis can apply to the same subject matter. While Hanson's basic position remained that there exists a logical appraisal of untested hypotheses, the examples he gives of strategies used in discovery – arguments from analogy, simplicity, aesthetic elegance, and explanatory fertility – all seem too remote from deductive logic to be analyzed down into anything commanding the respect of philosophers. Perhaps, Hanson would have been better off extending his accounts of good inductive reasons and his exploration of cross-type inference to have shed some light on the inference patterns so often active in discovery. Hanson's conceptual arsenal was rich enough to mount such an attack, but he seemed resigned to progressively limit his notion of the logic of discovery. Perhaps in this area more than any other might, we have expected the mercurial Hanson to have changed his mind once more and to have offered up an argument for a stronger notion of the logic of discovery had he lived.

We see in the essays on religious belief (Part V) Hanson's propensity to speak his mind. For all of his outspoken unbelief on the subject of religion, Hanson enjoyed many friendships with devout believers and was not one to shy away from discussing such higher matters. The longer of the two essays, "What I Don't Believe," was solicited by Hanson's friend Edward MacKinnon, at that time a Catholic priest. The piece was meant to be the first installment in an exchange between Hanson and MacKinnon concerning the rationality of religious belief, but Hanson was killed as MacKinnon's private response was making its way through the mail.

Even on the subject of religion, seemingly far removed from philosophy of science, we can find significant traces of Hanson's thought on science. In fact, though it may pass by unnoticed upon first reading these essays on religion, they encapsulate much of Hanson's mature philosophy of science. If we regard theistic existential claims as factual claims, as Hanson hurriedly argues we must, then we must treat the claims in accord with their logical type, whether we are confirming or disconfirming them. Despite the wonderful reputation for neutrality the agnostic enjoys, the agnostic is actually guilty of some logical double dealing. When considering the factual claim that God exists, the agnostic cannot confirm the claim due to the lack of evidence in its favor. As Hanson vividly portrays in both essays, he knows exactly what kind of evidence would convince him of the existence of God – the sky could open up, and the "Michelangeloid" God could show Himself, letting it be known how little He cares for Hanson's theological quibbling into the bargain. Since phenomena of this type, and others less dramatic, have not been observed, Hanson believes there is no evidence to support the claim that God exists.

Since there is no evidence for God's existence, the case against God ought to be closed. However, the agnostic just won't listen to reason. Instead of considering "God exists" as being disconfirmed in the same way that "the Loch Ness Monster exists" is disconfirmed (viz., by the absolute paucity of confirmatory evidence), the agnostic shifts ground and claims that no evidence – or, better put, lack of evidence – could ever disconfirm God's existence. This is not fair dealing: if the claim is regarded as confirmable, then it must also be disconfirmable. If the evidence does not support the existence of x, then the evidence disconfirms x.

Judgments of the quality of Hanson's writing on religious belief no doubt hinge on the reader's religious convictions. Even if one doesn't like Hanson's beliefs, it is hard to criticize his general strategy of moving the discussion from the otherworldly back to the shared commitments of scientific inquiry. The focus on the nature of religious belief and the logic of evidence appears to be a fruitful path, both toward a mutual understanding of one another's world views and toward exorcizing the pernicious subjectivity that hides behind the idea that differences in religious belief always come down to a *Weltanschauungskampf*, a battle between eternally incommensurable worldviews.

In this expanded edition of *What I Do Not Believe, and Other Essays*, we have been able to include two additional pieces. The first is Hanson's enigmatic essay "Observation and Explanation: A Guide to Philosophy of Science," which was published as a free standing book by Harper and Row. Sadly, practically nothing is known about when this short piece was composed or what its relation was to Hanson's other posthumously published textbook in philosophy of science, *Perception and Discovery*. The work does, however, seem to have been composed near the end of Hanson's life and expresses his mature philosophy of science. In it, he counters popular objections to his earlier published views, though in his usual indirect way. Here, as elsewhere in Hanson's work of the 1960s, we find a muted impatience with the "vogue" status of the theory-laden observation thesis. For Hanson, theory-laden observation marked just one aspect of the conceptual structure of science – a *significant* aspect, to be sure, but one whose full significance could only be appraised after studying it alongside the other concepts at the epistemological core of science. For Hanson, science represents a concerted attempt to render the world intelligible, and the various concepts central to that attempt are interdependent and cannot operate, or even be fully understood, independently. Concepts like fact, discovery, explanation, and cause are just as weighty as the concept of observation, though their perplexities are not as dramatically revealed as those of observation.

In "Observation and Explanation," Hanson calls for moderation and argues for a *via media* between the extremes of "dustbowl" empiricism and formalism. In this essay, Hanson launches once more his own distinctive philosophy, this time not portrayed as an overdue philosophical analysis of science as in *Patterns of Discovery: An Inquiry into the Conceptual Foundations of Science* but as the only way past seductive false philosophies of science. At the same time, though, the study of the extreme positions is indispensable for finding the elusive middle course. The essay's style is fresh and engaging, and it is rife with Hanson's aphoristic brevity. One well-acquainted with Hanson's work might accuse him of passing off his own philosophy as *the* philosophy of science; however, Hanson is probably no more guilty of that crime than were Hempel, Nagel, or other writers of introductory books in philosophy of science. It is fairer to Hanson to emphasize the substantial effort he had put into the philosophical education of science students. From the beginning of his career at Cambridge through his years at Indiana, Hanson acted as a philosophical ambassador to science, and this essay, along with *Perception and Discovery*, represented his final contributions to the pedagogy of philosophy of science.

The second new addition, Hanson's essay "The Trial of Galileo," is something of a "lost" work – it was published in a small run by the now defunct Hartford College for Women and was left out of the published lists of Hanson's works. The published version of the essay was put together by Stephen Toulmin from an audio recording of the lecture. Toulmin knew Hanson's literary style so well that the published version is indistinguishable from one of Hanson's self-edited works. Hanson's lecture was one of the six sponsored by the Hartford College for Women on the theme of trials where justice and the law came into conflict. What better topic to exemplify the decaying regard for authoritative institutions in the latter half of the 1960s? This setting for the lecture explains great deal about the essay's goals and its direct, and somewhat didactic, style. Hanson, ever the loud and pugnacious advocate for freedom of inquiry and expression, saw much in Galileo's story that reflected the problems of Cold War America.

Hanson's closeness with some of the best Galileo scholars of his day is evident in the piece, as Hanson expertly lays out the rich medieval ferment in physical and theological thought that set the stage for Galileo. Hanson shows how the courageous Galileo, with his unexampled powers of debate and irrepressibly sharp tongue, was bound to clash with small men and the Mother Church that emboldened them. Hanson himself was clearly able to see many of his own struggles reflected in the mighty travails of Galileo. Hanson was a vituperous advocate for freedom of thought, speech, and religion; his firebranding certainly earned him some recognition, not all of which was positive. Like Galileo, Hanson was something of a member of the Catholic Church's loyal opposition. Galileo, of course, remained a devout Catholic all his life, but he sought to moderate its dogmatic position on natural philosophical inquiry; Galileo was concerned not just for the future of natural philosophy (science) but for the Church itself, regarding it a tragic outcome should the new knowledge not issue from Catholic soil. Hanson, though baptized and raised Catholic, "converted" to atheism in adulthood. Nonetheless, he retained a great deal of respect for the Catholic intellectual tradition and especially loved the musical and artistic expressions of Catholicism. In short, Hanson was charmed enough by Catholicism to feel the profound tensions that must have animated Galileo in his fateful struggle. Hanson, ever the polemicist, ends the piece with a warning about the unchanging weakness of human nature and the necessity for those who respect truth to resist dogmatism and institutionalized thinking.

Even as we reach the 50th anniversary of Hanson's death, it is impossible to read the essays in this volume without feeling remorse at how much was lost in the plane crash that took his life – so much talent, humor, boldness, passion, and humanity and so many more intellectual vistas to have been taken in. Hanson's remark about Galileo that "intellectual gadflies are rarely stationary" (164) applies to himself as well – sadly, some intellectual gadflies don't live long either.

[A note on the text: since Hanson's career was split between the English and American academic worlds, his works were published in both British and American styles. The styles of the original publications have been retained to reflect their places of origin.]

Department of Philosophy and Religion Studies   Matthew D. Lund
Rowan University
Glassboro, NJ, USA

References

Hanson, Norwood Russell. [1958] 2010. *Patterns of discovery: An inquiry into the conceptual foundations of science.* Cambridge: Cambridge University Press.
Lund, Matthew D. 2010. *N.R. Hanson: Observation, discovery, and scientific change.* Amherst: Humanity Books.

Contents

Part I
Philosophy of Science

Chapter 1
A Picture Theory of Theory-Meaning

> His (Kepler's) admirable method of thinking consisted in forming in his mind a diagrammatic or outline representation of the entangled state of things before him, omitting all that was accidental, observing suggestive relations between the parts of his diagram, performing divers experiments upon it, or upon the natural objects, and noting the results. –C.S. Peirce, *Values in a Universe of Chance*

Perplexities concerning Scientific Theories persist because the usual 'singled valued' philosophical analyses cannot do justice to the problematic features of so complex a semantical entity. The components of theories are like law statements, and like models and hypotheses, being conceptual entities which are used in a variety of ways – not all of these being always compatible with the others. Thus many physicists characterize the classical laws of motion, as if they functioned in a definitional way.[1] But sometimes these laws seem remarkably empirical.[2] Others characterize such laws as 'conventional'; they shape entire disciplines much as the rules shape the game of chess.[3] Law statements are not exclusively any one of these – definitions, factual claims or conventions. They are *all* these things.

Consider: "The sun rises in the east". It is impossible from only hearing or seeing these words in isolation to know whether this claim is functioning in a definitional way or in a descriptive way. Thus if tomorrow the sun parts the horizon 90° from where it arose this morning, it might still be rising in the east *if* one treats "east" as the *name* of that place where the sun rises (wherever that may be). If one defines "east" in the terms of celestial coordinates though, it will be an empirical/factual/synthetic claim that the sun rises in the east. So the very meaning of "The sun rises in the east" is elusive until one comprehends this assertion's local use in a specific context. This latter is quite free to change.

Much this same diversity and flexibility should mark our understanding of scientific theories. What a scientific theory *is* cannot be finally determined – for theories

---

[1] Kolin, and sometimes Poincaré, for example.

[2] Mach and Broad frame the Second Law as fundamentally a factual statement based on experience.

[3] Reichenbach, and Poincaré again, are cases in point.

© Springer Nature B.V. 2020
N. R. Hanson, *What I Do Not Believe, and Other Essays*, Synthese Library 38,
https://doi.org/10.1007/978-94-024-1739-5_1

are context-dependent instruments of conceptualization. Tomorrow's enquiries can transform yesterday's scientific theories into semantical structures different from what today's philosophers pronounce them to be. (Who now reads Mach for insights into contemporary Quantum Theory? or even Schrödinger?)

Let us look at theories in a way different from those which dominate discussions in philosophy of science. Think of theories not as ideal deductive systems, as precise languages, or as convenient empirical shorthands. That is, they are used sometimes as if they were definitional/analytical/calculational systems; sometimes as if they were ideal languages (well-chiseled logicians' Esperanto); sometimes as if they were elegant compendia of factual information. Theories are all these things – but they are more too. Explore yet another facet of scientific theories – one which disappears in the glare of the analytical spotlight.

How can theories enable us to *understand* a subject matter? What is the difference between a heap, or a list, of descriptive assertions and a theory –which is itself largely constituted of those same descriptions? These questions recall the contrast between a mere generalization (e.g., that all white, blue-eyed, tom cats are deaf), as against a law of nature (e.g., that all bird's wings have a convex top-side). If the generalization is imagined refuted, we are required only to effect a quantitative readjustment; we may have to say that 99% of all white, male, blue-eyed cats are deaf, rather than all of them. We will still know what cats are, however. No conceptual readjustment is forced on us by a feline counterinstance. With a law of nature, such as that wings of birds have convex top-sides – if one were to encounter a counterinstance of this, *conceptual* difficulties would ensue at once. The full concept of bird *flight* requires a wing imagined so shaped. Faced (*per impossibile*) with a bird wing curved otherwise, one might come to doubt what a bird wing *is,* and what role it plays in flight – doubts which do *not* now punctuate the thinking of aerodynamicists and ornithologists. It is as if one imagined an exception to: *all unsupported bodies in terrestrial space move toward the center of the earth.* An exception to this would have to be a body in a state of levitation or 'negative gravity', either of which possibilities raises doubts as to what *bodies* were in the first place.

It is sometimes said that a Law of Nature explains its subject matter, helps us to understand it, makes it more intelligible and comprehensible -as against a generalization which only correlates observables via actuarial techniques; these observables may concern 'unrelata' like the simultaneous occurrence of sun spots and wheat failures, where no conceptual link binds such phenomena. Analogously, a scientific theory entices philosophers because it somehow explains its subject matter; it helps us understand 'interconceptions' between phenomena.

What does all this mean? What is it in a theory such that before it was formulated all the data, the descriptions, the initial conditions – however accurately recorded – did not compose into a coherent and intelligible subject matter, whereas after the theory has been generated and coupled with observations one can *comprehend* the subject matter?

Consider theories *pro tem* as conceptual entities located at the crossroads between epistemology and philosophical psychology. Think no more, for now, of the logical and the semantical aspects of scientific theories; everyone always talks about that.

Fig. 1.1

Fig. 1.2

Let us view theories as instruments of intelligibility. Ask with me: "How does the conceptual structure of a theory make understanding possible?"

Reflect on those picture-puzzles, dear to learning theorists and Gestalt psychologists. The sheep-in-the-tree (Fig. 1.1), and the figure on page 14 of *Patterns of Discovery* (Hanson [1958] 2010) (reproduced here as Fig. 1.2).

These constellations of lines cohere dramatically when once it is signaled what they are. The cluster of dots and blobs and shapes set out just above (in Fig. 1.2) can be seen as a medieval Christ-like representation. Often this appears as an unintelligible chaos of patches, and lines – before it constitutes a picture of any significance. How is it that a conceptual structure, a pattern in imagination, can give meaning to gaggles of dots, shapes, lines and points? How is scientific observation possible?

Consider Fig. 1.3: When labelled 'a Mexican on a bicycle (seen from above)' something happens within the perceptual field. The experience now is qualitatively different from what it had been before when this was a mere configuration of lines. How so?

How doesn't matter (the problem is philosophical, not psychological; conceptual, not factual). *That* patterns affect the significance of lines, dots, shapes, and patches – which might have been in perceptual turbulence otherwise – this is our fundamental datum. It has profound epistemological consequences. Knowledge is

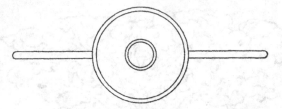

Fig. 1.3

a function of how our experiences cohere. Observations made *before* the perceptual pattern is appreciated, are epistemically distinct from the observations, (and
their descriptions), made *after* that pattern has cast them into intelligible constellations – although the observations and descriptions, those before and those after,
might be 'congruent'.[4] The descriptive terms, the assertions, the observations
themselves, when considered in terms of repeatability and what is 'written on the
page', these might be identical both before and after the pattern is appreciated. The
lines in the drawing above did not shift geometrically when the caption was
assigned. Yet there is an epistemic distinction between the earlier and later encounters, a distinction of deep importance.

Clearly, talk about *patterns* differs in type from talk of lines, shapes and dots.
Patterns do not fill the same logical space as do the lines and dots being patterned: a
pattern – e.g. of this Mexican atop a bicycle – is not itself detectable or visible or
drawable, not as the shapes and the lines are. This is not to say that they are not
detectable or visible at all. How else should we come to know them? Patterns are
detectable and can be made visible to those who cannot see them – but not necessarily by adding more lines. Describing this encounter differs from speaking of objects
of sensation as appreciated by all normal observers. 20-20 vision is no guarantee of
seeing the Mexican on the bicycle. Patterns are not *elements* in an epistemic configuration. Rather, the pattern is the configuration itself. By analogy, the plot of a novel is
not another cluster of words; the form of a sonata is not just another cluster of notes;
the planform design of a building is not merely more bricks and beams; the aerodynamic structure of an aircraft wing – its airfoil section – isn't just more ribs and skin
plates; indeed, the meaning of a proposition isn't only another articulated term!

Much as the level of 'pattern talk' differs conceptually from that on which talk of
dots, shapes, lines and patches obtains – so also theoretical talk differs conceptually
from observational and descriptive talk. The more comprehensive suggestion is this:
that just as perceptual pattern recognition at once gives significance to elements
perceived and yet differs from any perception of dots, shapes and lines – so also
conceptual pattern recognition at once gives significance to the observational elements within a theory and yet differs from any awareness of those elements *vis-à-vis*
their primary relationship to events and objects. The ways in which theories, conceptual structures, are meaningful with respect to the observation statements is
qualitatively a different type of concern from that involved in discussions of how
observation statements are meaningful with respect to things.

---

[4] The temporal references, 'before' and 'after' are inessential. This exposition would not suffer
were 'independently of' and 'dependent upon' introjected.

Fig. 1.4

At this juncture, some parenthetic autobiography. A psychological fact: there are moments when I find myself confronting a cluster of symbols, or observed anomalies, such that after having come to view these through the appropriate *scientific theory* they configure, cohere and collapse into meaningful patterns within a unified intellectual experience. This, to me, seems not unrelated to what is involved when I appreciate dots and lines in a qualitatively different way after having mastered the perceptual pattern structuring those marks.[5] Consider Boyle's law as understood in 1662, then simply a stack of statistical correlations; Boyle didn't extract that famous generalization himself, his followers did. That law – that correlation considered *before* the advent of kinetic theory and before classical statistical mechanics – resembles the dots without the pattern, the observations without the theory, the descriptions without the explanations. Boyle's Law began life as the *merest* correlation. It explained nothing. Only when general gas theory and the kinetic hypothesis caught up with it, did Boyle's generalization come to function as Laws of Nature are reputed to do. Bracket with this example the historical problem concerning the anomalous motions of Saturn and Jupiter. This was a descriptive thorn in the side of astronomical explanation, B.L. ('Before Laplace'). Laplace undertook to set out a conceptual framework for mechanical ideas, a Stability Proof in terms of which this anomaly – the apparently secular aberrations in the motions of Saturn and Jupiter – could be regarded as but local irregularities in what was really a 900 year cycle – a periodic, repetitive 'aberration'. It is a little like what one should expect in a microcinematographic film of meshing gears in a fine clock: crude and lopsided in fine scale, but precise and perfectly periodic at the macrochronometric level. Descriptions of Saturn and Jupiter B.L. were independent, unrelated and unsynchronized, whereas these same descriptions A.L. constituted almost different subjects for one's attention.

Please permit me to spell out this primitive analogy in more detail. Consider the concept of a *scene*. More specifically, think of a dawn seen from a hillside. There sits a landscape painter, busily conveying to his canvas a configuration like Figs. 1.4 and 1.5. Some passersby may say of this painting that it is 'true to life' (Fig. 1.4), that it captures what is significant 'out there' (Fig. 1.5). Painting is an activity of the

---

[5] Cf. the earlier illustrations.

Fig. 1.5

appropriate type *to* capture features of the original – the tree, the hill and the other landscape objects 'out there' and 'committable to' canvas. There is a structural identity between what can be seen by the painter from the hillside and what can be seen on the canvas he has painted. And this is just as important for his painting's being 'true to life' as is the identity at the color – shape – line level. *Of course* the tree should be painted green, as it is, and not pink, or silver. But no less important is it that it should be depicted as to the left of the sun – and not stretched horizontally above it. Something, which I shall designate 'the scene', is 'out there' for inspection; one can stand on the hillside and survey the scene to the east.

One can also describe what the artist has put on canvas as 'the scene he has painted'. The scene on his canvas and the scene 'out there' are structurally so related that it is meaningful to speak of the former as constituting a replication of the latter, something one cannot claim of sounds, textures or tastes, no ingenious combination of which can replicate the scene at dawn; the scene-as-paintable eludes the powers of music, of tactile sensation and even of cookery. Thus the term 'scene', from a conceptual point of view, is specific yet Janus-faced. It alludes to an objective subject matter 'out there', and it also refers to one's plastic representation of that subject matter.[6] The *same* scene can be both 'out there' and also on canvas.

That the artist has put the same scene on canvas as obtains 'out there' is pertinent to whether his rendition is veridical.

I don't want to refer to the scene *per se* as if it were an 'interim designatum'. That would proliferate entities, since Antiquity a philosophically suspect practice. Nonetheless, aspects of subject matters are reproducible in this way *because* of their possible structural identity with aspects of the reproduction – this is all I wish to remind you of.

---

[6]Yet the designation is specific in that it excludes myriad other kinds of representations of the world. The real steak's possession of properties which can induce gustatory delight in me is no part of any *scene* of that steak. The nightingale's song is replicable, but not because it is part of a scene. The scrape I endure may be due to the icy, rough surface of the granite I clamber upon, but the scrape is not a replication of any part of the granite, whereas my visual memory of the granite may indeed have properties of the granite block itself – such as those an artist could commit to canvas in a painting of that block.

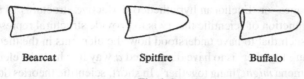

Bearcat   Spitfire   Buffalo

Fig. 1.6

Many terms do similar work; 'landscape' has the same mirrorlike semantical quality. The landscape is something tended by a gardener, something one can view from a distance. It is also what is capturable on canvas by a draftsman, or painter. Again, the subject matter and its representations can share something of considerable conceptual importance. Were this not so, the subject matter would not be representable at all.

There are myriad such 'bipartite' terms. The 'plan-form' of a bird's wing, as referred to by ornithologists and aerodynamicists, makes reference to such geometrical relationships as the chord-span ratio, the angular sweep back of the leading edge, the relative root-to-tip rate of narrowing, and the contour shape of the wing (elliptical?, rectangular?, triangular?) The wing's 'aspect ratio' is another such term – this is the relative thickness of the 'fuselage' as against the length of the wing, as viewed directly forwards or aft. The tip configuration of the wing, whether blunt or pointed or round, will also be part of the understood designatum of 'plan-form' (Fig. 1.6). The plan-form and the aspect ratio of a bird or an air craft can be drawn out on a piece of drafting paper, *and* it can also be inspected in the 3-D wing itself, as found on the living bird, or the operational aircraft. Wittgenstein's point about the structure of the bird's song as being something which is *in* the song itself, and also *in* the gramophone recording of the song – and also *in* the musical score which captures that song in notes (*à la* Delius) – this point is close to what I am groping for. The song, its recording and its score – share a common structure. The plan-form on paper and in the actual bird wing share a common structure, as is true also of the tip-configuration, the aspect ratio, the dihedral, etc. The landscape, the scene, *is* the common structure shared by objects-in-configuration 'out there', and color-patches-in-configuration on the painter's canvas. [Aside: *Facts* are the common structure shared by events 'out there' (as when they are 'hard', 'stubborn' and must 'be faced') and by the truth as stated about those events (as when we 'state the facts', 'list' them and base theories on them).]

My suggestion will be that, analogously, states of affairs, that is, constellations of phenomena, are often rendered understandable and intelligible and comprehensible *because* some objective, structural component of those phenomena is duplicated in a corresponding structural component within some scientific theory. Scientifically understanding phenomena x, y and z consists in perceiving what *kinds* of phenomena they are – how they relate each to the other within some larger epistemic context, how they are dependent upon, or interfere with, each other. Insights into such relations 'out there' are generable within our perceptions of the structures of theories; these theoretical structures function *vis-à-vis* our linguistic references to x, y and z in a way analogous to how the *scene* stands to the tree-and-hill 'out there', and also to the painted patches on canvas. Thus, in contrast to the delineation of theories

as 'ideal languages' or 'Euclidean hypothetico-deductive structures', I suggest that
the important function of scientific theory is to provide structural representations of
phenomena – such that to have understood how the elements in the theoretical representation 'hang together' is to have discovered *a* way in which the elements of the
original phenomena *might* 'hang together'. In short, scientific theories do not always
argue us into the truth; they do not always demonstrate deductively and forcefully
what is the case. Often they *show* what could be the case with perplexing phenomena, by relating representations of those phenomena in ways which are themselves
possible representations of relationships obtaining 'out there'. Theories provide patterns for ordering phenomena. This, just as much as they provide inference-channels
through which to argue towards descriptions of phenomena.

Before proceeding, consider some classical objections to the so-called 'picture
theory of meaning'. Clearly, if one takes all forms of representation to be fundamentally *iconic*, as one would in a landscape painting, then the painter will be felt to
represent elements in the original 3-D configuration by way of *iconic* tokens in the
copy configuration (2-D). That is, his tree here will share some properties of the tree
out there, (perspectively considered). Its shape, for example, oriented with respect
to the sun, and the hill, will display ratios in relative height, width, and color, analogous to what obtains in the original. *The* sun and *his* sun will have a common geometry both internal, with respect to its discoid design and coherence, and external,
with respect to its relations to tree and hill. A color transparency, e.g. of the Kodak
variety, could be moved from its superposition on the scene out there, to superposition of the scene on canvas, and it would be logically possible for there to be shape-
congruence and 'color-congruence' all the way through, both in superposition I and
in superposition II. And so that representation on canvas will stand to the original
(3-D) in a way which is designated as "iconic". This is proved by the Kodak transparency's congruence with each.

Now, *vis-à-vis* scientific theories, where the mode of representation (if there is
one) is linguistic and descriptive, it is obvious that this is not any crudely iconic
representation. Theories are not simple pictures. The word "tree" has nothing iconically in common with what this word may designate, namely some actual tree.
(There is nothing arboreal about "tree"!) Similarly the word "sun" is not *iconically*
connected with any perceptual object or any physical object. Words represent not
because of property-sharing. They have no property in common with what they
represent – save for onomatopoetics like "toot", "crash", "smooth" and "short".
[These seem to me relatively unimportant, semantically; they certainly constitute no
paradigm of word-object meaning]. It will be the *conventional correlation* of words
with objects which holds our attention here. Consider a term well-known in analytical mechanics – *"syzygy"*. This word does not represent iconically any rectilinear
configuration of moon, earth and sun (which is what the word means). It is not due
to any iconic relationship with objects in the Solar system that this linguistic term
means what it does – although you will perceive that there is something about these
designations ('y', 'y', 'y') which seems to tie in with the three bodied problem
involved; sun, earth and moon. Nonetheless, *'syzygy'* is related to planets as a painting may be related to trees. To hear it for the first time, is not to know (simply from

Fig. 1.7

the configuration of the sounds and symbols) that it connects semantically with moon, earth and sun – in the way in which "toot" might connect semantically with a passing train, or "buzz" with a passing saw. In other words, statements, paradigmatically, designate; then they characterize their designata as being of this or that type, or as having these or those properties. Thus, "The moon is a pocked sphere" – where "The moon" is the designation of an astronomical object, and "is a pocked sphere" characterizes that object, that designatum. Pictures represent in a non-designatory way; they are non-specific with respect to the attention-directing they may stimulate. Does Fig. 1.7 designate the moon, its sphericity, its discoidity, its pock-marks, its yellow color... or what? Statements place one's attention precisely on particular designata, and then they discriminate between, and select from the appropriate alternative characterizations of that designatum. Thus, of all the things that it may be true to say of the moon as depicted above – e.g. that it is spherical, that it appears as discoid, that it is pock-marked,... etc., – the statement "The moon is a pocked sphere" selects one of these specific data as its unique and direct message, and articulates it pointedly. That is why it is true that one picture is worth a thousand words; a picture is a thousand times less specific than a short sharp statement. But, by the same token one word is worth a thousand pictures; a statement can supply a focus for the attention quite different in type from anything generable *via* confrontation with a picture.

These objections to the picture theory are well-known, and yet I am going to suggest something sometimes suggested by others – that all this critical carping on the distinctions between originals and icons, as against originals and statements, really misses the profound point of the picture theory of meaning. Objections concerning the non-iconic ways in which words and statements represent, these really deal with the hyper-fine structure of discourse *versus* pictorial representation. These are directed to the ways in which words like "moon" are, or are not, correctable in function with line configurations such as shown in Fig. 1.8. Aside from such hyperfine structural differences, statements and drawings remain deeply analogous *vis-à-vis* representational features to be discussed in a moment. Thus the objections to the picture theory advanced by such people as Edna Daitz and Irving Copi concern just the minute superficialities of word tokens and claim tokens. What else could be the point of noting that "cat" does not look feline and that "moon" sheds no light?

However, let us attend rather to the structure of discursive knowledge in more general terms, and not restrict our interest to the indivisible tokens through which that structure is conveyed. Consider the structure of discourse itself, and the corresponding structure of representational knowledge. These different kinds of struc-

Fig. 1.8

Fig. 1.9

tures can perhaps convey insight and information about the structures of the originals in much the same way – so much so that there is yet more to be said for the classical picture theory of meaning with respect to how it helps us understand linguistic meaning. A claim such as "The sun rises to the right of the juniper" does indeed have something in common structurally with Fig. 1.9 (a configuration an artist might put on canvas). The structure common to both the claim and the sketch makes it possible to learn from both to what extent they might be veridical. Certainly the claim and the sketch, because of some common structure, stand or fall together; either both of them have the structure of the original (i.e. "sun-to-the-right-of-juniper"), or neither of them has. The picture theory, then, as articulated at high speed in the *Tractatus* (Wittgenstein 1922) and at very slow speed in Wisdom's articles[7] on "Logical Constructions", may be articulated improperly in both contexts. For these celebrated expositions dwell overmuch on language token-physical object correspondences, and not enough on structural correspondences. This latter undertaking will constitute the philosophical burden of this essay, a burden relieved somewhat, I hope, by special illustrations from Fluid Mechanics which will serve as a typical scientific theory in the analysis to follow.

Our paradigm should not be the interconnections and resemblances between paintings and their subject matter (or between photos and their subject matter), but rather between such a thing as a map and its subject matter. Even better, for our purposes, will be the logical linkages between a chart and its subject matter, or between a highly schematic diagram and its subject matter. Let's begin with maps, our first approximation. A map must share some structure with the original terrain to

---

[7] The articles to which Hanson refers are collected in Wisdom (1969). –*MDL*.

be useful at all. This much stresses the iconic, and might even incline one to think of maps as if they were stylized paintings from above (high above) ground, or even vast aerial photographs; they are neither of these things. Still maps do resemble paintings and photos in this – they must share some representational structure with the original terrain mapped or else they would not be at all reliable or informative. A map has got to indicate to us, for example, that the Greater Pittsburgh Airport is northwest by west of the Golden Point, that Allegheny Airport is roughly south, and that Pittsburgh-Wilkinson Airport is roughly east. These must be fairly stable and veridical representations; if the representation fails in this respect then it simply won't be doing for us what a map is expected to do, to wit, provide us with a representation of the geographical structure of Pittsburgh such that if we can 'locate ourselves within' the representation we have located ourselves within Pittsburgh. Still, noting this requirement of cartographic verisimilitude is compatible with recognizing also that there is an extensive conventional vocabulary within any map – so much so that a painting or a photo of Pittsburgh from above just cannot serve as a map of Pittsburgh. Thus, one has in the map, a legend (see Fig. 1.10) and by appreciating how one designates tall towers, state capitols, railway lines, airports, state parks etc., one can make one's way through the urban jungle with the aid of this graphic, but necessarily stylized, representation of that jungle. Such reference points cannot be expected in an aerial photograph (see Fig. 1.11), of course, anymore than one can expect them to stand

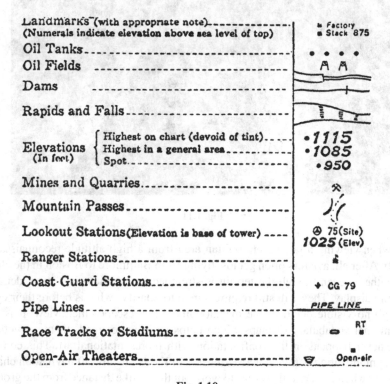

Fig. 1.10

Fig. 1.11

forth when viewing a large metropolitan area from a high altitude reconnaissance aircraft. After all, aviators often get lost flying from Baltimore to Boston in the clearest weather – even though the terrain is stretched below them in a most detailed dioramic display. They will still require a map to 'clarify' what is before their eyes, although no visible detail of the megalopolis below escapes their view. It is this conventional vocabulary of maps which helps us learn to 'read' them, something which never happens in the confrontation with representational art. This conventional symbolization is much more extensive in En Route aircraft navigation charts, especially when one actually begins to work out things like distance from the ground,

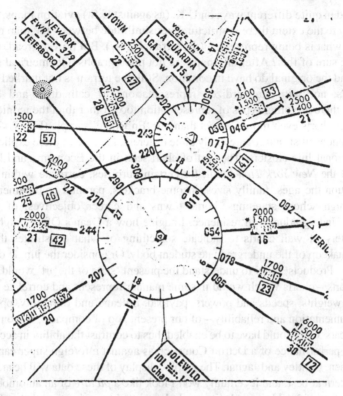

Fig. 1.12

and one's distance from appropriate objects in the air space and terrain ahead, by extraordinarily stylized and conventional blobs and shapes of color which, for a novice, would simply be unintelligible. [Notice how, in the En Route aircraft navigation map reproduced here as Fig. 1.12, the outlines of famous land masses have almost disappeared, in favor of the much more stylized depiction of 'airways' – so much so that 'map' almost seems less appropriate a word for this rendition than does 'chart']. These features of aircraft maps and navigation charts are valuable not because they represent iconically, but rather because they have a fixed and widely applicable significance; this makes any attempt to comprehend the geographic complexities of Pittsburgh in principle like the comprehending of any other urban-geographic subject – Harrisburg, Philadelphia, New York or Boston. Maps must be *read*, as pictures need not be. They require training, legends, glossaries and vocabularies. In just this they differ from country to country. An aviator in Great Britain finds that local maps contrast markedly with American versions of the same locale – not in the iconic details, of course, but in all those other features with respect to which an aerial chart is supposed to serve as his guide. He soon realizes that there is much to learn, and to comprehend, merely to understand what the 'message' is in those maps. For cities, towers, airports, train tracks, bridges, monuments... etc. are symbolized in quite different ways on U.K. (as against U.S.) aviation maps. Both are useful only to the extent there is structural verisimilitude between the map representation and what is being represented (the terrain below). But how to 'read' the chart so as to be sure of this? After all, maps are not photographs. Geometrical structure the map and the original do have in common. Still, the terrain is not marked with As, Es and Cs, nor with standardized representations of cathedrals and canyons. Moreover, there are other types of representation too – other than the iconic, that is.

Consider charts even more abstract than our 'En Route' sample (Fig. 1.12) above. The correlations of structures are really quite different, in some of these explicitly non-representational line-clusters. The sort of chart seen in the Economic and Financial sections of the *New York Times* will, for certain purposes, correlate within a single representation the ages, family sizes, income brackets, parental employment, major subjects, high school standings, home towns and career objectives, of (say) all Pittsburgh Undergraduates in residence. Imagine how a Dean's Office might have a whole battery of wall charts to indicate something of what constitutes the cross-sectional makeup of the undergraduate student body. Or consider the jungle of Detroit Automotive Products. Fully to understand the present 'state of the art' would virtually require charts – charts which would in some manner represent and correlate (say) the respective weights, speeds and powers, payload, mileage and durability, operational ease, instrumentation and reliability – of our present zoo of compacts, wagons, sedans and sportscars. One would have to be enabled thus to contrast the ability in acceleration or braking performance of a Detroit Compact, as against a foreign super car, like the Saab, Citroen, Bentley and Jaguar. The graphic display of these data will be an intricate rococo pattern at best; and it certainly won't look like a *picture* of an automobile.

Charts of considerable complexity might be needed to understand the performance properties of some small internal combustion engine. One would have to relate within the same curvilinear configuration a potpourri of parameters concerning things like compression differentials, average fuel flow, the brake mean effective pressure, the revolutions per minute of the crank shaft, the generator shaft, the drive shaft – the oil temperature after 5 minutes of idle running, the mixture ratios of fuel and air, the lubricant's viscosity and the coolant's efficiency – all of these things as considered in this one small engine at a moment, perhaps 5 minutes after having been fired up. From such a graphic display of parameters one could then delineate with accuracy the power-plant's total performance after having run for 10 minutes, 15 minutes, or any time whatever. Those lines and their interrelationships will indicate changes in the numerical values of variables whose functions really constitute what is the essence of the machine in question. The R2800-30 W Pratt and Whitney aircraft engine is as shown in Fig. 1.13.

This is all of its constituent dynamic parameters – their waxings, wanings, influx; representational description of that powerplant's performance. [This chart and the engine have much in common – structurally, that is!]

The chart concept gets quite complex and abstract now as we address a modern, high-performance aircraft wing. A wing *is* its coefficient of lift, plus its coefficient of drag, its frontal resistance and its skin friction; it is the eddying turbulence at the trailing edge, the vorticity and the wingtip configuration; it is the stalling point-where the boundary layer lifts off the upper surface – and the starting vortex.

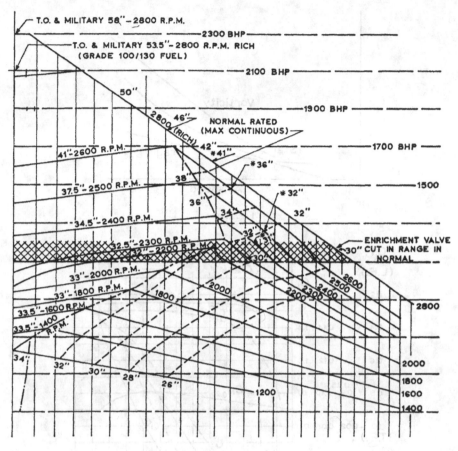

Fig. 1.13

Understanding a wing, appreciating what it does, what it is – *is* just being aware of how such lines would slope and intersect when representing the aerodynamics of that shape. To contrast the S-shaped airfoil section envisaged by Richard von Mises (Fig. 1.14) and a quite distinct airfoil section known as NACA 2412 (Fig. 1.15) is to appreciate intuitively how the appropriate graph-representations of the aerodynamic parameters of these wings will differ, or will be the same. Aerodynamicists are really discussing the properties of such *shapes* when they discuss stagnation points, laminarity at given angles of attack, the stability of burble zones in the boundary layer, the vorticity of the downwash and the induced drag and form drag of a particular wing-section's shape – none of these being photographable (or at least not primarily photographable). These locutions really concern the structural interrelationships between these functions which describe the slopes of the performance lines on aerodynamic airfoil charts. No two airfoils have the same characteristic charts; when two charts are identical they designate one and the same airfoil shape – something known *without* actually introducing the shape as a third bit of evidence.

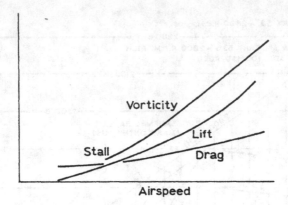

Fig. 1.14

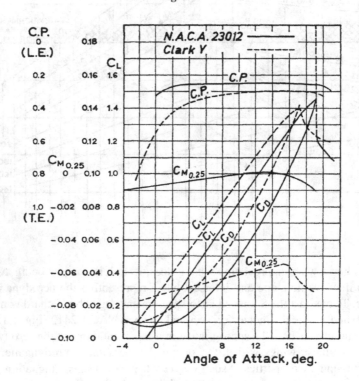

Fig. 1.15

The charts *are* the wings; they are everything aerodynamically significant about the wings. Yet they are certainly not pictures. They *are* structural representations of aspects of the wing, however. Such charts, diagrams and graphs as I've described are veridical in a large number of instances. They are informative because they share structures with the actual wings in question – dynamical structures, not geometrical structures. They provide a pattern, through which the multiform and chaotic manifestations of the original appear as correlated parameters. These patterns provide conceptual *gestalts* which allow inferences from one parameter to another parameter throughout a charted system of data-lines. Thus from knowledge of what in fact is the numerical value of x (the respective angle of attack) supposed to obtain with NACA 2412, plus a knowledge of the airspeed, one can infer to a value for 2412's coefficient of lift, coefficient of drag, and its trailing edge turbulence. These data tumble right out of this representational approach, since to know the shape of the general parameter-configuration, and the value of one parameter, is to be positioned for inference to all the other parameters which describe 2412 [the parameter-configuration as graphed *describes* NACA 2412, as a compass describes a circle; 2412 is completely delineated *via* the drawn parametric interactions].

Once a constellation of parameters is captured in a graph, simple Cartesian transformations can render them algebraic, and less obviously pictorial and geometrical. Any cluster of criss-crossed lines in a plane becomes totally comprehensible when transformed into a cluster of algebraic formulae, each one of which 'programs' where every point on a data-line will fall. Thus co-ordinate geometers easily transform a horizontal straight line into $y = k$; a vertical one into $x = k$; a sloping straight line (through x_1, y_1 with slope m) into $y - y_1 = m(x - x_1)$; a circle (radius r, center at a, b) into $(x - a)^2 + (y - b)^2 = r^2$; a parabola (vertex at o, focus at a, 0) into $y^2 = 4ax$; an hyperbola (center at o, foci at c, 0 and $-c$, 0; transverse axis $2a$) into $x^2\!\big/a^2 - y^2\!\big/(c^2 - a^2) = 1$ ... and so forth. Familiar exponential and logarithmic curves are $y = e^x$, $y = e^{-x}$, $y = log_e x$ ... etc. The Gaussian distribution so important in laboratory work is $y = e^{-kx^2}$. In general, one can find an algebraic descriptive equation for any locus of points defined geometrically. Hence all lines, or segments thereof, are completely described algebraically. For physical curves like projectile paths, light rays, celestial orbits... etc., and for physical surfaces like ship hulls, airfoil sections, gas flow lines... etc., this analytical technique is obviously very powerful, for the full battery of algebraic procedures is at once made available to the physical understanding of any 2-D curve, or 3-D curve-set. The situation is identical for the analysis of datagraphs, the observation-point lines on which are interrelated simply by algebraically interlinking the equations of these lines. The result is a very much fuller understanding of the dynamical properties of the subject-matter partially described by each observation-line; this is signally true in studies of airfoils, engines, markets and societies. The sequence is always from observations, to numerical descriptions (of these observations), to point location (on a graph), to curve construction (out of the observation points), to algebraic description (of the parametric curves), to functional interrelation of the algebraic descriptions. The end result is an algebraic structure which *is* (or is at least analogous to) the dynamical structure of the actual airfoil, the engine, the market or the society. [As the algebra exfoliates in logical space, so does the airfoil, or the engine, perform in actual space-time.]

This natural development from discrete measurements to algebraic formula-clusters is open to an insight advanced by Wittgenstein where he says "language sprang from hieroglyphics, but the essence of representation remains".[8] Analogously, I will argue that after curvilinear data graphs are rendered into algebraic and function-theoretic form, the essence of structure remains as between the processes in the subject matter and the processes in the algebra. If the graphs were informative as structural representations of the aero-dynamics of a wing or the thermodynamics of an internal combustion engine – and if there is no difference that makes any difference between the graphed lines and the algebraically-symbolized lines – then the algebra is *eo ipso* informative as a structural representation of the original subject matter, wing or engine. Physical theory, I submit, can be thought of as a result of compiling, meshing and unifying many such charts. I don't mean this genetically or factually, of course; it is not a historical remark. But reflect on the conceptual possibilities here. If a physical theory is construed *pro tem* as a result of data chart compiling, each graph is structurally related to complexes of phenomenal processes and, if the data-graph compilation grows towards theorizing by being transformed into generalized algebra from which one then undertakes to detect still higher orders of formal and inferential connection – if one allows all this to stand, it will suggest how phenomena and theories relate, how measurements and algebra connect, how the world and our ideas of it are linked. And in this view of the rise of theory the essence of representation is not lost. The graph lines represent data-structure; where the data ascend or descend numerically, when they oscillate or spread randomly, the graph lines do the same. They share structures to that extent. But the algebra does this to no less extent; the algebraic descriptions are thus also representative of processes in the phenomena. But complex and inelegant algebraic expressions can be traced to simple, powerful and elegant higher order algebraic claims. Equations fuse, collapse into one another, reveal themselves as but special cases of much wider and abstract mathematics. Most sets of equations can be shown to be functions of other sets, given sufficient time, ingenuity (and computers). In fact, for any particular body of descriptive equations it can usually be shown that there is an indefinitely large number of functional relationships which bind the equations. It is the special office of theoretical insight to opt for *one* of these functional reticules as against others equally faithful to the data. Thus, that a theory should square with the facts is a necessary condition for its acceptance. But it is not a sufficient condition. That is, after threading data into alternative abstract functional relationships – there always will be alternatives – further considerations must be weighed in the court of scientific theory.

The sophisticated complexities of theoretical language should never obscure the fact that much of the understanding such language-systems provide issues from structural insights they afford into the 3-D behavior of the original phenomena.

Before developing an example or two of what I am getting at (and, as usual, the examples may prove richer than the analysis), let me discuss some advantages of our analogy. Remember, the analogy concerns theories considered as structural patterns; or perhaps 'structuring patternings' would be more felicitous. Theories as blueprints are our primary interest; theories as narrative will barely concern us. One immediate fallout from this way of viewing theories is that it becomes difficult to speak of them

---

[8] Hanson's free translation of part of Wittgenstein (1922, 4.016). –*MDL*.

as being descriptively true or false in the manner that observation-statements are descriptively true or false. Indeed it becomes hard to think of them as being true or false in *any* simple sense. This is compatible with many arguments by many philosophers of science; 'of theories ask not whether they are true or false, but only whether they apply or do not apply'; 'theories are not simple conjunctions of observation-statements'… etc. So, just as patterns suggested within a pictorial configuration are not true or false, but rather are effective or ineffective *vis-à-vis* the color patches they are intended to relate, so also patterns set out within a theoretical representation are not true or false, but rather are such that they do function effectively *vis-à-vis* the observed phenomena, or do not function effectively *vis-à-vis* those phenomena. For theoretical patterns to 'function effectively' *vis-à-vis* phenomena, they must provide conceptual structures which permit one to move inferentially from descriptions of one phenomenon to descriptions of other phenomena – much as a pictorial pattern permits the eye to move smoothly from one color patch to another within a larger picture.

These patches, Figs. 1.16, 1.17, and 1.18, are 'given meaning' in one way through the pictorial pattern shown in Fig. 1.17 and are 'given meaning' in quite a different way in the pictorial pattern shown in Fig. 1.18.

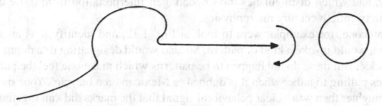

Fig. 1.16

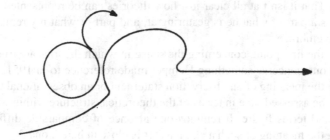

Fig. 1.17

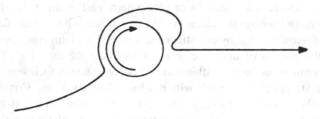

Fig. 1.18

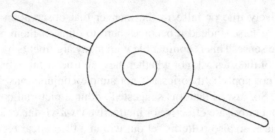

Fig. 1.19

It is the same with theories, wherein the inference-patterns of one will structure observation statements in *this* way, while the inference-patterns of another will structure them in *that* way. It was not the observed data, but the patterns which distinguished geocentrism from heliocentrism in sixteenth century astronomy, which distinguished wave theory from corpuscle theory in seventeenth century optics, which distinguished phlogistication from oxidation in eighteenth century combustion theory, which distinguished vitalism from mechanism in nineteenth century biology, and which distinguishes the Copenhagen Interpretation from those of its critics in twentieth century microphysics.

If someone, for example, were to look at Fig. 1.19, and identify it as an *x*-ray tube, that would invoke a kind of pattern. So also would designating it 'a donut with toothpicks'. But these do not happen to be patterns which make one feel the patches and lines pulling together when it is dubbed 'a Mexican on a bicycle'. Your spontaneous laughter then was a clear behavioral signal that the marks did knit together for you in a dramatic way. This is part of what Ernest Nagel had in mind, when he urged that in his view theoretical-predicates-as-variables constitutes a healthy philosophical attitude. That it isn't at all clear just how theories can be represented as true or false – this is a part of what he is gesturing at, and part of what my remarks so far are meant to embrace.

That was the first point, concerning the sense in which theories are true or false. The second one concerns something Hempel made reference to in 1951, namely – determining the meaning of an observation statement or an observational term. This cannot even be assessed save in terms of the theoretical structure within which such statements and terms figure. It reminds one at once of comparable difficulties in determining the meaning of such a claim as "It is close in here"; one does not know whether such words are uttered in a small, smoky room, or in a large hall filled with university officials, or in an embarrassing situation, or what. Similarly with "The sun rises in the east"; were this spoken by one for whom 'east' is the *name* of where the sun rises, no matter where, the claim couldn't be false. Where 'east' designates a direction, as determined e.g. by celestial coordinates, the claim could be false. One must know the context of utterance in order to know the meaning of the claim uttered. Observation statements within a theory derive much of their semantical content from the structural framework within which they do figure. Consider a pronouncement like 'red now' in an astrophysical context, wherein it may be a highly technical reference to such things as photospectrometric tabulations of the red shift and consequent confirmation or discontinuation of rarified cosmological theories. [This is not unrelated to Hoyle's recent abandonment of the Steady State Theory.] In an ophthalmological context, of course, 'red now' will be semantically charged in a different way. In a philosophical (phenomenalistic) context it will be different again. In a chemical context (involving titrations) it may be different yet again... etc. This connects with a point that Wilfrid Sellars makes concerning how it is that theories indicate why particulars fall under the empirical laws they do fall under. It seems to me remarkably analogous to understanding why it is that certain dots and shapes and lines cluster within a pictorial configuration as they do, given a 'significance pattern'. Of course, this also ties in with the view that theories, in general, stand or fall *en bloc*. Observation-statements that won't fit into the overall theoretical conception are treated as anomalous, and one's reactions to the anomaly *vis-à-vis* the theory's structure are of the greatest importance here. The advance of the perihelion of Mercury wrecked classical celestial mechanics *in toto*[9]; the non-conservation of parity did not wreck quantum mechanics. Different responses to anomalies indicate that different meanings attached to the structural principles of the theory in question.

Understanding how phenomena are sometimes felt to become comprehensible when viewed through a particular theory is somewhat analogous to the 'Gestalt click', and this is itself instantiated *via* the pictorial illustrations which support the analogy I have been shaping. It also relates to another point that Sellars has been zealous in making: to wit, that theoretical terms themselves might well be construed as meaningful in the restricted sense that they are *structurally effective*. This point pierces the classical positivistic and hypothetico-deductive positions which urge that the *real* meaning to be found in theoretical structures is all imported 'upwards' from the observational statement level (Braithwaite's semantical zip-fastener); the rest is syntactical veneer – all form and no content. But significance, and meaning, is often purely a matter of form, and not of content. One may have theoretical terms like ψ in quantum mechanics, or i in classical thermodynamics, 'structural terms' which are important for forming the framework in terms of which the observational details hang together; these terms, and operations on them, are what make that structure apparent and effective, and as such they constitute an important part of 'the meaning' dimension of these terms. Yet they are in no obvious sense connected with factual observations. The meaning of i is not 'zipped' up into it from the ostensive correlation level of observation statements. ψ is not a semantic composite of all the special coordinating definitions which strap quantum mechanics to laboratory facts. Rather, a theory and its constituent abstract terms can be considered a sort of conceptual gestalt for observation statements. There are, in the history of science, several examples of mere unstructured inference, lacking any appropriate accompanying gestalt. This was the criticism of Leibniz and Poleni and Bernoulli of Newton's *Principia* when first it appeared. They treated it as being but a mere algorithm – one which failed altogether to link up the sort of causal explanations and speculative understanding so important in their kind of natural philosophy. Leibniz regarded

---

[9] At least this is true of the assumption that classical celestial mechanics could be applied in principle to *any* astronomical phenomena, any where and any how in the universe.

Newton's theory simply as an algorithm which, almost *per accidens*, seemed incredibly effective at grinding out numbers. But this wasn't really natural philosophy in Leibniz' sense. Thus their criticisms. Parts of quantum field theory today are describable in much this same way; consider the mathematical divergencies which result immediately from the technique of renormalization. Non-Hermitean S-matrices, negative probabilities, 'ghost' states... etc. are the ghastly issue of this inelegant manner of forcing calculation even after understanding has departed. One's comprehension of the subject matter of microphysical radiation is almost to be distinguished here, from the degree to which numbers can be successfully churned out. Analytical mechanics required no stacks of statistics, successive approximations, confirmed predictions... etc., because the full understanding of classical physics was embodied within a terse, powerful, rich and elegant system of inferential patterns. Nothing like that obtains in microphysics today, and undigested statistics have taken the place of natural philosophy – which may not matter since computers have taken the place of natural philosophers. A further example of this business of drawing mere inferences *sans* any appropriate gestalt can be seen in the many aerodynamic recipes used in the late nineteenth century by aeronauts, and by physicists, in order even crudely to approximate to this incredibly multiparametric, turbulent subject matter. This is to be contrasted with theories which, at rosy moments in the history of science, are felt to be imbued with an intrinsic gestalt – a key to the intelligibility of the phenomenon in question. This was the attitude of Euler and Helmholtz towards Newtonian mechanics itself. Euler at one stage asked: "What is it to explain any phenomenon in nature? It is simply this, to 'reduce' that perplexing macrophenomenon to a mechanical analysis of its microconstituents, their energies, motions and positions". This was carrying the argument of the *Principia* all the way. To explain was to 'apply Newton'. Similarly aerodynamic theory today, following comments of von Karman and von Mises, does provide a conceptual gestalt which functions with respect to successive predictions in the way in which the conceptual gestalt functions (Fig. 1.20) with respect to the lines rendered within its subject matter (Fig. 1.21).

Many moments in the history of analytical mechanics, especially those involving planetary theory, fall within the plot just delineated. The contrast between geocentric and heliocentric theory is clearly representable in these structural terms: the facts were equally accessible to protagonist and antagonist. The *arrangement* of those facts, however, is what generates the *dramatis personae* within the history of astronomy. The arguments which terminate in an hypothesis positing the existence of some trans-Uranic object, the planet Neptune, and the structurally identical arguments which forced Leverrier to urge the existence of an intra-Mercurial planet, the

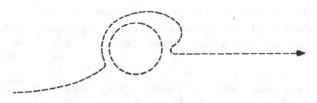

Fig. 1.20

Fig. 1.21

planet 'Vulcan', to explain the precessional aberrations of our 'innermost' Solar system neighbor – these arguments are formally one and the same. They run: (1) Newtonian mechanics is true; (2) Newtonian mechanics requires planet P to move in exactly this manner, $x, y, z,...;$ (3) but P does not move à *la* $x, y, z;$ (4) so either (a) there exists some as-yet-unobserved object, $o,$ or (b) Newtonian mechanics is false. (5) (4b) contradicts (1), so (4a) is true – there exists some as-yet-undetected body which will put everything right again between observations and theory. The variable 'o' took the value 'Neptune' in the former case; it took the value 'Vulcan' in the latter case. And these insertions constituted the zenith and the nadir of classical celestial mechanics – for Neptune *does* exist, while Vulcan does not.[10] Some of Kepler's arguments leading to his First Law are also relevant here; Tycho Brahe had the data, Kepler divined their structures. And there are many other historical instances of this recognizing-of-patterns-in-data which highlight the evolution of science.

Now Book Two of Newton's *Principia* was designed to represent a difficult and turbulent subject matter, namely, the dynamics of fluids. Newton makes the important suggestion that all undulatory behavior within a fluid subject matter can be represented in punctiform terms. He urges that the laminar flow of fluids, their turbulence, their vorticities, their currents, viscosities, resistances, eddies, densities, and so forth – all of this can be treated simply as a mathematical manifestation-in-the-large of what is fundamentally appreciable in terms of those punctiform interactions articulated in Book One of the *Principia*. By the time Newton composed his *Principia*, certain empirical facts were well known. Galileo had concerned himself with the resistance a fluid exerts against a given object, as had Leonardo before him. Both reckoned this resistance to be directly proportional to the velocity of the fluid itself, or at least to the relative velocity of the fluid as against a moving object. Huygens perceived the matter more clearly and set up a little experiment – the end result of which was his discovery that the fluid's resistance to any object moving through it was proportional to the *square* of that relative velocity. This is the figure we use today. Thus, whether it be a fish swimming through a still stream, or rapids rushing against a rock – the resistance felt by fish or rock is proportional to V^2, i.e. the square of the fluid's velocity relative to the motion of the fish, or rock, through it.

Now one of the interesting moves made in Book Two of the *Principia*, a move which had a remarkable effect on subsequent history of science, consists precisely in a pictorial type of representation (again, where 'pictorial' is in inverted commas;

---

[10] Cf. Hanson, 'Leverrier: the zenith and nadir of Newtonian mechanics', Chapter 6, this volume.

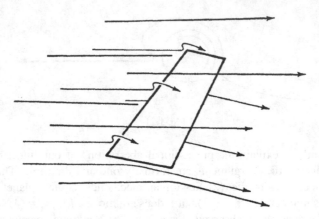

Fig. 1.22

Newton's is not a presentation of pictures in any graphic sense). His is, rather, a presentation of structures in the sense more difficult to comprehend – more difficult, but essential to our understanding of what *this* scientific theory (fluid mechanics) does by way of 'explaining' its subject matter. Newton concerns himself with the resistance that will obtain between a fluid flow which is assumed to be laminar (without internal rotations or turbulence), and a given flat plate. The latter can be thought of as being simply two dimensional (of zero thickness) for the purposes of this enquiry (Fig. 1.22). [I'll have to tilt it so we can view it from the port quarter.] Newton construes the force exerted by this fluid, or, rather, by the indefinitely large number of particles of which this fluid is constituted, upon this particular plane plate – as being proportionally related to a number of things. One of these, of course, will be the *density* of the fluid. In liquid mercury the force upon the plate will be greater, for a given relative velocity, than it will in ordinary water. Moreover, this force will vary with the *relative velocity* with which the fluid moves against the underside of the plate; when it passes slowly, the force manifested will be different from when it is moving very quickly (as every water-skier knows). This force is also a function of the *area* of this particular plane plate; a postcard will not receive the hydrodynamic shock experienced by a barn door – other things being equal.

Finally, and here is the joker within Newton's deck, the force the liquid exerts upon the plate will be a function of the *sine squared* of the angle of inclination of the plate to the liquid. That is the representation Newton gives (Fig. 1.23). In short, it should be pointed out that $F \propto \rho V^2 S sin^2 \alpha$ within the 'mechanics of perfect fluids'; this law is a simple deduction from having assumed our fluid to be *inviscid* – i.e. there are no internal interactions, or frictions, between the particles of the fluid, *irrotational* – i.e. the particles of this fluid are restricted *vis-à-vis* their degrees of freedom in motion; they do not rotate, no vortices are generable within the fluid. And finally, this ideal fluid is *incompressible* – i.e. calculations do not depend on the *springiness* or 'squeeze' of the fluid. These assumptions, although profoundly counterfactual, make the 'perfect fluid' game go beautifully. The developed algorithm of Euler and Bernoulli is a magnificent and elegant symbol-system to comprehend. The mere fact that it applies to nothing should not be taken too seriously, perhaps.

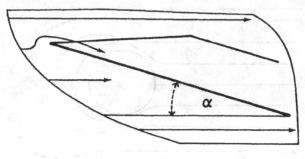

Fig. 1.23

The same is true of many 'idealized' physical theories. And, of course, *this* theory is fundamentally punctiform.

We were discussing one of the consequences of this particularity when one represents the fluid resistance on the underside of a flat plate in terms of the sine squared of the angle of incidence. This particulate model of such an ideal fluid comports well with such a sine square law of resistance. Applied to *practical* cases within fluid mechanics, however, the model and its associated 'laws' are less than wholly satisfactory. How does the air support a bird – like an albatross or even a hummingbird – which beast always displaces a volume of air which weighs but a small fraction of its own total weight? The condor seemingly soars eternally – supported upon a substance, the air, which (volume for volume) weighs less than one thousandth what the bird weighs. Were a similar situation to obtain with water and ships, the oceans' floors would be strewn with wreckage in less than 5 minutes. Archimedes' law, then, conveys little for our understanding of flight. Nor do the analyses of Leonardo and the influential Newton. Suppose, *à la* Leonardo and Newton, that the only thing that holds birds aloft is the pressure upon the underside of the wing; then a bird such as an albatross (moving at an airspeed of 15 miles on hour) – in order to achieve the value for F compatible with these other known parameters, – must angle its wings' incidence up to something like 60°! A Boeing 707, to move as it does at take off or cruising speed, would also have to tilt up its wings to about 65° – given the sine squared law! This would force calculations for the associated drag which would be totally out of the question. Birds and planes could not soar and glide as they do with wings tilted up like snow plough blades! The sine squared law requires that we'd have to minimize the angle of incidence (α) in order to keep the drag factor within the bounds of conceivability. This would necessitate enlarging the value for the wing area, S, such that the actual area of 707's wing would have to be about the size of two football fields. Either that, or boost the value of V^2, by fantastic increases in propulsive power – the result of which will be fantastic increases in gross weight. Reflections such as these forced many to conclude that, by mechanical means alone, birds *couldn't* fly. The only effective manner of calculating such parameters derived from Newton's *Principia* – this is what constituted 'analysis of natural phenomena' in the eighteenth century. But this analysis (the sine squared law in particular), was quite incompatible with what one observed in the case of bird flight. Therefore, there must be something extra-physical or extra-natural about bird

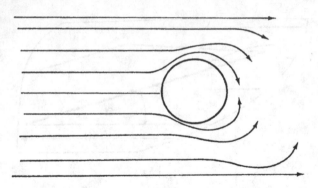

Fig. 1.24

flight – something beyond Newton, which came to mean 'beyond science'; many simply attributed this to divine intervention, occult qualities, mysterious miracles… etc. The extant literature thus makes the special problem of flight extraordinarily interesting from the point of view of the History of Science.

As *we* know, there is another natural effect primarily responsible for flight. Newton didn't realize (how could he?) that this further effect contributes to the flight of a bird at least five times more upward force, or 'lift', than is generated by way of direct impact of the fluid molecules on the underside of a wing. Newton's picture of material collisions, although not negligible in the phenomenon of flight, is minor indeed when compared with the 'suction' operative on the topside of the wing.[11] The magnitude of this lifting force is a function of the shape of the wing, but in some cases it can be 50 times greater than anything calculable *via* the classical sine squared law.

Consider the so-called Magnus Effect – so-called because a Professor Magnus, in Germany, worked with it in the early nineteenth century; it is actually stated quite clearly in Newton's *Principia* (Fig. 1.24). A cylindrical rod is here seen on end. Laminar (non-turbulent) airflow comes from left to right. The d'Alembert paradox is to the effect that on either side of the stagnation point air will flow symmetrically around the rod, curling in behind it, and impinging on the aftermost point (diametrically opposite the stagnation point). How like Aristotle and the *antiperistasis* theory articulated in Book IV of the *Physica*.[12] Newton entertains a rotation being given to the submerged cylinder or sphere itself. This rotation of the cylindrical rod, or sphere affects the liquid flow across the rod, as shown in Fig. 1.25. Indeed, the rod itself now moves across the flow, a phenomenon for which there was no expectation in 'perfect' fluid mechanics. At the topside, indeed, a further effect also operates, the Bernoulli Effect. This instructs us that, in laminar liquid flow, where velocity increases relative pressure decreases, and where velocity decreases pressure

---

[11] 'Suction' is just a quick way of designating 'pressure differential' – which moves the wing from the high pressure region below it, toward the low pressure region above. So this is, in effect, the same phenomenon Newton had in mind with his sine squared law. Only, the actual difference in pressure is many times greater than could be calculated by that law.

[12] Cf. Hanson (1965). See Part III of this volume as well.

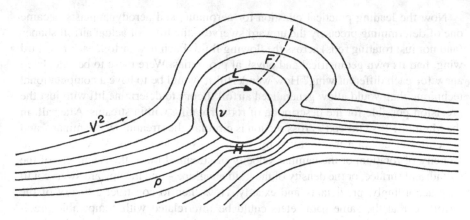

Fig. 1.25

increases; also, where pressure decreases velocity increases and where pressure increases velocity decreases. $\varrho V^2 + P$ = cons. At the rod's topside V^2 increases, so P decreases – and the rod lifts. This is what is actually operating: (1) a positive pressure beneath the rod, due to particle friction between rod surface and fluid flow and (2) a negative pressure above (due to the Bernoulli Effect). Newton notices the translation of the rod across the stream line, and he remarks the force (F) of that translation. The translation will be a function of the velocity of this fluid's motion across the rod (V_{fl}^2), and of the angular velocity of the rod's rotation (V_r^2); $V_{fl}^2 \pm V_r^2 r2 = V^2$. When fl and r have the same sense of motion their velocities are added; when their senses are opposed the rotatory velocity is subtracted from the flow velocity. The resultant variation in V^2 will determine a corresponding variation in P (since they both add up to a constant figure for a given fluid of a given density ρ). From knowledge of this variation in topside pressure one can *calculate* the degree of translation of the rotating cylindrical object across the laminar liquid flow. Increase this rotatory velocity and you increase the displacement across the laminar flow.

It became clear, with the failure of the sine square law, that some other force was operative – some force beyond the positive pressure exerted on the bottom side of an object inclined to a fluid flow. It was the ornithologists (not the physicists), who tumbled first to this conclusion. After all, *contra* the sine square law, birds could fly even with the tiny wings they have – they do! Some professional physicists appreciated the situation and joined the ornithologists *pro tem*. Thus Lord Rayleigh made a number of serious contributions to our understanding of what it was in the *actual* motion of the swan's wing which made it aerodynamically effective during take off and landing. It slowly dawned on biologists, fledgling aeronauts, and finally upon physicists, that there was *something* about the convex camber on the upper surface of the airfoil section of a swan's wing – or that of a condor, an albatross or a gull – which was definitely responsible for the dramatic lifting effect operative in, and responsible for, bird flight.

Now the leading practical question for aeronauts and aerodynamicists became one of determining precisely the upward swerve – the lift – of actual airfoil shapes (and not just rotating rods) across the flowing fluid. Each new airfoil, each new bird wing, had its own geometrical and physical properties. Were these to be calculated anew for each different wing? How wonderful it would be to have a computational technique that would allow generalized airfoil theory to determine lift with just the precision possible for the description of rods, rotating within streams. After all, in the Magnus Effect the rod's cross section is circular, the streamflow is laminar – and the entire calculation is structured by symmetry considerations. Vary the angular velocity of rotation, or the laminar velocity of stream flow, or the roughness of the cylindrical surface, or the density of the fluid envelope – and the upward swerve will vary accordingly, predictably and exactly. How then to construct a theory of the airfoil so that the same parameters could be interrelated with comparable precision – as against the early exposures to aerodynamic phenomena, wherein the physical properties of each airfoil had to be determined separately through observation?

It was the triumph of Kutta and Joukowsky to have discovered precisely such a calculational technique. Their researches paved the way for a computational aid which makes the determination of an airfoil's lift no more difficult for us today than was the determination of a rotating cylinder's upward swerve through a fluid difficult for latter day Newtonians. I will delineate this final example in some detail, since it is clearly a case of perplexing phenomena having been rendered intelligible through a theory which *shows* the structure of those phenomena in a representational way.

To review: an ideal fluid impinging broadside upon a cylindrical rod will curl around behind the rod in symmetrical fashion, turning in against the trailing edge with a force of impact exactly equal to the force of impact at the stagnation point forward. The paradoxical result would be that such a rod would experience, and exert, no effective resistance within such an ideal fluid. Now rotate the rod rapidly. Because of friction and turbulence below, as against the lack of such effects above – plus an increased fluid velocity (V^2) above (and further decreased pressure) – the rod will lift across the flowlines with a force (F) that is proportional to the density of the fluid (ϱ), its laminar velocity (V^2), and the velocity of the rod's rotation (v).

When this much was known about cylindrical rods and their rotatory behavior within non-ideal fluids, airfoil theory was *still* in a primitive experimental state. To learn how a given airfoil shape would 'lift' across laminar flowlines, one could not analyze and calculate – it was necessary to experiment and observe. To that extent, full understanding of the physical phenomenon involved was lacking.

Bird wings lift – but no one knew how or why. Rotating cylindrical rods lift – and this seemed to be fairly well understood. F. W. Lanchester had *the* physical insight within unnumbered millennia of wonderment about flight. Just as the rod's rotation is itself a kind of *circulation* within a laminar flowstream, so also an airfoil may also be a circulation – indeed, a simple deformation of the kind of circulation one sees operative in the Magnus Effect. Now, the Magnus Effect is equally well instantiated in the complete absence of a physical rod. If, across a flowstream, one can induce a vortical motion within the fluid (as, e.g., by rotating vanes on the walls of the tank

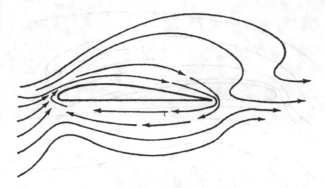

Fig. 1.26

or tunnel) that very vortex will itself swerve upwards just as calculated in the case of the rod. And the magnitude of that 'lift' across the flow will itself be determinable as a function of the fluid flow velocity, the fluid density, and the angular velocity of the vortex rotation.

Suppose now, muses Lanchester, that an airfoil – e.g. a bird's wing – is uniquely suited to induce vortical circulation within the fluid flowstream (Fig. 1.26). Suppose, that is, that the main function of a wing was not itself physically to support a flying object, but rather to induce around itself a vortex, a circulation, whose behavior within the flowstream will be in principle identical to the behavior of a vortex manifesting the Magnus Effect. On this view the first function of an airfoil shape would thus be to generate vortical circulations around itself, the properties of which circulations will differ in accordance with differences in the airfoil shape itself. Thus some airfoil shapes will decrease pressure above for slower flowstreams than others will – but they will pay for this by generating greater trailing edge turbulence at increased relative velocities. Other shapes will move and lift effectively through flowstreams of higher velocity, but will be almost like flat plates when the flowstream is more leisurely.

The primitive representation of an airfoil-induced circulation, then, would be something like in Fig. 1.27. Lanchester then argued that the tendency of the flowstream to curl beneath the trailing edge of the wing and to proceed forward – this tendency would be obliterated by increased velocity in the flowstream itself, which would 'wash' this turbulence far astern. This physical insight immediately collapses a number of striking observations together, to form one impressive pattern of 'explanation'.

What makes the airfoil *start* to lift? What is it about the *shape* of an airfoil that induces within the flow around it an effect comparable to what is generated by the rotation of a submerged rod? Lanchester argues that the air flows around a symmetrical shape in a way analogous to fluid flowing around a non-rotating cylinder. Particles of air separated at the leading edge will arrive at the trailing edge at the same time and, with but a modicum of turbulence, will proceed astern – with no observable effect on the airfoil itself. If, however, the upper surface is cambered much more radically than the lower surface, then two particles of air separated at the leading

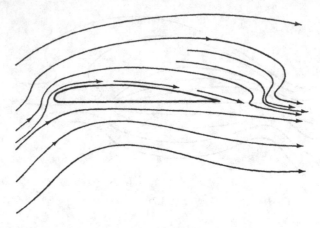

Fig. 1.27

edge will not reach the trailing edge at the same time. And, because of the slight angle of inclination of the shape to the streamflow, there will be greater positive pressure below the wing than above it, and hence the particle arriving at the trailing edge before its 'twin' which traveled over the upper surface – that lower particle will tend to move upward immediately after passing the trailing edge. Indeed, a vortex will be generated at the trailing edge of the wing, the immediate effect of which is to institute a small Magnus generator within the flowstream above. This 'starting vortex' will pull the particles of fluid across the upper surface of the airfoil much more rapidly. This will further increase the velocity of air flow above, which will in turn further decrease the pressure within the fluid above the wing. As the relative velocity of the wing through the flowstream increases, Lanchester argues, this 'starting vortex' will suffer the same fate as did the earlier particles which sought to curl around over the trailing edge and proceed forward again – i.e., they will be swept astern. But what has happened above the upper surface of the airfoil shape is, in every physical way, analogous to what happened at the upper surface of the rotating rod in the Magnus Effect example. That is, if there is *any* relative velocity of the flowstream over the airfoil, a primary circulation will be instituted (Fig. 1.28). This immediately gives way to the starting vortex astern – which, in turn, draws the fluid over the wing so as to reduce most dramatically the effective pressure above, and to increase most dramatically the difference between the effective pressure below and that above.

How then, by way of this model, does the whole aircraft (bird, airplane) fly? Lanchester's representation is as given in Fig. 1.29.

In short, by analogy with the kind of circulation one witnesses in the Magnus Effect, there is instituted around a wing an initial 'vortex loop'. Increase in relative velocity of airfoil through flowstream washes the starting vortex astern, leaving a 'vortex hoop'. So, around any aircraft moving at constant speed, in constant attitude, there is to be found a *vortex horseshoe*. As soon as any parameter is varied – that is, if forward speed is increased, or the angle of attack increased, or the density of the fluid increased – a new starting vortex appears, with its resultant vortex hoop.

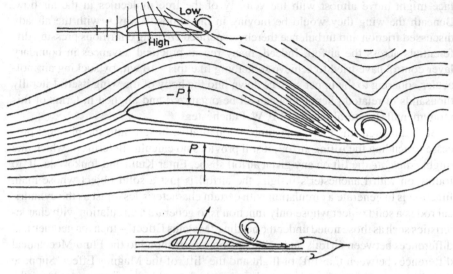

Fig. 1.28

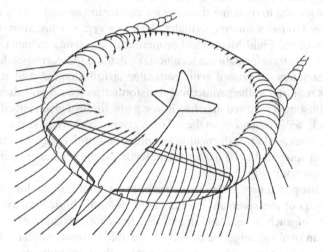

Fig. 1.29

And again, this at once gives way to the vortex horseshoe – all by close analogy with what obtains in the Magnus Effect.

Every other complex feature of aeronautical design can be at once related to this fundamental insight within aerodynamic theory. For it was Lanchester who perceived in the phenomenon of circulation the key to the ancient problem of mechanical flight. In fact, were it not technologically so formidable a thing to achieve, a wing with a surface which moved *à la* the cylindrical rod of our Magnus Effect case – would be very efficient indeed. Boundary layer problems on the topside would be drastically reduced; because the air molecules adhering to the wing surface might move almost with the velocity of the free molecules in the air flow. Beneath the wing they would be moving in the opposite way – with the already discussed friction and turbulence thereby generated creating enormous pressure differentials across the airfoil. Indeed, many recent practical advances in boundary layer control have tried to achieve something like this – to wit, by sucking air molecules adhering to the upper surface down into the body of the wing itself. Literally thousands of related phenomena can all be organized and aligned in terms of this monumental structural insight of F. W. Lanchester.

Still, although this much constitutes a qualitative insight into the physical processes attending the lifting airfoil, it still provides no calculus in terms of which one might *calculate* the lift on any given airfoil shape. Enter Kutta and Joukowski. If, to follow on with Lanchester's insight, the airfoil is just a solid object whose main function is to generate a circulation with certain characteristics – just as the cylindrical rod is a solid object whose only function is to generate a circulation with characteristics such as those noted under the heading 'Magnus Effect' – then the geometrical differences between airfoils and cylinders must correspond to the Fluid Mechanical differences between the 'lift' of flight and the 'lift' of the Magnus Effect. Suppose that there were a transformation technique which could allow one to 'reduce' the complex shapes of airfoils to the simple shapes of cylindrical rods. This would, in effect, be tantamount to reducing the complex circulation around a wing to the simple circulation around a rotating cylinder. And *vice versa*. Thus, from the simple determination of the Fluid Mechanical properties of a rotating cylinder, one could (by way of such a transformational technique) calculate the correspondingly more complex parameters associated with particular airfoil shapes. And the specific deformations required in the geometrical transformation would give the clue to the numerical adjustment required in noting how e.g. the lift on a given airfoil, at a given angle of attack, will compare with the lift on a given circulating cylinder.

The Kutta-Joukowski transformation technique is intuitively quite simple. Begin with a cylinder imagined to be made of a perfectly plastic substance. We imagine it to deform as shown in Fig. 1.30.

What has happened here is that there has been no deformation along the vertical axis, while units of measurement have doubled along the horizontal axis. And one would expect of such a resultant shape, that the characteristics of a circulation around such an oval rod might be directly calculable in some similar way. Thus, if the total bottomside turbulence of the rod were x, the total bottomside turbulence beneath the oval would be $2x$.[13]

Infinite variations are possible, of course. A circle can be deformed (according to rules) in an indefinitely large number of ways. Thus, suppose we place our original cylindrical section eccentrically upon our reference lines. And suppose that we opt for a trebling of each of the four areas now apparent, under the restriction that the boundary line remains smooth, i.e. still described by a continuous function. Such a

---

[13] Of course, the increase will not always be linear in this way. The relationship will be very complex, from a functional point of view, in most cases. But there will be *some* functional connection between the geometrical deformation, and the corresponding fluid mechanical deformation.

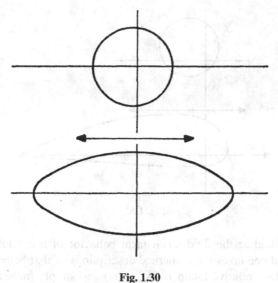

Fig. 1.30

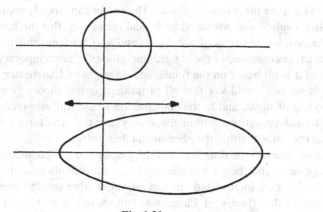

Fig. 1.31

deformation might be as shown in Fig. 1.31. Associated with this would be an identically-shaped vortical circulation, the properties of which will be to the original 'circular' circulation as is the new shape to the original cylinder.

The way is now clear for the generation of almost any shape one can envisage. Thus Fig. 1.32, and given the fundamental intuition of Lanchester to the effect that it is not the object of such a shape, but rather the fluid mechanical characteristics of a vortex circulating around such a shape – it is this which generates the specific dynamical characteristics one should be able now to correlate with any shape within actual tests.

Thus, take any bird wing. Determine the plane geometry of a representative airfoil section of that wing – and then determine to what degree this is a deformation of a corresponding circle. This will have been indirectly to have determined what will be the corresponding aerodynamical properties of this airfoil section. All one

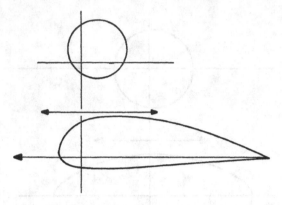

Fig. 1.32

must do is to calculate the fluid mechanical behavior of a cylindrical vortex as undistorted. Then one takes the numerical descriptions of that behavior (for given velocities, densities, relative laminarities… etc.) and simply makes distortions of these numbers analogous to the geometrical distortions that the given airfoil section constitutes *vis-à-vis* the original cylinder. The reader can already appreciate that this will require a subtle and advanced analytical technique. But the basic principle is simple. Indeed, in the hands of von Mises and von Karman, this transformational technique has become one of the algorithmic glories of contemporary aerodynamic theory. And it is all based on the fundamental insight of Lanchester. To the degree that that is so, one could say that all of modern airfoil theory is effective to the understanding of flight, and to the construction of aircraft, to precisely the degree that the formal operations within the theory are the structural analogue of the dynamical operations within the phenomena themselves.

Within no time, all the drawings and the graphs and the geometrical representations disappear – and their Cartesian algebraic equivalents take their place. Then these equivalents are interrelated through infinitely more subtle 'internal relations', well known to the Theory of Functions. But, notwithstanding the unbelievable advances in calculation and theoretical transformation made available to the physicists and aerodynamicists by this Cartesian development, this approach is effective precisely because of the *original* structural felicity of the basic representational intuition, e.g. that of Lanchester, and the *de facto* phenomenon itself.

We have come a long way round to Nirvana in this exposition of *A Picture Theory of Theory Meaning*. After our initial feinting with the picture theory of the *Tractatus* and Wisdom's *Logical Constructions*, we considered how observational data – measurement numbers, position-time event registrations, process description… etc. – were all map-able on data graphs the parametric representations on which have a structural identity with the dynamical features of the phenomena themselves. Thus the identifying structure of a physical process can be represented on a graph, just as the 'essence' of a given powerplant can be completely encapsulated on a chart or graph appropriate to it. So there is something in Section 3 of the *Tractatus* which remains not fully explored by philosophers to date. *Data charts are representations of physical processes in virtue of the fact that they have the same structure as the process itself. They are structural pictures of the dynamical reality.*

Although this much may not seem to constitute the familiar treatment of the nature of scientific theories, I submit that it is but a small step from the structural representations of data graphs, to the most sophisticated algebraic treatment of complex physical processes – complete with some insight into what we mean by 'explanations *via* theory'. For, a curve of any slope on a data graph is representable (in principle) algebraically, by way of a Cartesian transformation. 'And the essence of representation is not lost thereby'[14] (Wittgenstein 1922, 4.016). Whatever structure the curves on the data graph had, the corresponding algebraic formulations have precisely the *same* structure. Therefore, if the charts and the phenomena were related by structural identities, the algebra and the phenomena are also related in this way. Indeed, the algebra *is* the structure of the physical process.

At this point the algebraist assumes command. Any list of algebraic statements will be such that there may be an indefinitely large number of ways of interconnecting them, by inventing functions powerful enough so that one can infer from any one expression to any other one. The result, an exploration within the Theory of Functions, permits the perception of, and creation of, ever higher-order conceptual patterns. And it is within these that our paradigm examples of physical *explanation* are to be traced. This is also where scientific creativity and theoretical insight assume command. For, in general, there will be indefinitely many possible connections between the algebraicized data – from which the scientist selects.

That all this constitutes a much more plausible account of the nature of scientific theories than can be provided by the 'ideal language' approach – this is made clear within a living, contemporary discipline which is as yet a long way from being 'ideal'; its problems are largely unresolved and one's efforts can hardly henceforth be devoted to axiomatic elegance – which is the case in our present reflections concerning classical mechanics (the discipline which provides the model for the 'ideal language' approach to theory). Fluid mechanics, and its exciting contemporary offspring *aerodynamics*, is a multi-parametric nightmare. All major advances seem to have been made through the uses of the theorist's imagination; it is there that intuitions of structural interconnections within the phenomena are instantiated within a structural model which becomes no less representational after all the initial diagrams are destroyed and replaced by the most sophisticated mathematical analyses, transformations and higher-order theoretical connections.

Scientific theories enable us to understand perplexing phenomena precisely because they enable us to *see* on the page some of the same structures which are *there* in the phenomena themselves. The theory allows us to comprehend what makes things 'go' – and to work our ways into the phenomena, along the dynamical structures (as it were) by way of inferences through the algebra which itself has the

---

[14] The English translation given here appears to be Hanson's own. The full sentence Hanson paraphrases is as follows, "Und aus ihr wurde die Buchstabenschrift, ohne das Wesentliche der Abbildung zu verlieren." *–MDL.*

same structure as the phenomena – or at least a structure compatible with the phenomena.

Perhaps the best name for this function of theories *vis-à-vis* their subject matters should make no mention of 'pictures' at all. Perhaps our title might more justifiably have been *Structural Representation Theory of what Theories do*. This approach, as recommended within the foregoing, will not only let us make sense of the advancing frontier of science, with all its brow-breaking perplexities, it will also shed light on the perennial philosophical problems concerning the meaning of theories, the semantical status of theoretical terms, the interrelationship between laws and generalizations and measurements and observations-and finally, it will be *scientific understanding* itself of which we may yet have some satisfactory explanation.

References

Hanson, Norwood Russell. [1958] 2010. *Patterns of discovery: An inquiry into the conceptual foundations of science.* Cambridge: Cambridge University Press.
———. 1965. Aristotle (and others) on motion through air. *Review of Metaphysics* 19: 133–147.
Wisdom, John. 1969. *Logical constructions.* New York: Random House.
Wittgenstein, Ludwig. 1922. *Tractatus logico-philosophicus.* Trans. C.K. Ogden, with German original (*Logisch-philosophische Abhandlung*). London: Routledge and Kegan Paul Ltd.

Chapter 2
On Elementary Particle Theory

Abstract In 30 years the science of elementary particles has made few achievements compared with its unsuccessful essays. The recent works of Schwinger, Tomonaga, Feynman and Dyson, however, have had some success (Particularly with relativistic phenomena and the relativistic subtraction). We have here a hint that progress is being made on the formal side of the discipline – though even this work is profoundly disturbing in some of its purely mathematical aspects (Cf. also Schwarz (1954). In a recent lecture at Cambridge Heisenberg remarked how the technique of 'renormalization', – an important formal innovation due largely to H. Bethe, – leads to the introduction of non-Hermitean operators which ruin the unitary character of the scattering matrices. This is really a fundamental change in the theory, not simply an ingenious bit of repair-work. It is objectionable because it leads to oddities like negative probabilities and experimentally vacuous 'ghost' states). There could be no better time to review the situation from a physical and philosophical standpoint, even if this proves to be an over-ambitious undertaking.

2.1 Introduction

The accounts offered of the complexities of nature by chemistry, astronomy and the biological sciences are often quite timid, and even a little naïve. Workers in these fields appreciate the inadequacies of the pictures of the world they offer. Nonetheless these pictures can be explained to any non-specialist patient enough to learn. This can be done because studies of the structure of molecules, of the stellar universe and of living things – all of these clearly reflect, and are reflected in, some aspect of the ordinary world in which we live.

By contrast, elementary particle physics progresses from formalism to formalism. Much of this is indispensable. But it is a pity that fundamental physics seems to be leaving behind all but the most able mathematicians. And of these many seem to forget that physics is still an inductive science, directed to the exploration of the real world. In some ways elementary particle theory has been in the tradition of rational mechanics, and the symbol-games of Hertz.

A revised version of an article appearing in Scientia, March, 1956.

© Springer Nature B.V. 2020   39
N. R. Hanson, *What I Do Not Believe, and Other Essays*, Synthese Library 38,
https://doi.org/10.1007/978-94-024-1739-5_2

Worse, the formalist rarely presents his invention for what it actually is. Abstractions multiply while relatively little thought is given to our conceptions of the stuff of the physical world, (though I daresay that most mathematical physicists would say that this is just what they are doing; – that is what makes micro-physics so extremely complex).

For the experimental physicist, on the other hand, electrons (+ and −), protons, γ-rays, neutrons and mesons are things in the physical world, to be treated like anything else he has learned about from experience.[1]

2.2   The Aim of the Theory

The theory purports to describe the interactions between elementary particles. It began, with De Broglie and Schrödinger, from this conception: that the formal relation of photons to an electromagnetic radiation field is typical for all other fields and all other particles. Each sort of particle is just the quantum aspect of a corresponding continuous field. The field is described by means of a system of functions of space and time. This system forms either a scalar, a vector or a tensor, (or a pseudo-vector or pseudo-scalar), or is transformed in some other way. All the simple possibilities of form have been explored (Rosenfeld (1948), Wentzel (1949)). The connexions of the dynamical magnitudes of the field with the wave functions which specify it have also been established. (This last has proceeded according to a programme that is in effect little more than a transcription and development of the Maxwell-Lorentz treatment of the classical electromagnetic field.)

After its invention by Heisenberg, Born, Jordan, Dirac and Pauli 20 years ago, this formalism developed in several ways.

Experiments on the interaction of high energy electrons and matter made it clear that quantum electrodynamics would have to take relativistic effects into account. The Compton effect, the production of electron pairs by γ-rays, of *Bremsstrahlung* by electrons and other charged particles (not to mention the multiple processes revealed in cosmic ray research) – all these had yet to find their proper theoretical description. This was supplied by Heitler and others. They used perturbation theory to obtain formulae for 'cross-sections' of the various processes. Their method was known to be unsound mathematically (divergence difficulties were their undoing). Nevertheless, forms were found that 'meshed with' the results of many experiments.

---

[1] For an admirable expression of this attitude, see P. E. Hodgson (1954). Hodgson remarks that "… the discovery of the nature and properties of the fundamental particles is almost entirely a matter of pure experimental research. There are no comprehensive theories which could aid the experimentalist in his investigations. Practically the only theoretical laws which can be used with a high degree of confidence are such fundamental ones as the conservation of mass-energy, charge and spin" (52). This is a little hard, but it does indicate how little elementary particle theory serves the experimenter at the present time.

Recent work has succeeded in formulating a systematic and unambiguous way of deriving the probabilities of higher order processes. It has also removed other discrepancies in the theory, — discrepancies resulting e.g. from the presentation of the sources of the fields as point singularities. This recent theorizing has made it possible to allow for the coupling of the electron (in a hydrogen atom) with the electromagnetic field, even when no photons are present initially.[2] One effect of this is to change slightly the energy levels of hydrogen.[3] Indeed, the calculated value of the splitting is very close to the value obtained in the micro-wave experiments performed by Lamb and Rutherford, as well as the results of recent spectroscopic determinations. And it is very much to the credit of the theory that it explains the anomalous g-factors for electrons; in this it shows good agreement with experiment. Best of all, the theory in its present phase goes very far towards resolving discrepancies, e.g. those between the calculations of the earlier theory and the precise experimental measurements of the ratio of the resonance frequencies for nuclear induction in hydrogen and deuterium.

Thus, at least until the very recent appraisal of Dyson (1949) on the theory's validity, it appeared that a promising formulation of the quantum electrodynamics of electrons and photons had been found.

2.3   The Structure of the Theory

Initially, photons, electrons, nucleons and the meson fields must be represented kinematically. Then one proceeds to a dynamical model by assuming a form for a Hamiltonian density in the field. The energy densities of each of the fields is thus represented as such. Represented also are the interaction densities for the fields with each other, as well as with electric, magnetic and radiation fields (of which the radiation field vectors, interpreted in terms of photons, can be measured in the laboratory).[4]

Symbols for the wave functions of the fields appear as operators for the system as a whole. They represent the creation and annihilation of the particles to which they correspond. Only the interaction terms come into the calculations made in recent work. Self-energies are removed by covariant transformation (thereby satisfying relativity requirements).

A method for calculating cross-sections was introduced by Heisenberg a decade ago. Consider a system of incident waves of characteristic types: to each there corresponds an emergent wave system in which the characteristic types are excited in a way determined by the nature of the scattering, receiving or radiating antenna. A matrix made up of reflexion and coupling co-efficients represents the connexion

---

[2] Cf. also Schiff (1949, 325).

[3] Removing, e.g., the degeneracy between the $2S_{1/2}$ and $2P_{1/2}$ levels.

[4] The electric and magnetic vectors cannot be measured in the laboratory.

between the incident and emergent wave amplitudes. The quantum mechanical analogue of this wave mechanical situation is the Heisenberg S-matrix.[5]

The matrix calculation proceeds by a formal expansion in powers of either the fine structure constant (for electrodynamics), or the interaction constant (for the meson field). The divergencies of the coefficients in the series are now eliminated; the convergence of the infinite series seems unlikely however.

I cannot set out here the mathematical details of these conceptions. Nor can I do any more than mention the diagrammatic innovations of Feynman and Dyson by the use of which the complexities of computation can be ordered and greatly simplified.[6] Formally, the theory is of absorbing interest. The variety of its methods constitutes a remarkable contribution to the literature of mathematical physics.

What is to be said, however, of the physical conceptions with which this cluster of theoretical conceptions works?

2.4   The Physical Significance of the Theory

For 20 years Dirac has looked to classical physics to suggest models on which to base new inventions within quantum theory. Pauli's book (1946) exhibits the same thing. Pauli regularly translates the models of classical mechanics into quantum physical terms. There are many obvious cases of classical conceptions suggesting research which may prove important for the future of quantum theory, (perhaps 'field' is itself a case in point). What a remarkable commentary on the fruitfulness of present forms of quantum mechanics for modern physical thinking. And in precisely this respect some current views of the field theory of elementary particles are open to criticism.[7]

---

[5] This can be derived from the Hamiltonian density referred to above, although Heisenberg's aim was to evade the Hamiltonian formalism.

[6] Perhaps 'diagrammatic innovations' is not quite the right way of putting it. It may be argued that Feynman and Dyson have here a real physical idea, and not just a formal gadget.

[7] It is of course true that Feynman has constructed transition matrices in quantum electrodynamics that take into account processes of successively higher orders. E.g. one electron may interact with another by emitting a virtual photon which is absorbed by the second electron. The probability amplitude for the first electron (I) going from A to B, and the second (II) from C to D, is conceived as follows: it is the sum of the amplitude that they proceed freely (the possibility of exchange allowed for) and of the series of amplitudes corresponding (a) to interactions between the electrons and the photon fields, (b) to the creation of virtual pairs, (c) to the annihilation of created positrons with one or both of the original electrons,... and so on. The terms in this series are structured so as to exhibit the steps in each process. Thus (I) may be propagated from A to P, emit a photon, and then go on to B; while (II) is propagated from C to X, absorbs the photon (from P) and then goes on to D. The total amplitude due to this interaction is obtained by summing over all possible pairs of points P and X. Formally, this resembles the calculation of probabilities in the statistics of a stochastic process. But the quantity represented is not itself a probability – it is a probability amplitude. So that although the processes can be thought to occur as per the formalism, the propagation of probability does not follow classical statistics (because the interference of wave-mechanical

In classical statistics we deal with the combination of independent probabilities for alternative routes in a stochastic process. But in wave mechanics the probabilities are not independent. The law for their combination depends on the accumulated phase difference along alternative routes. Or, in the language of Dirac's transformation calculus, the operators of quantum mechanics are non-commutative, while those of classical mechanics are, in principle, completely commutative. This should warn us that, attractive as Feynman's graphs are as an aid in managing physical theory, they require supplementation (by e.g. concepts of the existence and motion of the particles logically connected with the substitution of probability amplitudes for probabilities).

When an alternating electromagnetic field is surveyed, a map of the field can be made. In this manner the field can be represented as a physical existent for the radiophysicist. True, his picture of the situation is not perfect. And this is not only because of limitations with regard to matters of fact, like the size of the antenna, the sensitivity of the detector... etc. The experimenter is limited in principle by the (reciprocal) relations connecting timing measures and frequency measures, or locating measures and wave number measures or in general, the Uncertainty Relations. But these limitations may indicate no more about the way the world is, than do the necessary distortions and limitations of an ordinary cartographic projection.[8] It is the representations, i.e. the notations of quantum theory, which may be deficient, but even then only to the extent to which they can depict the world. As to the existence of entities whose representation requires a formal technique of a finer scale than either sort of map (the electromagnetic field or the cartographic) can supply, no reference whatever is made, nor can it be made within most systems of present day quantum mechanics. It is with a modicum of scepticism, therefore, that we should hark to the claims of formalists regarding what does or does not exist in the world. A lunar map is not itself a moon. $\nabla \psi + \left[\dfrac{8\pi^2 m}{h^2} \right] (E - U) \psi = 0$ is not itself a moving elementary particle.

Particle fields are not even like electromagnetic fields in this respect, – they cannot be surveyed or mapped. It is the particles themselves, the colliding, scintillating entities of the laboratory, that constitute the physical reality.[9] The apparatus of fields

---

probabilities plays an essential role – and operators for the creation and annihilation of particles are represented by complex numbers).

[8] The further you move from the 'equator' on a standard Mercator global projection the more uncertain you become as to the true character of the area being mapped. This distortion is a feature of the projection, not of the Polar Regions themselves. This is, in effect, the argument of D. Bohm (1952). He reconsiders the possibility of there being 'hidden parameters' in quantum physical phenomena, our ignorance of which requires the conceptual limitations set out in the uncertainty relations. For the purposes of this article, the author also espouses this attitude. In other contexts however, he would wish seriously to quarrel with Bohm's thesis, as well as that of De Broglie, Einstein, Podolsky, Rosen and Jeffreys.

[9] There are, of course, physicists who would quarrel with this statement, Schrödinger and March, for example. But the strict continuity of these 'classical' versions of wave mechanics are quite useless for many-body problems, as is well-known. Thus, for any N interacting particles, the theoretician requires a $3N$ dimensional phase space for his calculations. The three-body problem must

is not a representation of reality on the particle scale – despite an assertion to the contrary by March (1951). This is merely a device for calculating the modes of the appearance of particles in any system endowed with a certain amount of energy. These modes are not 'classical' except in a very general way.

Classical physics does not treat of entities that can be created and annihilated: it deals with continuous motions, not successions of particulate births and deaths. This is not creation in any ordinary sense, of course. For there is a specific mechanism for creation and annihilation at the quantum level (it is operative always in pairs, conservation of change and of spin are always manifest). Only this mechanism makes this idea of creation acceptable.

Despite all this, the essentially classical idea of a particle field is threaded into the formalism of elementary particle theory, a subject which purports to treat only of what is accessible to experiments. And that the field idea is classical must be clear from its assumption of uniformity. Are not the discontinuities of quantum physics incompatible with this assumption? Does not the new physics require the field to be quantised? This importation of classical forms, this forcing of potential energy (and other representations of dynamical interaction) to depend on space and time coordinates, indicates a reliance on analogy of a remarkably naïve variety.

2.5   Some Remarks

The understanding of elementary particle phenomena requires a physical insight deeper than is shown in many theoretical papers. Recipes, e.g. for the elimination of the convergence of integrals that upset the evolution of the algorithm, belong to the engineer's handbook – not to natural philosophy. A singularity, we remember, is only a limiting conception within mathematics. What explanation can there be for such a conception entering a physical theory which purports to spring from and incorporate the 'experimental' discovery that continuity is not the law of nature on the microscale? Why not, as Dirac appears to be considering, adopt a theory that leaves no place at all for singularities?

Such questions must be met if atomic physics, – which is rather more than the mathematical theory of atomic phenomena, – is to shake free of its reliance on formalisms which, from the standpoint of natural philosophy, are indigested and perhaps indigestible. The placid contention that the mechanics of atoms transcends physical explanations of what goes on in the world, and the docile acquiescence to this contention by many physicists (who are too busy learning the latest formal

---

be worked out in a nine dimensional configuration space, which (as a literal account of a physical situation) is unintelligible to the experimenter. So the wave functions must be interpreted statistically, i.e. as giving the density of electrons within some 'classical' volume element, or the probability of finding a particular particle within such a volume element. So we agree with Born that the particle is the physical reality, as it would in any event be natural to suppose from the most cursory consideration of collision behaviour.

drills to perceive what is happening to their science) – this situation cannot continue indefinitely. For the aim of physical science is to gain an understanding of the physical universe. This platitude looms in importance against the attitude which regards any question unanswerable by some current formalism as 'mere theology' (to use an expression now well known in Cambridge). Has not the history of the calculus shown us that an algorithm is not enough? Has not the history of theories of light taught us of the consequences of forcing a mathematics based on the wrong physical idea?

We shall apparently have to endure these hard lessons again. For, so long as the algorithm supplies correct answers, it will be unfashionable to question either the possibilities of which the formalism treats, or the actualities with which experiment purports to deal. But whether or not it suits the formalist the physicist inherits a tradition dedicated to the observation of nature. He must imagine processes as actually taking place. The average theoretician does not worry in the slightest about questions of interpreting the ψ function; he leaves 'that sort of thing to the philosopher of science'. But this is the real chink in the armor, like it or not.

2.6 Conclusion

I cannot conclude without some suggestion about how this requirement of the scientific imagination might be fulfilled.

We must, of course, form a conception of the modes of the existence of elementary particles. That they should (occasionally) be locatable in space and time is essential for our saying of them that they are 'physical existents'. But how in principle this may be done is left open (Hodgson 1954). The idea of continuity must, of course, be abandoned. And we must think of the particles as persisting by successive creation and annihilation. This is not asking too much of us. The cinema picture is a quite suitable model. Such motion is discontinuous and to be treated formally as a stochastic process – remembering, of course, that the statistical algorithm is not the usual one, but that of wave mechanics with its interfering probabilities.

Why must we treat the stochastic process in this way? Fully to answer this question would require a reappraisal of our conception of physical existence. We may need to re-learn the lessons that communication theory and computing machines teach about representing the world. Suppose e.g. that Köhler's apes built a computer. Would we rely on its results just because it turned up the right answers to all the test problems fed into it? No, we should wish to understand that it gave the correct answers because the machine had the proper structure. Reliance on a method of representing physical phenomena does not depend merely on whether or not it grinds out acceptable answers. It must help us better to understand nature – the connexions between the formal picture and the physical world must stand out intelligibly. That is, given two formalisms, both equally accurate in producing answers to questions, we should still be able to ask questions about which of the two provides the greater insight into the workings of nature (questions that are not 'theology', but physics – natural philosophy).

The atomic theory, for example, explains how sub-atomic space is not like the space between molar objects. The theory does not destroy our ordinary conceptions about macrophysical objects – it explains them. Likewise a theory of elementary particles should help us to see how we and they belong to the same world.

References

Amaldi, E., C.D. Anderson, P.M.S. Blackett, W.B. Fretter, L. Leprince-Ringuet, B. Peters, C.F. Powell, G.D. Rochester, B. Rossi, and R.W. Thompson. 1954. Symbols for fundamental particles. *Nature* 173: 123.

Bethe, H.A. 1947. The electromagnetic shift of energy levels. *Physical Review* 72: 339–341.

Bohm, David. 1951. *Quantum theory*. New York: Prentice Hall.

———. 1952. A suggested interpretation of the quantum theory in terms of "hidden" variables, parts I and II. *Physical Review* 85: 166–193.

De Broglie, Louis. [1924] 1963. *Recherches sur la théorie des quanta* (thesis). Paris: Masson.

Dirac, P.A.M. 1930. *Principles of quantum mechanics*. Oxford: Clarendon Press.

Dyson, F.J. 1949. Radiation theories of Tomonaga, Schwinger, and Feynman. *Physical Review* 75: 486–502.

Feynman, R.P. 1948. Relativistic cut-off for quantum electrodynamics, parts I and II. *Physical Review* 74: 939–946. 1430–1438.

Heisenberg, W. 1927. Über den anschaulichen Inhalt der quantentheoretischen Kinematik und Mechanik. *Zeitschrift für Physik* 43: 172–198.

Heitler, Walter. 1944. *Quantum theory of radiation*. 2nd ed. Oxford: Oxford University Press.

Hodgson, P.E. 1954. Fundamental particles: The present position. *Science News* 31: 49–68.

Kanesawa, Suteo, and Sin-Itiro Tomonaga. 1948. On a relativistically invariant formulation of the quantum theory of wave fields. V: Case of interacting electromagnetic and meson fields. *Progress of Theoretical Physics* 3 (1): 1–13.

Koba, Zirô, Takao Tati, and Sin-Itiro Tomonaga. 1947a. On a relativistically invariant formulation of the quantum theory of wave fields. II: Case of interacting electromagnetic and electron fields. *Progress of Theoretical Physics* 2 (3): 101–116.

———. 1947b. On a relativistically invariant formulation of the quantum theory of wave fields. III: Case of interacting electromagnetic and electron fields. *Progress of Theoretical Physics* 2 (4): 198–208.

March, Arthur. 1951. *Quantum mechanics of particles and wave fields*. New York: Wiley.

Pauli, Wolfgang. 1946. *Meson theory of nuclear forces*. New York: Interscience Publishers.

Powell, C.F. 1954. Excited nucleons. *Nature* 173: 469–471.

Rosenfeld, Leon. 1948. *Nuclear forces*. New York: Interscience Publishers.

Schiff, Leonard. 1949. *Quantum mechanics*. New York: McGraw-Hill Book Co.

Schrödinger, Erwin. 1927. *Abhandlungen zur Wellenmechanik*. Leipzig: Barth.

Schwartz, Laurent. 1954. Sur l'impossibilité de la multiplication des distributions. *Comptes Rendus de L'Academie des Sciences* 239: 847–848.

Schwinger, Julian. 1948. On quantum-electrodynamics and the magnetic moment of the electron. *Physical Review* 73: 416–417.

Tomonaga, Sin-Itiro. 1946. On a relativistically invariant formulation of the quantum theory of wave fields. *Progress of Theoretical Physics* 1 (2): 27–42.

———. 1947. On the effect of the field reactions on the interaction of mesotrons and nuclear particles. III. *Progress of Theoretical Physics* 2 (1): 6–24.

————. 1948. On infinite field reactions in quantum field theory. *Physical Review* 74: 224–225.

Van Hove, Léon. 1951. Sur le problème des relations entre les transformations unitaires de la mécanique quantique et les transformations canoniques de la mécanique classique. *Bulletins de l'Académie royale des sciences, des lettres et des beaux-arts de Belgique. Classe des sciences* 37: 610–620.

————. 1952. Les difficultés de divergences pour un modèle particulier de champ quantifié. *Physica* 18: 145–159.

Wentzel, Gregor. 1949. *Quantum Theory of Fields*. Trans. C. Houtermans and J. M. Jauch. New York: Interscience Publishers.

Chapter 3
Some Philosophical Aspects
of Contemporary Cosmologies

A subtitle for this paper might have been 'Creation, Conservation and the Cloak of Night' – for it is these semantical knots in the reticulum of contemporary cosmology which I shall seek to unravel, or at least to re-knot. Clearly, much of the fabric of scientific cosmology is rent and torn by sharp breaks in exposition and by jagged misunderstandings of philosophical principle. I may come to grief on these same reefs, but the course must be sailed.

I am going to make an *assumption* – something no philosopher or cosmologist should undertake lightly – to the effect that the reader knows a fair-to-middling share about modern astrophysics and cosmology, or at least about their more stable subject matters. For *my* problem of exposition is that I cannot both set out the structure of so much that has been written so well by Gamow, Hubble, Shapley, Baade, Hoyle, Bondi, Gold, McCrea, Milne, Whitrow, and many others, *and* also snipe away at certain weak theoretical junctions and conceptual targets within that structure. My self-set task in this paper concerns only just such a sniping expedition, to which I now turn gingerly, addressing the title caption 'The Creation of All There Is'.

3.1 The Creation of All There Is

References to "creation" bulk large in contemporary Cosmological theories. For "Big Bang" cosmologists the creation of the universe consisted in some "Primeval Atom" having reached a state of infinite internal density. The present state of our universe is thus but one moment in the explosive expansion out of that initial "Atom". The Steady State cosmologists, on the other hand, argue for Continuous Creation – through the auspices of which the background material within our observable universe (whose diameter is about four billion light-years) is maintained

© Springer Nature B.V. 2020
N. R. Hanson, *What I Do Not Believe, and Other Essays*, Synthese Library 38,
https://doi.org/10.1007/978-94-024-1739-5_3

at a constant and uniform level. Indeed, these ideas (the Big Bang and the Steady State) are polar concepts on a semantical spectrum concerned with "the physical problem of the creation of the universe." But when Big Bangers and Continuous Creators argue with each other about creation, they are too often at semantical cross purposes. Thus Gamow discusses "fundamental questions, such as whether or not our universe had a beginning in time" (Gamow 1952, vii), without any apparent awareness that there are semantical difficulties and conceptual perplexities in the very idea of the universe *in toto* having had a beginning in time at all – as against the myriad processes within the universe which certainly have had such beginnings.[1] There are deep differences between questions concerned with empirical temporal beginnings – e.g., whether some positron-negatron pair 'materialized' in a cosmic ray bombardment at time *t* – and questions of a quite different kind, such as that concerned with "the creation of all there is".

Questions about the moment of a 'pair creation', the age of our galaxy, or of our sun – or our earth – these make good sense within the appropriate, scientifically respectable empirical contexts. Furthermore, our present factual information about recession velocities, rates of material condensation, relative hydrogen densities, and stellar spectral compositions – these are obviously germane to questions about the *de facto* age of the universe; at least they can refute estimates which make the universe too young. But these may relate only to what should be called "the current phase of universal evolution". What went on *before* this phase began? (Do you even understand the question?) What can we reliably say of the "*pre*-galactic past"? Gamow answers

...we have no information about that era, which could have lasted from minus infinity of time to about three billion years ago... (137)

Well, if we have no information about that era, then it *could* have been just an earlier stage in a beginningless, yet ever-continuing universal cycle, within which our present configuration and expansion constitutes but one further sinusoidal "phase" or periodic pulse. And what in the world does "the minus infinity of time" really mean? Indeed, when it comes to that, what does "the temporal origin of the universe" mean? Surely, we must be able to answer such questions in a semantically satisfying manner (i.e., intelligibly) before quantitative determinations can be assessed?

No *unique* semantical or conceptual perplexities attend questions about the age of our earth, or of our sun, or of our galaxy. Radiogenic lead isotope ratios, spectral analyses of the intrasolar distribution of the elements, determinations of our galaxy's population of white dwarfs – these constitute evidence germane to answering such "when, and how long" questions. Similarly, rates of the dispersion of stellar clusters, descriptions of the colligation of interstellar dust, calculations of energy distributions within our galaxy – these also constitute data relevant to queries like "could the universe be only one million years old?", "... only ten million years old?", "... only 50 million years old?", etc. These stabs at "age guessing" are shown

---

[1] Cf. Scriven (1954).

by the available evidence to underestimate the antiquity of our universe. But what questions lie beyond?

Indeed, Big Bangers and Continuous Creators all agree that the evidence suggests our universe, in its very early youth, was considerably different in constitution and appearance from what it is now. Yet most Big Bang speculations about "the creation" consist in "backward extrapolations" within which our exploding universe is theoretically imagined to implode through 'reverse time', as it were, collapsing onto an initial point – a "Primeval Atom" of infinitely great density. But this Disneyoid picture presupposes that all intra-universal processes are constant during the entire extrapolation and the corresponding theoretical intrapolation; this assumption is incompatible with the general recognition just cited (of Big Bangers and Continuous Creators alike) that the universe must at one time have been different from what it is now. When the fact that nuclear forces are greater than gravitational forces made a *cosmological* difference, and when temperatures and pressures obtained for which there is not even an analog anywhere in our present understanding of physical processes – is it not conceivable that our present conceptions of extrapolated processes might not have applied? Can we even be sure that the "pre-expansion" stages of our universe were also themselves expansive? Contractive? Steady? What can we be sure about in this context?

In short, were the universe *very* different just after its "initial moment" from what it is now, its constituent processes would also have been very different from anything we now know. The physics of these processes might also have been vastly different from anything we know now (*if* it is meaningful even to suggest such a thing). But if the physics of our very young universe differed too markedly from universal physics now, there can be no secure intellectual basis for conceptually "rewinding" the universe back to its birthday, a debating point which pierces Continuous Creation as well as the Big Bang approach.

Because, if we don't know what we are theoretically winding the universe back *through*, we cannot possibly comprehend what we are winding it back *to*!

Thus the question "when did the present expansion begin?" has a profoundly different logic from "when did the universe begin?" – assuming this latter to be physically significant at all. It is semantically acceptable to suggest that universal expansion began 10^{10} years before its present state came to obtain – and 10^{10} years *after* it lay quiescent. We have no particularly good reason, on the present evidence, for saying anything like this. But neither have we good reason for denying it! We are in no position to have adopted an intellectual stance on the matter at all! As Gamow said "we have no information about that era".

From all of which it must appear that the analogies between Theological discussion and Cosmological discussion cut deeply against the latter – particularly where 'creation' talk is invoked. Just as medieval theologians, following Aristotle, perceived that it made perfectly good sense to suppose that the universe never had a beginning in time at all – and that the logic of talk about particular creative processes was quite unlike the logic of talk about the creation of *everything there is* – so similarly, contemporary cosmologists should not use expressions like "the creation of the universe" as if they 'unpacked' just like expressions concerned with the creation of processes *within* the universe. Remember, 'creation' is a time-dependent word, one which we understand in terms of considerations like "How long did Beethoven's creation of the Ninth Symphony take?" and "When was Einstein half through creating the Theory of General Relativity?" The creation of the universe is, in any physically intelligible context, tantamount to the creation of Time – since in the absence of physical processes there is (simply and dogmatically) no such thing as Time. And how long did *that* universal creation take? When was the creation of Time half finished? The questions are completely senseless. Yet they are the semantical legacy of careless 'Creation' talk. *Ergo*, Cosmologists no less than Theologians, must be doubly wary of references to 'The Creation of All There Is' – if only for no better reason than that scholars whose business is the analysis of *meaning*, and its sundry perplexities, will not be able to understand such references if they are left completely unguarded and unqualified.

Let us turn now to the second of my self-set targets. I have called it: "Conservation Beyond the Edge".

3.2   Conservation 'Beyond the Edge'

It is a simple consequence of all this that the total amount of energy that can be observed at any one time must be equal to the amount observed at any other time. This means that energy is conserved. So continuous creation does not lead to non-conservation of energy as one or two critics have suggested. The reverse is the case for without continuous creation the total energy observed must decrease with time. – Fred Hoyle, *The Nature of the Universe*

The argument here turns on the fact that the 'edge' of the observable universe must be set (empirically) at the critical distance of 2,000,000,000 light years. This is twice as far as the distances expected to be reached through observations with the giant 200-inch telescope at Mount Palomar. Beyond that 'edge' the light emitted towards us by receding galaxies "neither gains ground nor loses". The velocity of the recessions of these galaxies is beginning to get so close to that of light itself that no signal sent from such fleeing sources, from such a distance, can *ever* reach us here on earth. Hence, the "knowable" universe all lies on 'this side' of the two billion light-year boundary. One can theoretically extrapolate beyond, of course. But no physical meaning whatever can attach to such extrapolations, since they are in no way vulnerable to empirical disconfirmation. No signals relevant to such extrapolations can reach us as a matter of principle; the extended reflections, therefore, can be neither confirmed nor denied.

On the "Big Bang" theories, the galactic population of the "knowable" universe would steadily diminish, as is well known. Since, as the Hubble-Humason "red shift" (1929–1931) indicates, all extra-galactic objects are moving away from us at incredible velocities, the furthest known and yet-knowable objects (just on this side of 'the edge') will shortly go beyond the "2 billion light-year limit" – and hence disappear into the 'forever unknowable'.

As Hoyle's remark signals, he regards such a dynamic state of affairs as a threat to the principle of the conservation of energy. The "Continuous Creation" theory, however, of which Hoyle is a provocative spokesman, seems to uphold the conservation principle. "So continuous creation does not lead to non-conservation of energy…", "… without continuous creation the total energy observed must decrease with time".

This conclusion is not wholly warranted. Hoyle does not deny that the "red shift" reveals galaxies continually to be disappearing into the great unknown beyond the epistemic 'edge'; this is as recognizable and incontrovertible a datum for "Continuous Creation" theories as it is for the "Big Bang" theories. So, the same phenomenon – universal expansion – is accepted by both of these rival theories as an initial condition. However, according to Hoyle this single datum constitutes a non-conservation of energy for the "Big Bang" theories. Whereas it apparently poses no problem at all for a "Continuous Creation" theory, which presupposes that H atoms are being continually created *ex nihilo* as background material.

This last point has caused considerable consternation within discussions of the New Cosmology. But Hoyle, Bondi, and Gold are certainly correct in arguing that this 'continuous creation' assumption is on exactly the same logical footing as the seldom-stated, but much more orthodox, presupposition that *all* matter was "originally created" in the initial "Big Bang". When placed side by side, both commitments make the brain reel. But that does not matter. What does matter is whether, and how much, such higher order assumptions are useful in generating lower order observation statements about, e.g., the universal prevalence of H and its relative sparsity here on earth, the darkness of moonless nights (as against the 'glow' Olbers showed we should expect); indeed, the red shift itself… etc. So it is not the 'top of the page' starting points of cosmological theories which should detain us, because of their semantical novelties or even their incomprehensibility, but rather the inferential techniques from there 'downwards' inside the theory; until at last the observation statements tumble out at the 'bottom of the page'. Hoyle's references to energy conservation constitute just such a 'top of the page' commitment: What does it do for us further down in the argument? What does the Big Bang theory with *its* version of 'conservation' do down there? Continuous creation or the instantaneous creation of all there is – these are both unexamined anchors for cosmological theories. So what? What is *secured* by these strange links and hooks?

Hoyle's argument turns almost on a "play of words". What his critics had in mind when *they* assailed the failure of 'New Cosmologists' to conserve energy led directly to their rejection of creation of hydrogen *ex nihilo* – *the* fundamental tenet of the Hoyle-Bondi-Gold cosmology. Generating something out of nothing is as much a cause for perplexity as would be the production of nothing out of something – the standard worry of Conservationists. The Big Bangers generate nothing out of something. The Continuous Creators generate something out of nothing. Conservative questions arise in both contexts: *'ex nihilo' et 'ad nihilo'; ergo cogita.* Hoyle's argument does not counter this uneasiness. Rather, he compounds it; he refers exclusively to the properties of the "observable universe" – that world enclosed within the 'surface' of a sphere-of-observation whose diameter is four billion light years – with

our galaxy as 'the center' (a remarkably 'pre-Copernican' attitude, but wholly justifiable as the *meaning* of "the observable universe"). Hoyle remarks what is certainly true: that on the "Big Bang" theory the total energy within this
knowable-observable universe is steadily decreasing. Whereas, on the "Steady
State-Continuous Creation" view, the aggregate energy level within the knowable-
observable universe remains constant, creating within as much as flees without.
Thus, given conservation as an operational necessity in Cosmological theory, the
red shift forces the Steady State theorist to keep the energy level within the observable universe *constant* in the only way imaginable, by continuous creation. Whereas
the operational impermissibility of Continuous Creation is itself what makes the
Big Banger grant an energy falloff within the observable universe (thanks to the red
shift) only to be finally 'conserved' in the 'great Beyond'.

But how can *this* move of Hoyle's be relevant to consternation concerning conservation? Whatever critics of the New Cosmology were upset about in the conception of Continuous Creation, they should still be worried about it, and are, after
having heard Hoyle. Creation of hydrogen *ex nihilo* wrinkles any smoothly orthodox picture. And Hoyle's substitution of "the knowable universe" for (the more
usual) "universe", while operationally respectable, nonetheless constitutes a semantical shift of some magnitude. The orthodox Big Bang theorists could easily counter
by saying that all the energy there ever was, or has been, or will be, is the result of
universal expansion from the initial "primeval atom" of infinite density. (The usual,
but again quite 'metaphysical' presupposition.) Our *observations* of much of this
matter-and-energy-in motion are limited by the galactic population having expanded
beyond our capacity any longer to observe – namely, two billion light years. Hoyle
does not deny that matter-and-energy phenomena are continually fading into the
unobservable distances beyond the reach of our finest instrumental probes and
detectors. But if *that* is what he calls "non-conservation of energy" then the
Continuous Creation theory is just as vulnerable as the Big Bang theory at this
point. *Both* theories accede to the fact of super-observable recessions. If this is a
failure of conservation in one case, it must *also* be that in the other – whatever *additional* moves may be made to keep things at a 'steady state'. The red shift cannot
constitute a failure of conservation for Big Bangers and not for Continuous Creators,
however many theoretical appendages one may hook on to the initial description.
What's sauce for the one must garnish the other as well.

Hoyle exploits the semantical question. He argues, as a good "Operationalist",
that it is senseless to refer to matter-and-energy phenomena beyond the limits of all
possible observations; all one can significantly discuss are those happenings within
our sphere of observation. What has slipped beyond is for us like what color discrimination tests would be for the blind; we *are* blind to what we can never observe,
and our discussions of such are thus operationally meaningless – quite beyond verification or falsification. Within that sphere of observation, however, given all the
matter-and-energy resident therein at this moment, some of it will assuredly soon
recede out beyond the two billion light-year edge. This is just what Hoyle designates as "non-conservation of energy" in the Big Bang case. But his own Continuous
Creation account is, so far, absolutely identical to the Big Bang story. The difference

shows only when Hoyle patches his operational version of the conservation principle with the additional hypothesis of the continual creation of the background material. But now *this* is exactly what his earlier critics had singled out as non-conservation!
Thus Hoyle harried the Big Bangers because the energy level in their *observable*
universe is continually falling – as we infer from the red shift. And the Big Bangers
badger the Continuous Creators because of their desire to maintain a constant
energy level in the observable universe – by way of a creation *ex nihilo*. A falling
energy level spells failure of conservation to the Steady Staters. Creation *ex nihilo*
spells failure of conservation to the Big Bangers.

Clearly, these arguments are at cross purposes. Different concepts and different
inferences are being used by Big Bangers and Continuous Creators, *as if* they generated conclusions which were not only relevant to each other, but actually contradictory to each other. This is not so. The Big Bangers suppose there to be a total sum
of energy in the universe – all resulting from the incomprehensibly explosive expansion of the Primeval Atom; our present observations of that still-continuing expansion are limited in principle by the two billion light year limit. The matter-and-energy
itself, however, *continues* expanding to infinity, even though passing beyond our
powers of observation – thus runs the Big Banger's argument. After all, nature does
not stop where our telescopes do. That is how the Big Bangers imagine energy to be
conserved: the initial store of it remains undiminished. But the recession of *observabilia* ensures that we are increasingly cognizant of less and less. And quite a traditional view this is.

Hoyle, as we saw, finds all this operationally indiscrete (as in some sense it certainly is). To talk thus of phenomena beyond our powers of observation, especially
when these are invoked as 'hidden parameters' to save abstract theoretical commitments, could soon lead to gibberish. Hoyle thus restricts the physical sense of the
expression "the universe" to that of "the observable universe". Rather than conserve
matter-and-energy as do the Big Bangers, by imagining it mushrooming through yet
further unknowable spaces, Hoyle (the Operationalist) chooses to keep all of his
explicantia on *this* side of the 'edge'! He will not even talk about energy conservation in the unknown beyond; rather, he conserves it here, where it matters observationally, by having new matter-and-energy created continually within the sphere of
observation.

So Continuous Creators and Big Bangers agree that matter-and-energy expands
beyond observable limits. But while Big Bangers conserve energy 'out there'
beyond the fringe, Hoyle denies meaning to such talk. Hoyle conserves energy *via*
continuous creation 'in here'. The Big Bangers counter by denying any meaning
to *that*!

Moreover, as all New Cosmologists admit, the *process* of the continuous creation
of hydrogen is itself unobservable in principle: just as conservation 'beyond the
fringe' is unobservable in principle.

It is thus difficult to see any logical advantage, one way or the other, attaching
to this small detail of argument within the more comprehensive cosmological
position advanced elsewhere by Hoyle, Bondi, and Gold. At this dialectical juncture no triumph and no loss can be scored by a Conservation Corps. In fact, philo-

sophical confusion is the only result of raising the issue at all. Because the Big Bangers and the Continuous Creators mean different things by "universe", by "conservation of energy", and by "physically meaningful claim". This cannot but lead to a rather benighted dialectical exchange. Which brings me to the third of this trinity of philosophical perplexities about contemporary cosmology – "Why Is It Dark at Night?"

3.3   Why Is It Dark at Night?

Newton once argued as follows:

> The light of ye Sun is about [9 • 10^8] times greater in our region then that of one of ye brightest of ye fixt stars called stars of ye first magnitude. And if we were twice as far from ye Sun his light would be four times, if thrice as far it would be nine times less, if four or five or six or 7 times as far it would be 16, or 25, or 36 or 49 times less & if we were [3 • 10^4] times as far his light would be [9 • 10^8] times less & by consequence equal to that of a star of ye first magnitude.[2] So then the nearest fixt stars are about [3 • 10^4] times further from us than the Sun. And so far as ye nearest fixt stars are from our Sun, so far we may account ye fixt stars distant from one another. Yet this is to be understood with some liberty of recconning. For we are not to account all the fixt stars exactly equal to one another, nor placed at distances exactly equal nor all regions of the heavens equally replenished with them.
>
> For some parts of the heavens are more replenished with fixt stars then as the constellation of Orion with greater or nearer stars & the milky way with smaller or remoter ones. For ye milky way being viewed through a good Telescope appears very full of very small fixt stars & is nothing else then ye confused light of these stars. And so ye fixt clouds & cloudy stars are nothing else then heaps of stars so small & close together that without a Telescope they are not seen appart, but appear blended together like a cloud.
>
> Were all the fixt stars equal & placed at equal distances from one another, the number of the stars next about us would be 12 or 13, those next about them 50, those next about them 110, those next about them 200 [those next about them 300, those next about them 450] or thereabouts. And tho their magnitudes & distances be not equal yet this affords ye true reason why the smallest stars are the most numerous. For there are about 15 stars of ye first magnitude 50 or 60 of ye second, 200 of ye 3d, 300 or 400 of ye 4th.[3] (Newton et al. 1962, 375, 376; MS. Add. 4005, fols. 21–22[4])

In this singular passage Newton notes two things: (1) that in increasing spherical volume elements of evenly distributed stars the stellar populations will increase as r^2, and (2) that the light from our sun and all other stars will decrease in intensity as $1/r^2$. On this latter matter, Newton's knowledge was quite general:

---

[2] *Radiation falloff* = $1/r^2$.

[3] Thus, at radius 2, population = 12×2^2 or *48*; at radius 3, population = 12×3^2 or *108*; at $r = 4$ it is 12×4^2 or *192*: thus when $r = n$, the stellar population at r is $12 \times n^2$ – *thus population increases as* r^2.

[4] Since Hanson had studied the Portsmouth collection during his years at Cambridge, he lists the manuscript references alongside the page numbers from the Hall volume, which was published after Hanson had moved to Indiana. –*MDL*

...And since in removing from a lucid body the light thereof decreases in a duplicate proportion of the distance,...(381)

Indeed, that light intensity falls off as the inverse square of the source's distance
from us, was already well known in the optical works of Kepler. The volumetric
increase in stellar populations, however, is an insight due entirely to Newton. This
idea was developed by Halley and Herschel into a cornerstone of modern astrophysics and cosmology.

Thus, Newton was well aware that, if we do live in a uniform universe, then the
number of luminous objects at distance r goes up like r^2, and aware also of the fact
that the intensity of light received from each of them goes down like $1/r^2$. Now,
compare these brilliant Newtonian observations with Hermann Bondi's summary
presentation of Olbers' Paradox:

... I have had a bee in my bonnet about the darkness of the night sky – that the simple fact
that it is dark at night seems to me to give a considerable clue to the structure of the universe. Many of you will be familiar with Olbers' old argument in which you simply suggest
that, if we do live in a uniform universe, then the number of objects at distance r goes up
like r^2, the intensity of light received from each of them goes down like $1/r^2$, and therefore
from every thin spherical shell we get the same amount of light, which should add up to an
infinite amount, or, if we are rather more cautious, to a very large finite amount. The main
way out, nowadays, is to ascribe the actual result, the darkness of the sky, to the expansion
of the universe; the distant sources move away from us at a speed so high that it materially
diminishes the intensity of light that we receive from them, so that instead of diverging, or
converging to a very large sum, the amount of light from distant matter in fact converges to
a very small sum. (Bondi 1962, 135-136)[5]

From all this one can perceive 'The Cosmological Problem' lurking in *The
Principia*, as revealed by Olbers, and engaged by contemporary cosmology and
astrophysics. Let us probe a little further into the historical background of Olbers'
work – to see more clearly what his revelation consisted in.

As in all History of Science, the very first stroke of such a probe uncovers the
orthodox, received story as untrue. As the astronomer Otto Struve (1963) has
pointed out, Olbers' Problem had been formulated 80 years earlier by the Swiss
astronomer Philippe Loys de Chéseaux (1744).[6] In the 1881 edition of the
Encyclopaedia Britannica, P. G. Tait wrote:

Chéseaux and Olbers endeavoured to show that because the sky is not all over as bright as
the sun, there is absorption in interstellar space...an idea ingeniously developed by Struve.[7]

---

[5] The Halley Lecture at Oxford University, May 16, 1962.

[6] Bode's *Jahrbuch* was an ephemeris that marked out in advance the positions of celestial bodies
for the year in question. The *Jahrbuch* for 1826 was published in 1823, and it contained Olbers
(1823). Chéseaux's work was based on comet observations from December 1743. Hence, Hanson's
remark that Olbers' problem had been formulated 80 years before Olbers' article. –*MDL*

[7] The quote is taken from Struve (1963, 141), who slightly alters Tait's original wording. The
Struve referred to by Tait is Friedrich Georg Wilhelm von Struve (1793–1864). Tait's entry is to be
found in the ninth ed., s.v. "light." –*MDL*

But Tait took the paradox as evidence for a finite universe, a fact noted by Struve. In the appendix to Chéseaux' 1744 memoir we find the title 'On The Force of Light, Its Propagation Through the Ether, And On the Distance of the Fixed Stars'. The discussion resembles Olbers' in every significant way. To account for the darkness of the night sky, Chéseaux assumed the existence of an 'interstellar fluid' of unknown composition, whose transparency is 3.3×10^{17} times greater than that of water – a computation not too difficult, given both $1/r^2$ and the darkness of night.

Chéseaux compared the brightness of Mars and Saturn (near opposition) with those of first magnitude stars. Assuming that these stars were of the same intrinsic brightness as the sun, he estimated that they were about 240,000 astronomical units distant. The apparent diameter of such a star would then be 1/125 of a second of arc, *if* it were the same size as the sun. All the first magnitude stars together would occupy only a tiny fraction of the entire celestial sphere. A second set of stars, appearing a quarter as bright, would be twice as far, and each of them 1/250 second in diameter. But since there would be four times as many of them, they would occupy just as much sky area as the disks of the first magnitude stars. Intensity fall-off is always equal to stellar increase – so each new spherical shell must add continually to the celestial brightness.

By including stars at greater and greater distances, Chéseaux concluded that the universe of stars need not even be infinite for the sky to be as bright as the sun. If the fixed stars were uniformly spread in space out to a *distance equal to the cube of 760,000,000,000 astronomical units*, every point on the celestial sphere would be covered by a star disk, and we would receive from the hemisphere of the sky above the horizon nearly 100,000 times as much light and heat as the sun gives us.

Olbers uses virtually the same argument to place the first magnitude stars about 350,000 astronomical units distant, so that Procyon, for example, would have a parallax of 0.6 of a second of arc. The diameter of a typical first magnitude star would then be 0.005 second.

To explain the darkness of the night sky, Olbers assumes that

space is transparent only to this extent, that of 800 rays emitted by Sirius 799 reach to a distance equal to our distance from Sirius. (Olbers 1826, 146)[8]

This, incidentally, is another 'top of the page' assumption determined by the observations cited at the 'bottom of the page' – in this case the darkness of night. In this, Olbers' words are on a par with the Primeval Atom's Big Bang for he who wishes to 'explain' the red shift, and on a par also with the assumption of Continuous Creation for he who wishes to conserve energy in the Observable Universe. Olbers then computes that stars 554 times as remote as Sirius lose half their light by interstellar absorption before we see them. At 5500 times as far, the surface brightness is reduced to 0.001 that of Sirius. Olbers concludes that stars more than 30,000 times the distance of the Dog Star do not make any contribution at all to the brightness of the night sky.

---

[8] The page reference given here is to the English translation of 1826; however, Struve seems to have supplied his own translation (1963, 142), and that is what Hanson actually quotes. –*MDL*

Of course, the hypothesis of interstellar absorption completely fails to resolve the paradox. Because if all that much radiation were being stopped in absorptive material of any kind the latter would soon be in a state of boiling incandescence-irradiating as much at us as it stopped from coming to us from beyond. Bondi beautifully analysed Olbers' assumptions, including such implicit ones as the neglect of the motions of very remote stars or galaxies. The amount of light we receive from an object that is, say, 500 million light years distant does not depend upon the present luminosity of the object, but on its luminosity 500 million years ago, when the rays we now see were being emitted.

So Olbers' unspoken assumptions can be and must be restated as follows:

(1) The average density of stars and their average luminosity do not vary through-out space. (Cosmological Principle.)
(2) The same quantities do not vary with time. [Perfect Cosmological Principle.]
(3) There are no large systematic movements of the stars.
(4) Space is Euclidean.
(5) The known laws of physics apply.

One of these premises must be incorrect, *if* we are both to reason validly on them and yet also avoid the factually false conclusion that the night sky is bright. Bondi suggests that (2) or (3), or both, may have to be dropped. If we drop (3), and accept the red shifts of the distant galaxies as detected by Hubble and Humason, then we can account for the darkness of the night sky without invoking interstellar absorption. For if distant galaxies are receding rapidly, the energy content of the radiation we receive from them is proportionately reduced.

Bondi's summary of his rather elaborate discussion is worth reading:

> This little argument may well serve as a prototype of scientific arguments. We start with a theory, the set of assumptions that Olbers made. We have deduced from them by a logical argument consequences that are susceptible to observation, namely, the brightness of the sky. We have found that the forecasts of the theory do not agree with observation, and thus the assumptions on which the theory is based must be wrong. We know, as a result of Olbers' work, that whatever may be going on in the depths of the universe, they cannot be constructed in accord with his assumptions. By this method of empirical disproof, we have discovered something about the universe, and so have made cosmology a science.... Thus the darkness of the night sky, the most obvious of all astronomical observations, leads us almost directly to the expansion of the universe, this remarkable and outstanding phenomenon discovered by modern astronomy. (quoted in Struve 1963, 142)[9]

Thus, almost any perplexity within contemporary cosmology turns out to have profound philosophical consequences. This is because of cosmology's concern for totalities: *all* there is, the *beginning* of it all, the *end* of it all. And the discussion of totalities often fares as do discussions of *limits:* One should never accord to the limit of a series properties appropriate only to members of that series. Infinity is not divis-

---

[9] Hanson's discussion closely follows Struve's, and he clearly got the Bondi reference from Struve. Struve gives no source for this quotation from Bondi, and it is not to be found in Bondi (1962). –MDL

ible by two in the way that any natural number is – although the series of natural numbers terminates in infinity. Mankind is not breathing, although many men are. The universe did not *begin* in just the way that its constituent processes began. *Its* total energy is not investigable as is that of its individual processes. Indeed, many of these processes, so easy to understand at first (like the night's darkness), turn out to reflect the greater complexities and perplexities which abound in any study of totalities, limits, and wholes. The darkness of night is a simulacrum of The Cosmological Problem – and those methodological wrinkles felt by Newton, Chéseaux, and Olbers are but small-scale versions of the philosophical tidal waves which inundate the researches of Hubble and Hoyle, Gamow and Gold, Baade and Bondi. The result is a *science* with infinite rewards for philosophers, reminiscent of Aristotle contemplating the spheres of Eudoxos.

References

Bondi, H. 1962. Physics and cosmology. *The Observatory* 82: 133–143.

Gamov, George. 1952. *The creation of the universe*. New York: Viking Press.

Hoyle, Fred. 1951. *The nature of the universe*. New York: Harper.

Loys de Chéseaux, Jean-Philippe. 1744. *Traité de la comète, qui a paru en Décembre 1743*. Lausanne: M.M. Bousquet.

Newton, Isaac, A. Rupert Hall, and Marie Boas Hall. 1962. *Unpublished scientific papers of Isaac Newton; A selection from the Portsmouth collection in the University Library, Cambridge*. Cambridge: Cambridge University Press.

Olbers H.W. 1823. Über die Durchsichtigkeit des Weltraumes. *Astronomisches Jahrbuch für das Jahr 1826*. Berlin: C.F.E. Spaethen. English translation: Olbers, H.W. 1826. On the transparency of space. *Edinburgh New Philosophical Journal* 1: 141–150.

Scriven, Michael. 1954. The age of the universe. *The British Journal for the Philosophy of Science* 5 (19): 181–190.

Struve, Otto. 1963. Some thoughts on Olbers' paradox. *Sky and Telescope* 25: 140–142.

Chapter 4
Stability Proofs and Consistency Proofs:
A Loose Analogy

Abstract A loose analogy relates the work of Laplace and Hilbert. These thinkers had roughly similar objectives. At a time when so much of our analytic effort goes to distinguishing mathematics and logic from physical theory, such an analogy can still be instructive, even though differences will always divide endeavors such as those of Laplace and Hilbert.

I

Philosophers of science make much of the distinction between physics and mathematics, so much that it seems perplexing to read of Galileo, Descartes, Newton, Laplace, Maxwell and Eddington as apparently *not* comprehending the differences. Granted, consistent claims which have consistent negations differ sharply from claims which do not: but in some respects physics and mathematics are analogous. By stressing one of these ways I hope to suggest how the physical world could have seemed to some to be only an instantiation of solid geometry and the differential calculus. *Are* the differences between mathematicians and mechanicians now known to be so great that any attempt to delineate partial analogies must fail? Let's see.

Lagrange and Laplace are positioned within Newtonian Mechanics analogously to the way Hilbert is located within the development of Elementary Number Theory.

Consider three theories, θ_1, θ_2, and θ_3. Suppose someone sought to ground these in a single discipline – θ'. Success would consist in demonstrating that θ_1, θ_2, and θ_3, is each one completely generable from θ'. Allow that θ_1, θ_2, and θ_3 were "properly made" to begin with; they are "well-formed". From a few primitive or unanalyzable concepts, and Formation Rules, Axioms are fabricated. By substituting values for the variable terms in these Axioms one can derive (*via* Transformation Rules), expressions which are "theoremic" – expressions which "unpack" from the Axioms.

Now, the *way* in which the values are determined distinguishes formal systems from scientific theories. Within the latter they are reached empirically: the "theorems" at the "bottom of a page" of scientific calculation are interpretable as "observation-statements".

© Springer Nature B.V. 2020   61
N. R. Hanson, *What I Do Not Believe, and Other Essays*, Synthese Library 38,
https://doi.org/10.1007/978-94-024-1739-5_4

θ_1, θ_2, and θ_3 are assumed, then, to be "well-formed". To render them as special cases of θ', one needs more general deductive techniques through whose use θ' can "envelop" θ_1, θ_2, and θ_3 – themselves being now 'theoremic' with respect to the Axioms of θ'. Let it transpire that from the primitives, Construction Rules, Axioms and Transformations of θ' all parts of θ_1, θ_2, and θ_3 can be derived – including what were originally 'primitives' and 'axioms' within these latter. This is what "reducing to θ'" means.

Naturally, θ' must be "well-made" too. Therein lies the rub, the nub and the hub of my position. Independent criteria are necessary for evaluating θ' – criteria additional to the mere fact that θ_1, θ_2, and θ_3 can be deduced from θ'. If θ' were inconsistent, the other three theories would follow by strict deduction. But so would their negations! Consequently, a first criterion might be that θ' cannot be accepted if it is demonstrably self-contradictory, for then θ_1, θ_2, and θ_3 "follow from" it only in a deductively degenerate sense. This is not the sense invoked when one speaks of theories (e.g., Huygens's Optics and Faraday's Electromagnetics) being "founded in" or "enveloped by" or "deducible from" some more basic theory (e.g., Maxwell's General Electrodynamics).

Granted, a self-contradictory θ' may stand on the same deductive ground as a consistent θ' so far as entailing θ_1, θ_2, and θ_3 goes. But one would like to show also that θ' is itself consistent, complete, non-redundant...etc., in short "well-made". *These* are the algorithmic criteria we wish any "more fundamental" theory to exhibit.

From this abstract account consider now the actual development of mathematics. It is known that 150 years ago the mathematical disciplines were several and unconnected. Some geometries had been "reduced" to general algebraic form. But the branches of higher Analysis, wherein $\sqrt{2}$, $\sqrt{-1}$, and π were deployed – these were not seen to have any direct logical connection with the principles of Elementary Number Theory. An "Arithmetization of Mathematics" was achieved by Gauss, Weierstrass, Dedekind and Peano. They showed the intricate concepts within Higher Analysis to be "reducible" to general notions of arithmetic – the ideas of *number, zero,* and *successor,* plus some set-theory (as when real numbers were identified, by Dedekind and Frege, with sequences of rationals). Peano went further; he formulated 'axiomatically' those basic propositions ("the Postulates") from which all of Elementary Number Theory, and hence all of higher mathematics, was generable.

But number-theorists perceived that in Peano's θ' the notions of *number, zero,* and *successor* were treated as primitive – hence, not further analyzable. Frege and Russell sought a more fundamental θ-'super prime' from which the θ-primes of Peano, Dedekind and Weierstrass could themselves be unpacked. The primitives in *Principia Mathematica* are just those of *negation* and *disjunction* (plus quantifiers of arbitrarily high order). Sheffer went still further; in *his* θ-'super-super prime' the *negation* and *disjunction* of *Principia Mathematica* themselves unpacked from the even more primitive "neither... nor".

Studies of fundamental arithmetic θ's loosely grouped to form a discipline: *Elementary Number Theory* (henceforth **ENT**). One of the great contributors to **ENT** was David Hilbert, whose 'program' is representable in terms of the foregoing.

If all specific mathematical theories, θ_1, θ_2, and θ_3 are themselves to be "arithmetized" (*à la* the θ's of Weierstrass, Dedekind and Peano) then some further things must be demonstrated about the ultimate θ' itself.

'Why?' Perhaps the inability to establish θ''s consistency and completeness, only reflects shortcomings in θ_1, θ_2, and θ_3? Were any of these undetectably incomplete no theory θ' (which encompasses them) could be complete either. Is it not enough that θ''s theorems and the joint set of those of θ_1, θ_2, and θ_3 are the same?

Not quite. It still remains that if **ENT** were self-contradictory, then all of mathematics would *necessarily* follow. But such an **ENT** would not then be an advance comparable to the "arithmetization of mathematics" in the nineteenth century – which is not to be characterized as the finding of evermore general contradictions from which mathematics might be unpacked! What Hilbert sought to show was that besides generating diverse disciplines (θ_1, θ_2, and θ_3), **ENT** (i.e. his θ') could independently be revealed as a consistent, a complete, and a well-made theory.

Classically, when seeking to establish an algorithm as consistent, it is undertaken first to demonstrate that it is *decidable*. A theory is decidable when any expression made in accordance with its Construction Rules is such that either it is, or it is not, a theorem (that is, these Rules determine what will be acceptable as 'well-formed formulae': the classical objective is then to demonstrate that there is no well-formed formula f of which it cannot be shown either that f is, or is not, a theorem of the theory θ'.) Thus, every expression "well-made" in θ' must be decidable within θ'; i.e., everything meaningful for θ' must be demonstrably true, or demonstrably false in θ'. That is the classical 'decidability' objective.

Only this can "close" an algorithm: ideally this should be established before seeking the algorithm's consistency. Because, if some expressions acceptable to θ''s Construction Rules were nonetheless *not* decidable – it could not be established that nothing inconsistent lurks within θ'. Perhaps some well-made but undecidable, expression of θ' has $(p \cdot \sim p)$ as a consequence, despite having been unpacked from well-formed prior expressions. One can ensure against this only by "closing" θ' to begin with – by providing a procedure through which everything well-made *via* θ''s Construction Rules is decidable in θ'.

An early part of Hilbert's objective, then – since he aspired for a general consistency proof – involved showing that his θ' (**ENT**) was *decidable throughout;* that everything well-made according to **ENT** is provably *part* of **ENT**. Transformation Rules alone would then reveal any acceptably-formed expression either as true within θ', or as definitely not so.

This sets the stage for Kurt Gödel's work of 1931. His theorem can be encapsulated thus: *No θ' can, if it is consistent, demonstrate its own consistency. Its non-completeness is thus a consequence* (Nagel and Newman 1958; Rosser 1939; Hanson 1961). That is, if any system which includes **ENT** is assumed to embody an effective decision procedure (i.e., a 'mechanical' technique for deciding of any well-made expression either that it is, or is not, a theorem) – then such a system can always be shown to generate inconsistencies. So no logical system which includes **ENT** embodies an effective decision procedure. *Ergo* no logical system which includes **ENT** can consistently demonstrate its own consistency. Any such system is *essentially* incomplete.

It is not my objective to explore the Hilbert program or the Gödel Theorem. I've sought only to suggest that when one undertakes to "axiomatize" apparently-independent disciplines, within one fundamental discipline, θ', further criteria concerning the suitability of θ' must be met. Hilbert sought mathematics' conceptual foundations. Gödel's Theorem shows that not all of what Hilbert seeks can be found, at least not in the unqualified manner typical of early programmatic statements of Formalism. For just assume that it can be found – that the decision-procedure required in a general consistency proof is itself a constituent of **ENT**: if **ENT** is then assumed to be complete it is inconsistent, and if it is assumed consistent it is incomplete.

So, simply enveloping θ_1, θ_2, and θ_3 within a more "fundamental" θ' may not be enough. It ought to be demonstrable that further things about θ' obtain – the things Hilbert sought to demonstrate of **ENT**, and which Gödel demonstrated could not be demonstrated.

A loosely analogous development structures the history of physics. This is not strange – given the intimate connections between pure mathematics and fundamental physics. Before the seventeenth century, physical science was a cluster of apparently independent disciplines. Benedetti, Galileo, Beeckman and Tartaglia had created *Mechanics* and *Ballistics*. Copernicus, Rheticus, Brahe and Kepler had given us *Celestial Mechanics.* Snell, Descartes and Huygens had formulated much of *Optics.* Crude theories of the Tides were available; some *Hydrodynamics* was known. Gilbert's work on *Magnetism* was familiar; and philosophical speculations on *Motion* abounded. But, as with nineteenth century formal science, these disciplines worked with independent concepts, principles and inferential frameworks. By the mid-seventeenth century, then, there were many independent scientific theories like our θ_1, θ_2, and θ_3. But there was no more fundamental discipline, θ', from the postulation of whose primitives and Axioms θ_1, θ_2, and θ_3 could be shown to follow as consequences. There was no *Elementary Physical Theory* (**EPT**), θ', from which e.g., Masspoint Mechanics, Ballistics and Celestial Mechanics could be generated.

Between 1665 and 1687 Isaac Newton created such an **EPT**. His θ' was presented as the *Principia Mathematica Philosophiae Naturalis*. Here (ideally) was a fundamental "Force-and-Gravitation-Theory" from which most of the previously independent theories of Mechanics, Ballistics, and Celestial Mechanics followed – as observational consequences.

This constitutes Newton's greatness. He was certainly no Planetary Theorist in the traditional sense; he was not a "coordinate-fixer" or an "orbit-shaper" as had been Ptolemy, Copernicus and Kepler. He was not, like Stevin, Torricelli, or Viviani, an experimentalist. He was not, like Brahe and (often) Boyle, a "data-recorder". He was primarily a systematizer; an architect, not a bricklayer. Newton discerned the axiomatic foundations of many of those apparently independent, physical disciplines extant when he was born. His θ' – "Force-and-Gravitation-Theory" – fused several $\theta_1 s$, $\theta_2 s$, and $\theta_3 s$... so as to form but theorem-subsets within his great **EPT** built on the Laws of Motion, of Central Forces, and of Universal Gravitation. As to his linking of mechanics and gravitation – some may question the legitimacy of this. Thus Truesdell notes that "90% of today's specialists in this area are as skeptical of Newton's theory of gravitation as Euler was. A working summary, yes, but a physical theory, no." This is a telling observation – but compatible with the *historical* recognition of Newton's theory as *the* great **EPT**, the θ' of all time. Insert the astronomical values for the variable terms, and Newton's fundamental formulae transformed into general propositions from which could be unpacked myriad predictions of the states of celestial bodies – as well as retrodictions of their past states! Similarly within Mass-point Mechanics and Ballistics.

The initial test of Newton's synthesis was to check whether the already verified observation statements within "derivative" physical disciplines (θ_1, θ_2, θ_3...) did really follow from calculation in Force-and-Gravitation-Theory, his θ'. As students of Lagrange, Clairaut, Euler, Laplace and Gauss know, the verification of θ' generated exciting moments within Analytical Mechanics. In crisis after crisis, Newton's theory ultimately triumphed, sometimes at the moment it was to be abandoned. True, delicate readjustments had continually to be made (and are still being made). Techniques were subtly refined; the observed values were 'plugged into' the theoretical variables in increasingly sophisticated ways. But Newton's θ' – as developed by Lagrange and Laplace, and especially by Euler and Gauss – proved itself, just as did the Weierstrass-Dedekind-Peano "arithmetization of mathematics". Newton attempted a corresponding "mechanization of physics". His *Principia* fused a diversity of physical theories into one well-made unitary discipline. From its assumptions, axioms, principles, rules and inferential techniques the descriptions and predictions of most apparently independent physical phenomena were generable. Newton had "reduced" gobs of physics to a single θ' – Force-and-Gravitation-Theory. Here was a 'Hilbertian Ideal' in part realized within the theoretical physics of the 18th and especially the nineteenth century.

Is there anything corresponding to a "Gödelian critique" within later developments of Newtonian mechanics? Not really, perhaps. But there are additional criteria used in assessing the suitability of Newton's θ' – criteria loosely comparable to the desired proofs of consistency and completeness within Hilbert's θ'. Obviously, these additional criteria for **EPT** cannot be identical with any of those for **ENT**. Proofs of consistency and completeness define the Hilbert ideal. Such demonstrations are impossible where *any* physical theory is concerned – although many philosophers and physicists continue to speak of orthodox quantum theory as being "incomplete" – as if it could be something else. A physical theory can never be 'closed' and rebuilt as a self-contained symbolic game, McKinsey and Suppes to the contrary notwithstanding. Further observations must always 'be to the point,' they must be relevant. All observation-statements unpackable from a physical theory must be synthetic – i.e., their negations must be consistent. Their acceptability thus cannot ever be a matter solely of their form. Since a physical theory conveys factual *information*, there *must* be other possible states of affairs which it claims do not obtain. Hence our certitude *vis-à-vis* anything 'at the bottom of the page' in **EPT** must be different from our corresponding certitude with respect to anything 'at the bottom of the page' in **ENT**. It is always possible that we made mistakes in our initial conditions, or in recording our observations. The suitability of our values for the variables within **EPT** is forever time-dependent: better techniques, sharper eyes, finer mathematics – might tomorrow require us to alter the numbers in our computations about e.g., lunar motions; (we now know of 'libration' and define 'π' differently from some of the ancients). In short, **EPT** must remain empirically vulnerable always. The suitability of our choices of values for variables within **ENT**, however, is not time-dependent in any important way. So, whatever may be the criteria with which we assess the further suitability of **EPT**, they cannot be identical with those Hilbert once envisaged for **ENT** – abstract proofs of consistency and completeness.

II

Our inquiry has reached this point: **ENT** and **EPT**, since they are like any θ' in that they encapsulate less 'general' disciplines, $\theta_1 \rightarrow \theta_n$, should meet additional conditions *independently* of their encompassing $\theta_1 \rightarrow \theta_n$. What are these? For **ENT** they would have consisted (ideally) in providing a general decision procedure, a self-administering consistency proof, and (derivatively) a completeness proof; this was the "Hilbert ideal". (Gödel, although initially in search of this same Hilbertian grail, ultimately exposed it as a mirage.) **EPT**, however, being logically different from **ENT**, cannot be subject to these criteria – a consistency proof and a completeness proof. What are *its* additional criteria, given that it can deductively generate θ'_1 (e.g. Terrestrial Mechanics) θ_2 (Ballistics) and θ_3 (Celestial Mechanics)?

The additional condition to be met by Force-and-Gravitation-Theory is the provision of a *stability proof.* What does this mean?

Consider any five 'boring' laps encountered during the Indianapolis Speedway "Big 500" Race. The laps between the 45th and 50th last May 30th were thus describable: once the big cars settled down they ran their circuits at even spaces with engines delivering less than full power. These are the "breathing" laps, the "coasters"! Suppose an ingenious mathematician designed a calculus faithful to the kinematics and dynamics of such laps. A "Keplerian" geometry of the moving cars could be generated; their angular velocities in the turns, their accelerations on the straightaways, their apparently retrograde motions as viewed from overtaking racing cars – all this might be captured within a rigorous symbolism by our theoretician. His calculus might even permit retrodictive and predictive remarks about the "middle" laps in Indianapolis races ten years ago, and ten years hence!

How could physicists be sure, from the *Principia* alone, that Newton's θ' was not merely a calculus like this? How could they be sure that the planets will not simply "shut down their engines", moving then into some celestial garage, after these few hundred present circuits are completed? What is it about the *system* of objects to which Newton directed attention which distinguishes it from that encountered during the "500"? An earth tremor could halt the latter. Would a "sky tremor" do the same to our solar system? Suppose a comet pierced our planetary array on its way from Polaris to the Magellanic Clouds – would not this disrupt the ordered gyrations of the system much as the "500" might be stopped by an enormous electromagnet at one of the turns? – one which killed magnetos, unbalanced camshafts, and even deflected the cars?

To this hypothetical taunt the history of astronomers would have answered:

no – our planetary system is stable. *Disturbances impinging from without result in perturbations which, owing to the inherent stability of our "quasi gyroscope", will be dampened and absorbed into the dynamic rhythm of the planetary machine.*

(It should be noted that since it might just have made sense to suppose (1) the solar system is a local instability in a stable universe, or (2) the solar system is locally stable in a generally unstable universe – the answer just given skirts a very large-scale commitment. Newton would have accepted that commitment; for him if the planetary system decomposed this would be a *feature* of the degradation of the universe as a whole. Historically, assumptions about the system's stability were wholly general – just as our local assumptions about the roles of the Second Law of Thermodynamics have vast astrophysical consequences.)

But the 'possibly unstable' taunt was haunting!, just as the 'possibly inconsistent' taunt haunted Hilbert. To *presuppose* that our system is stable – a presupposition as common to Aristotle, Ptolemy, and Sacrobosco, as to Copernicus, Brahe, Kepler, and Galileo – this is less satisfying than a proof; just as a *presupposition* of arithmetic consistency is less satisfying than what David Hilbert aspired for, although no mathematician has ever doubted that arithmetic is consistent. We know what unstable systems are like; a spinning top has an equilibrium unlike that of a stationary top balanced on its point. So the 'stability' supposition is not primitive to Newton's theory in the sense that alternatives were inconceivable. Our "Indianapolis 500" example might be generalized; what we observe in the heavens may be but a fleeting configuration of whirling bodies in but a temporary vortical phase of their interminable travels through space.

Newton perceived the problem. He felt obliged to meet those "additional requirements" for his θ', to which he had reduced the independent physical disciplines interlocked within the *Principia*. We considered the objections to reducing all of mathematics to an *inconsistent* θ'. Analogous objections would be lodged against **EPT** if apparently-distinct disciplines were reduced to a θ' which was not itself demonstrably 'stable'; a theory which descriptively encapsulated but a passing configuration of nearby celestial objects, and which was powerless to prove otherwise, would be a sorry θ'. No; Newton feels that his Force-and-Gravitation-Theory is fundamental to the "nature of things". This had to be demonstrated. His θ' had to be more than an ephemeral descriptive calculus; indeed it was soon felt (e.g. by Euler and Gauss) to provide a mechanical *understanding* of what kinds of objects planets, satellites, stars and comets really are! But it was not enough just to *say* this. One ought to prove it! The celestial mechanician should show that, because of the other dynamic properties of the system, it is *in its nature* to be stable; it will persist *ad indefinitum* despite myriad disturbances and shocks from without.

Demonstrating this is the "problem of the proof of planetary stability." It is loosely analogous with "the problem of the proof of arithmetic consistency" – since both proofs constitute "additional conditions" to be met by θ's which systematize less general, and apparently distinct θ's,$_{1,2,3}$. To stress this has been the objective of this paper. From here on, although analogies abound, divergencies mark developments of **ENT** and **EPT**. For although **EPT** has had its Hilberts, it has never yet had a Gödel – even though the ambitions of early Newtonians for **EPT** have required more complex, intricate and elusive techniques than the original bold conceptions would have suggested.

Newton could not piece together the required proof. He explicitly leaves it to God's infinite wisdom and omnipotence to keep planetary matters running aright. The Master Mechanician is expected to dampen disturbances so that they never enjoy an oscillatory growth sufficient to disrupt the dynamics of our solar system. Oscillations, remember, can begin with tiny deflections and expand to explosive proportions: slightly unbalanced camshafts have slowly decomposed massive engines. The makers of aircraft instruments devote most of their research to refining "dampening techniques". But in 1687 the theories of small oscillations and of dampening techniques, had not yet affected the astronomer's quiescence. (Besides, with the Mechanical Philosophy harassing Theology from all sides, what could have been more fitting for a devout physicist than to leave the stable maintenance of our planetary system to the Divine Engineer?)

This was like leaving it to God to ensure that arithmetic – since so much was based on it – was provably consistent and complete. I am told that some learned divines and despondent dramatists were as appalled by Gödel's discovery as John Donne had been with the Mechanical Philosophy. But, given the historical line from Weierstrass – through Dedekind – to Peano, formal theorists could not let the "arithmetization of mathematics" terminate in appeals to God's algorithmic stewardship. If **ENT** was to be the universal foundation stone, then *we* had to know that it was "well-made" for the task. Faith in God's consistency and completeness could never replace a proof of **ENT**'s completeness and consistency. Similarly with the historical sequence from the Parisian Impetus Theorists – through Galileo – to Newton; if the growing constellation of independent disciplines was really to be founded on Newton's θ', this also had to be demonstrably "well-made" for the task. Why 'reduce' θ_1, θ_2, and θ_3 to a fundamental θ' if one had then to assume again what was *already* assumed in θ_1, θ_2, and θ_3? Copernicus, Brahe, Kepler and Galileo all construed the planetary system as stable and conservative, in this reflecting their attitude towards the entire universe. Newton the 'Axiomatizer' initiated a program which, ideally, ought not to have had to assume the same things earlier theorists had to assume. So just as the 'arithmetization of mathematics' was thought by many to be unfinished without consistency and completeness proofs for **ENT** (*by* **ENT**), so similarly, the 'analytical mechanization of physics' was regarded as incomplete without a stability proof for **EPT** and its paradigm subject-matter – our planetary system – the universe *in simulacrum*.

Laplace's ideal is expressed thus:

> Astronomy is a great problem of mechanics in which the elements of the motions are the arbitrary constant quantities. The solution depends upon the accuracy of the observations,

and upon the perfection of the analysis. It is very important to reject every empirical process, and to complete the analysis, so that it shall not be necessary to derive from
observations any but indispensable data. ([1829–1839] 1966, 1.xxiii)

It is within this framework that Laplace's effort to write a Stability Proof must be
located.

The neglected demonstration occurs in *Mécanique Céleste* (Laplace 1829–1839,
vol. 1, bk. 2: §57). There Laplace argues (in effect) that *if* the planetary system were
multiply-periodic – if it were quasi-gyroscopic – then the orbits would tend towards
circularity, despite Kepler's discovery that they are all ellipses. Most of the *de facto*
orbits are but very small deviations from perfect circularity – deviations which
could, however, constitute just those small oscillations which grow progressively to
disrupt entire systematic configurations. What needs to be shown is that our system
is such that its constituent orbits grow more circular and, hence, the system gyroscopically more stable – rather than progressively more elliptical and hence disruptive; in 10,000 A.D. the orbit of Halley's comet (if it still be with us) should be less
eccentric than now *if* our solar system is really stable/is really multiply-periodic/is
really a Eudoxian-gyroscopic/is really such that all its apparent secular short-range
disturbances are ultimately periodic or repetitive over great spans of time. This is
the insight of Laplace.

Thus even Kepler noted something wrong with the "coupled motions" of Jupiter
and Saturn. Halley suggested an acceleration of the former and a retardation of the
latter: in one millennium Jupiter should have been displaced forward by 57′, Saturn
aft by 2° 19′. Here is the kind of secular-progressive displacement which might
threaten the stability of the entire planetary system. Within years such a gyrational
imbalance would begin to decompose it. (Imagine a helicopter's rotor-blades, one of
which slowly moved further away from another: the result will be a disastrous instability.) Could such progressive displacements be accommodated within Newton's
theory? Euler tried in 1748 and in 1752; at one time he even suggested the mechanical feasibility of secular perturbations – those which always 'progress' in the same
direction, without apparent resolution or periodicity. In 1763 Lagrange addressed
the Three-Body Problem – as first enunciated by Newton. He sought a *general* analytical technique for computing the displacements of any three objects due to gravitational interattractions amongst them. The usual method was to calculate this
perturbation initially for two of the bodies (*via* $F = Gm_1m_2/r^2$), and then to add "correction terms" *ad indefinitum* because of the displacement suffered by both due to
the attraction of the third body (but ignored in the initial computation). This general
and analytic solution has been as much a *sangreal* within **EPT** as has been a general
consistency proof within **ENT**. Lagrange generated novel methods and applied
them to Jupiter and Saturn. He *did* find a secular term for both. But it was too small.

Enter Laplace. He re-examined all the tiny higher-order terms neglected in earlier computations of the planets' mean motions. They all cancelled out – like the
printed border on the edge of a dinner plate, one which does not quite 'join up' the

first time around, but which would fit perfectly were the pattern drawn around the edge (say) another five times. The *continuous* dislocation of these planets thus had to be *nil* – just as the continuous "dislocation" of the plate-pattern is *nil* on the fifth circuit. Then Lagrange extended all this by demonstrating that the mutual attractions of all the planets *in toto* could not produce any secular progressive changes in their mean distances from the sun and the periods of revolution; these latter could be subject to periodic variations only. This was a monumental achievement. The road towards a general stability proof seemed now well-paved. For, Lagrange's division between secular and periodic motions had become fundamental to analytical mechanics. Sharply distinguished were those perturbations which were progressive (hence increasingly disruptive of a system's "quasi-gyroscopic" and multiply periodic attunement) – and those different perturbations which were ultimately repetitive and thus only complex additions to orthodox descriptions of planetary orbits. To reveal all apparently rhymeless disturbances in our planetary movements to be perturbations of the latter kind (aspects of higher-order dynamical patterns), to expose all secular aberrations as being ultimately periodic – this was the essential first step in any general, analytic stability proof. Lagrange made that step.

A successful Stability Proof would thus assume all individual planetary motions to be singly periodic; the motions of the system as a whole would then be demonstrated as multiply periodic, even though a complete cycle might take centuries. Secular disturbances, then, would be only short-range descriptions; the complete account being so multi-parametrically-periodic that the observed motions seem more like arbitrary cosmic whims. Imagine e.g. 6000 spinning roulette wheels with a phosphorescent point on each – all affixed to a lopsided Ferris Wheel, which is itself set upon a railway-turntable. On a dark night the ultimate rhythmicity of such an intricate and complex set of motions may well be disguised by sheer complexity. Who could predict how many times the turntable would spin before all the roulette wheels, and the Ferris Wheel, were simultaneously returned for an instant to their original starting positions? Yet if all these motions were *systematically* connected in some way, such prediction would have to be possible in principle.

With respect to Jupiter and Saturn, Laplace approached their observed accelerations and retardations as if they were just such phenomena – small segments of a long, intricate, periodic cycle. By 1784 he perceived that any five revolutions of Jupiter 'equaled' any two of Saturn; the planets resumed their coordinates with respect to each other, and with respect to the sun-and-fixed-stars, after each two revolutions of the latter or (what takes the same length of time) any five revolutions of the former. After each 59-year period the two planets will always meet again at almost the same place in the ecliptic. Two tiny terms of the third order (usually neglected in ordinary calculations) return in the positional descriptions for all computations involving 59-year repetitions, *and* these terms recur in precisely the same way. If analytically unperceived the continued cancellation of these terms grows into discrepancies between the (unadjusted) predictions, and the observations of planetary latitudes. The "conjunction position" of the planets shifts to other longitudes, as noticed by Kepler, Halley and Newton earlier. This (apparently) progressive effect reverses after each 450 years! Thus the perturbations, originally thought to be "secular", constitute a protracted rhythmic oscillation having a period of

900 years. With this important calculation all ancient and modern observations of Jupiter and Saturn were suddenly well represented. As Laplace said:

> The irregularities of the two planets appeared formerly to be inexplicable by the law of universal gravitation – they now form one of its most striking proofs. ([1829–1839] 1966, 1:324)

In effect, a *general* stability proof for celestial mechanics would demonstrate that every secular perturbation, even those initiated *de novo* by some energy-source from outer space, will ultimately be "dampened" within the repetitive rhythm of some periodic perturbation pattern – some of these requiring millennia for their full cycle.

Naturally, were a supernova to pierce our galaxy near the sun, the planetary system would decompose implosively. The compensatory mechanism Laplace envisages would be insufficient to dampen a disturbance of that magnitude. But all Laplace seeks is what every planetary theorist wanted – a demonstration that 'sufficiently' small disturbances to our planetary configuration will not surge in an oscillatory growth, thus

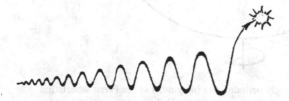

which would ultimately disrupt the entire system's dynamic geometry. Similarly, a well-designed aircraft instrument will incorporate a "dampening mechanism" such that external shocks of *sufficiently small* magnitude will be absorbed thus

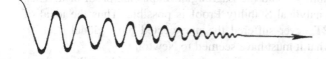

– within the device's energy-train without perceptible effects on the accuracy of the readings. The most perfect chronometer, or directional gyro, will cease to function under a steamroller. But the sudden dislocations such instruments encounter during 'normal' use will not magnify during transfer through the internal mechanism of these devices. This is all celestial mechanics, and Laplace, desired to demonstrate of our solar system – that it was a shock-proof machine, not also a catastrophe-proof creation.

This, then, is a major component within the "Laplacian program". Its comparison with a major component of the "Hilbertian program" – the consistency proof – has been delineated. Laplace's stability proof was originally thought to have been wholly successful – just as Hilbert never doubted that his objectives *vis-à-vis* a consistency proof for **ENT** would be achieved. Gödel formulated the absolute critique of the latter. There is no corresponding critique of the former – no sudden

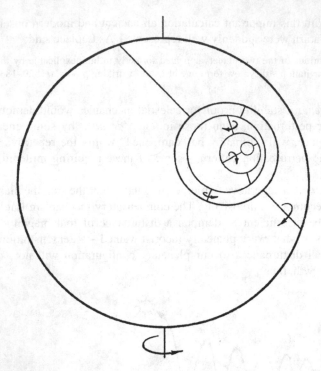

Fig. 4.1 A child's toy-gyroscope is singly periodic. But imagine, *à la* our Ferris wheel, that several such toys are superimposed, and supra-imposed (in the Eudoxian manner) to form one intricate conspiracy of motions. Our solar system, with its multiple rotations, revolutions, axes and satellites – is just such a conspiracy

death of an ideal! But a protracted history of unsuspected complexities has modified the Newtonian ideal of stability – and the Lagrangian-Laplacian proof of it. This, although in principle an analytical Stability Proof is possible. Thus the ideal of analytical security for **EPT** never suffered the unexpected blow Gödel dealt to **ENT**. But it is no longer what it must have seemed to Newton.

The Hilbertian program cannot, as a matter of principle, ever be fully realized; my limited understanding of Gentzen indicates no rebuttal to this. (He so broadens the *idea* of the Hilbertian program that it is far from clear that Hilbert would identify it with his own early pronouncements.) The Laplacean program can, in principle, be fulfilled, but then again only in the manner which invites saying that Democritus' ideals have also been fulfilled in the complex lists of Elementary Particles (at least 36 in number) we find today. The Democritean ideal, the Hilbertian ideal and the Laplacean ideal, if they survive at all now, do so in profoundly altered forms.

Consider what Laplace does in his "proof". I will use the expressions "multiply-periodic" and "quasi-gyroscopic" indiscriminately (even though not every multiply-periodic system is also quasi-gyroscopic). These are the visions of Stability Laplace invokes. A physical system S which is quasi-gyroscopic, or multiply-periodic, will be such that any external shock or disturbance, if less than some given magnitude ε

(the value of ε is a fundamental property of S) – this shock will be absorbed, or dissipated, within the periodicities and gyrations of the system.

A *general* stability proof for our solar system as conceived in the eighteenth century would consist in showing that it is multiply-periodic, or quasi-gyroscopic, in precisely this sense. How did Laplace proceed to set out such a proof?

Laplace simply *deduces* the consequences of the assumption:

The planetary system is multiply-periodic; it is a quasi-gyroscope in the Eudoxian sense.

It follows from this that small oscillations *will* be dampened, never to grow beyond the capacity of our planetary mechanism to contain them. Growing oscillations would never be observed, granting this assumption; all perturbations would slowly be absorbed into complex periodicities.

We do in fact observe all this. Aberrations originally thought to be initiating a limitless disruption – e.g., the Jupiter-Saturn problem – turn out to be resolvable in quasi-gyroscopic terms. Secularities are *observed* to be but short-range aspects of periodicities.

Do these observational facts confirm the original assumption? In some sense 'yes'. This is the "hypothetico-deductive" sense in which *any* assumption is (insofar) confirmed when its consequences are observationally verified.

The ideal case here is where *further* consequences of the assumption (perhaps initially unsuspected) also turn out to square with the facts. The assumption thus gains *independent* and *diversified* support. Thus the logical oddity of Laplace's undertaking: there is nothing unsuspected about the observed periodicities in planetary motions. These regularities are what first led to the conviction in all of Laplace's predecessors that the planetary system was stable! Hypothetico-deductively, then, Laplace's Stability Proof seems degenerate. Observations of those periodicities which led originally to the capital assumption that the system *is* stable – these themselves follow as the signal consequences of Laplace's assumption. In short, the assumed premise and the derived observational conclusion are so intimately connected (semantically) that the proof seems sterile.

Slightly more formally: to consider the solution of $y = -\sin y$, assume y is small. Then $y = -y$, $y = \alpha \cos t + \beta \sin t$
Hence:

$$y(t) \le \sqrt{[y(o)]^2 + [\dot{y}(o)]^2},$$

so that if $y(o)$ and $\dot{y}(o)$ are small, so is $y(t)$.

But every mathematician will remark this as wholly unrigorous. *This 'proof' assumes y to be small at all times in order to prove that in fact it remains small if it starts out small (and if ẏ starts out small too).*

Thus to say "the planetary system is stable" just *is* to say "disturbances from without will not in fact disrupt the system; periodicity is the pattern". Observing examples of this latter *will* support the assumption, just as observing men die will support "all men are mortal". But is the general assumption *proved* thereby? The question is particularly barbed when the observed examples unpacked from the assumption are the very ones which led to the assumption initially. Within the context so familiar to Lagrange and Laplace could anything so "merely empirical" ever conclude with *quod erat demonstrandum?*

There are other consequences of the stability assumption – but they were not readily verifiable in strictly astronomical terms. Hence, in the absence of *independent* support for the stability assumption (i.e., that y remains small at all times, and $\dot{y}$ too), Laplace's argument must be adjudged weak, deductively and hypothetico-deductively. Since his argument is to some extent inductive it cannot be a strict proof; continually confirming the observational consequences one unpacks from an assumption does not strictly prove that assumption. And since these same observational facts are both genetically responsible for, and also inferentially generable from the stability assumption, i.e. that y remains small, this cannot constitute a very confident example of empirical confirmation either.

This mildly negative appraisal of Laplace will be no surprise to anyone who knows eighteenth century mechanics. But the argument schematized above is accepted as almost a dogma of religion by most physicists (Cf. e.g. Jeffreys 1956). But it seems suspect in every way.

What would a less contestable form of Laplace's proof have been like? It would have shown analytically that our planetary system, with all its normal motions and unexpected perturbations, is representable not only *now* via the Force-and-Gravitation-Theory embodied in the Newton-Laplace θ' – the system must be proved always to be so representable. Not in the sense that unless it is θ' will no longer apply to it, but rather in the sense that the system is *in fact* such that θ' constitutes its most fundamental description and explanation. What is required is an analytical proof – i.e., an explicit demonstration expressed as analytic functions – that the solar system must be multiply-periodic, that it must be representable as a type of Eudoxian quasi-gyroscope. It must be demonstrated *as a consequence* of the *Principia*, or else as some needed addition, that nothing short of a celestial catastrophe could possibly disrupt the smoothly functioning mechanism analytically encapsulated within θ'. The bodies composing our planetary group must be shown to constitute a unified mechanism of which it can be proved that all disturbances to the system are 'absorbable' in Force-and-Gravitation-terms.

Consider now what this last request is a request for:

It seeks nothing short of a *general analytical solution to the multi-body problem.* What else could one be seeking here? What else other than a general proof that given groupings of bodies are not only describable *here and now* in Force-and-Gravitation terms, but that nothing *could* occur to make such terms inapplicable? Despite the multi-parametric influences each such body has on all the others (and *vice versa*) we should still be able analytically to generate all future state descriptions for the entire constellation: that is Laplace's quest. It is one thing to calculate the perturbational interactions between earth and moon in two-body terms, and then to figure in the sun's attraction on both – and then Mars' attraction on all three (each time correcting all earlier computations by redescribing the initial states). It is another thing to have a general technique through which all future states of planetary objects *en bloc* are describable so as exactly to underwrite an ideal stability proof.

Weierstrass and Poincaré gave their energies to this part of Laplace's program. They succeeded in refining the technique of "successive approximations". In finding values for the variable terms in ($F = G(Mm/r^2)$) one will first determine M and m as if these were masses of otherwise unperturbed bodies. But weighing M and m is accomplished only by assuming some value for F. As intimated, both M and m will be perturbed by gravitational influences of other objects, e.g., their own satellites, asteroids, planets, the sun, comets...etc. As one computes the influences of these other objects on our original M and m, assessing the interactions between these latter will have to be adjusted and readjusted *ad indefinitum*. But so will further adjustments follow when we consider the corresponding descriptive determinations involving the satellites, asteroids, planets, and comets, whose perturbational propensities we *assumed* in order to introduce corrections for M and m. There is no theoretical terminus to this. One can, however, get successively closer to the "true" perturbational situation within any multi-body complex *via* this technique.

This method is inductive. It cannot generate the Q.E.D. lusted after by Laplace. Although our solar system's being multiply-periodic may become plausible through this technique, such a conclusion can never be "deduced" from a series of successive approximations.

Laplace's "Proof of Stability in the Solar System" cannot be a proof within the orthodox mathematical tradition; not if he assumes on faith just what his algebraically less-sophisticated predecessors had also assumed on faith – viz., *that the solar system was stable.* To assume as he did that a system is multiply-periodic just *is* to assume its dynamical stability. That we can then unpack a cluster of observation-statements corresponding to the facts – this does not decide whether our observations of planetary motions are of an ephemeral conspiracy amongst 'celestial wanderers', or of the workings of the stable machine our planetary system actually may be. Since our astronomical observations are compatible with *either* hypothesis, they cannot *prove* one of them, as against the other – especially when one's "proof" begins with the very assumption one seeks to establish, viz., that $y(t)$ will remain small.

What then, since Laplace? Following Newton, he demanded that the planetary system be "multiply-periodic, or quasi-gyroscopic". As I've argued, he *assumed* this for θ' in order to prove it. Distinctly unconvincing. In Celestial Mechanics today this would be felt to be unrealistic. Just because Laplace used trigonometric functions in his long calculations this does not mean all his successors had to as well. They did not.

There are scores of Stability-concepts, somewhat weaker than what Laplace wanted (granted), whose proofs (for our solar system) are quite amenable to sound mathematics. With respect to the algebra set out earlier, mathematicians have long since found rigorous proofs which require no explicit solution, and apply to any system of the form:

$$\dot{y} = f(y,u)$$

$$\dot{u} = g(y,u)$$

in which f and g are analytic functions. In general, these proofs are such that *their solutions depend continuously on the initial values*. Indeed, a most elegant alternative to the technique of successive approximations is given by Moulton (1914, 268) – and there have been other attempts at analytical solutions to the n-body problem – both before and after Moulton.

It has sometimes been urged that there cannot be a strictly analytical solution to the n-body problem. There is extant, so far as I know, no demonstration of the *non*-existence of such a solution; work such as Moulton's drives one towards the opposite decision – even though the last word has not yet been said. As for the notorious Poincaré-Painlevé theorem – this refers to solutions analytic in the *masses*. The theorem reads: There is no strict solution to the n-body problem which is set out as analytic in the masses involved. But Siegel has remarked that this conclusion has no relevance for the existence of solutions for particular masses (like planets). Moreover, the Poincaré-Painlevé computations are purely formal – as is the more recent work of Cherry. Painlevé's effort is indeed no more than a programmic abstract. There is no sound proof known for this much-publicized theorem (Cf. Truesdell 1961, 21–36).

So, Laplace's ideals *vis-à-vis* stability in **EPT** have succeeded, or at least they have been constructively transformed – in a manner distinguishable from what transpired in **ENT**. Nonetheless, success or failure aside, the *occasion* for a Stability Proof in the former case and a Consistency Proof in the latter arose in that **EPT** and **ENT** functioned as θ' for other 'derivative' disciplines. Thus the loose analogy which structures this paper.

Appendix

Demonstrations of the stability of dynamical systems have been fraught with perplexities, despite all the elegance. If one has a set of simultaneous linear differential equations for n variables, and then eliminates these equations in succession so as to favor one variable, the process can terminate in a differential equation for that one. Substitute $e^{\delta t}$ for it: the result is then an equation for δ; if there is a repeated root then there will be a second solution for $te^{\delta t}$. Should such a thing happen in the theory of small oscillations it would transpire that a repeated value of δ will lead to terms of the form $t \cos Kt$, $t \sin Kt$ ($K = i\delta$) and, save for specially designed initial conditions, *a small oscillation will grow indefinitely*. This is another way of putting Newton's and Laplace's problem: the planetary system is always representable as such a set of simultaneous linear differential equations. What is to prevent a small external disturbance from provoking a small internal oscillation of this sort (as when a rock in the road is responsible for an oscillatory vibration in a car's speedometer-needle) such that the latter grows to disruptive proportions? This has never been found to happen in our system. But *could* it? Newton felt that the Deity would not allow it. But, if it did come about it would contradict the fundamental principle that if the planetary system's potential energy is a minimum in the position of equilibrium (the Lagrange-Laplace assumption), and the initial displacements and velocities were sufficiently small (but not zero!), then there will of necessity be a limit that no displacement can ever exceed. Otherwise engineers could, by clever constructions, 'feed' small oscillations into a suitably designed mechanism generating thereby *any* amount of oscillatory energy at the terminus of the power train – a conclusion which conflicts with every Conservation Principle known. Whatever it is that prevents the construction of a *perpetuum mobile* (of the First Type) also prevents extra-systematic disturbances from generating intra-systematic dislocations any greater than the original disturbance. But what then of the theory of small oscillations which allows for progressive enlargements of oscillatory amplitudes?

Laplace was deeply troubled by this. Routh ultimately found a resolution in 1877 (Routh 1877, and compare Heaviside 1902, 529). If the system in question is not dissipative and the roots are unequal, the zeros of the minor of any element in the leading diagonal separate those of the original determinant (Cf. Weierstrass 1858. Cf. also Lamb 1920, 222–226, and Bromwich 1906). If, then, the determinant Δ has a factor $(p^2 + \alpha^2)^k$, every first minor contains the factor $(p^2 + \alpha^2)^{k-1}$. When we evaluate the contribution from the initial conditions to the operational solution

$$\left[\text{i.e. } X_m = \frac{Erm}{\Delta} a_{rs} \left(p^2 u_s + p v_s \right) \right]$$

a factor $(p^2 + \alpha^2)^{k-1}$ will cancel out. We will be left only with a single factor $(p^2 + \alpha^2)$ in the denominator. This will happen with *every* repeated root; the interpretation will contain *only* terms of the forms cos αt and sin αt. Thus varying u_s and the v_s will alter ratios of the coefficients of these trigonometric factors for different coordinates; but they will never introduce terms like t cos αt or t sin αt – the trouble makers.

Consider now not a *general* initial disturbance, but one confined to the period equation. Suppose the equations of motion are these:

$$\ddot{x}_1 + f\,\dot{x}_1 + c_1 x_1 = b\dot{x}_2 = 0$$

$$\ddot{x}_2 + f\dot{x}_2 + c_2 x_2 + b\dot{x}_1 = 0$$

(where $c_1 < 0$, $c_2 < 0$, $f \geq 0$, and b is large enough to ensure stability when $f = 0$)

Assume solutions proportional to $e^{\delta t}$; δ must then satisfy

$$\delta^4 + 2f\delta^3 + \delta^2 \left(c_1 + c_2 + b^2 + f^2 \right) + f\delta \left(c_1 + c_2 \right) + c_1 c_2 = 0,$$

from which it follows at once that

$$\Sigma \delta = -2f \leq 0$$

and that

$$\Sigma \frac{1}{\delta} = -f\left(\frac{1}{c_1} + \frac{1}{c_2}\right) \geq 0.$$

The roots for $f = 0$ may now be chosen as $\pm in_1$, $\pm in_2$ [where $n_1 > n_2$]. For $f > 0$ (but very small) let the roots be $\pm in_1 - \alpha_1$, $-in_2 - \alpha_2$, to order f. Then:

$$\alpha_1 + \alpha_2 = f > 0$$

and

$$\frac{1}{in_1 - \alpha_1} + \frac{1}{-in_1 - \alpha_1} + \frac{1}{in_2 - \alpha_2} + \frac{1}{in_2 - \alpha_2} \doteq -\frac{2\alpha_1}{n_1^2} - \frac{2\alpha_2}{n_2^2} > 0$$

Such inequalities are consistent only if α_1 and α_2 have opposite signs: in fact, since $n_1 > n_2$, $\alpha_1 > 0$ and $\alpha_2 < 0$.

Hence any system which is kept stable by gyroscopic action *only* is such that small friction will always produce instability. The fastest free vibrations will be damped, but the slower ones will increase in amplitude with time (to achieve which the system's 'gyroscopic energy' must be rechanneled into the growing slow oscillations, upsetting thereby the system's 'gyroscopic stability').

If, in the case of our planetary system, we postulate a near-gyroscopic configuration, and neglect friction altogether, then all the roots are purely imaginary: hence the system is stable! If $c_{rs}x_rx_s \geq 0$, in the presence of very little friction $a_{rs}\dot{x}_r\dot{s}_s + c_{rs}x_rx_s$ will decrease and all oscillations will be gradually damped down! Such a system will then be called *secularly stable*: as with our planetary system, imbalance-oscillations will become periodic over long periods, and their amplitudes will decrease.

But if the quadratic form is not ≥ 0 and the system is kept stable only by the gyroscopic terms, the slower oscillations must increase in amplitude, leading ultimately to profound changes in the character of the overall motion. This kind of system will then be *ordinarily stable*, but *secularly unstable*.

Newton's worry consisted in having no argument against this latter contingency in the case of our solar system. Laplace's triumph consisted in his recognition of the form of the problem. To prove the planetary system stable, then, he had to assume $c_{rs}x_rx_s \geq 0$. And he had to postulate but very little friction in the system. If either of these suppositions are challenged, he must then infer that the stability of the solar system is not the result of gyroscopic terms merely – that forces other than gyroscopic determine the stable cohesion of the system.

This kind of argument is methodologically fascinating. In a way Laplace, to the degree that he reasoned thus, has not really proved anything about our planetary system as it exists *de facto*. Rather, he has deductively unpacked the semantical content of the suppositions: "Grant that the planetary system is quasi-gyroscope, in balanced 'vortical' motion around its own freely moving (but physically unoccupied) center of mass; and grant that very little friction is operative in the configuration; and grant that in its trigonometric description $c_r, x_r x_s \geq 0$". From this springboard one easily leaps to the conclusion that all oscillations will gradually be damped down, even those initiated by the sudden introduction into the system of quite violent and dislocating energy spasms from 'outside'. Any oscillation that does not thus dampen down was too large to begin with!

The reasoning is magnificent: *if* our system were as supposed, *then* it would be secularly stable!

But for what reasons ought we to grant such suppositions about our planetary system? If we simply argue that by doing so a proof of the system's stability follows, the reasoning is not only circular, but viciously so. This would give us no reason for being convinced of systematic stability other than a merely hypothetical kind – and no reason for making our 'gyroscopic' assumptions other than that the consequences are reassuring. Newton did as much by leaving it all to God!

No, we need some independent argument either in favor of the system's quasi-gyroscopic nature, or else in favor of the system's stability – from either of which we can then infer to the other! *But we cannot at once assume both what is to be proved and also that from which the proof is to proceed.*

What other reasons are there, therefore, for supposing the initial suppositions about our planetary system to be fulfilled? The answer is clear: we must find independent *theoretical* data from which it will appear that our planetary system's gross motion is comparable to something quasi-gyroscopic, e.g., something like 6000 epicyclically-mounted roulette wheels on a Ferris wheel, itself mounted on a railway turntable, which in its turn ... etc. But to show the latter really is gyroscopically stable (if it is) one must undertake an immensely complex computation, determining the force components of each roulette wheel, and the nature of the many dynamical systems it forms with other roulette wheels to which it is mechanically linked, and the relationships obtaining between the mass-centers of every such roulette-couple, and then the compositional effects of all these on the gross gyroscopic balance of the Ferris wheel, which in its turn must be ... etc. Clearly this is the multi-body problem in gyroscopic-engineering terms. But this is precisely the kind of thing for which Laplace must find a general analytical solution in gravitational-dynamical terms if he is ever to proceed with his further deductions.

To show the planetary system stable in such terms, then, one must first address the multi-body problem. Since a general analytical solution to this problem is not yet indubitably secured (however probable) the 'gyroscopic' springboard assumed in any practical demonstration must be pieced together by successive approximation technique – the ultimate result of which can never be a *quod erat demonstrandum* of the Euclidean type.

Ergo, the proof without which Newton's synthesis must be adjudged seriously incomplete may not yet have been finally given. One's convictions as to the stability of our solar system indeed, may not actually *be* founded on any formal demonstration. Ultimately it may rest on an immense accumulation of inductive considerations. These may be quite sufficient for planetary theory. But we must remark to what degree the logic of the situation, and our expectations withal, have changed.

References

Bromwich, Thomas John I'Anson. 1906. *Quadratic forms and their classification by means of invariant factors*. London: Cambridge University Press.

Hanson, Norwood Russell. 1961. The Gödel theorem: An informal exposition. *Notre Dame Journal of Formal Logic* 2: 94–110.

Heaviside, Oliver. 1902. *Electrical papers*. Vol. 1. London: Macmillan & Co.

Jeffreys, Harold, and Bertha Swirles Jeffreys. 1956. *Methods of mathematical physics*. Cambridge: Cambridge University Press.

Lamb, Horace. 1920. *Higher mechanics*. Cambridge: Cambridge University Press.

Laplace, Pierre-Simon. [1829–1839] 1966. *Celestial Mechanics*. Trans. N. Bowditch. Bronx: Chelsea Publishing Company.

Moulton, Forest Ray. 1914. *An introduction to celestial mechanics*, 2nd ed., revised. New York: Macmillan.

Nagel, Ernest, and James Roy Newman. 1958. *Gödel's proof*. New York: New York University Press.

Rosser, Barkley. 1939. An informal exposition of proofs of Gödel's theorems and Church's theorem. *Journal of Symbolic Logic* 4 (2): 53–60.

Routh, Edward John. 1877. *A treatise on the stability of a given state of motion, particularly steady motion*. London: Macmillan and Co.

Truesdell, Clifford. 1961. Ergodic theory in classical statistical mechanics. In *Rendiconti della Scuola Internazionale di Fisica, Enrico Fermi, XIV Corso: Teorie ergodiche*, ed. P. Caldirola, 21–56. New York/London: Academic.

Weierstrass, Karl. 1858. Über ein die homogenen Functionen zweiten Grades betreffendes Theorem, nebst Anwendung desselben auf die Theorie der kleinen Schwingungen. *Monatsberichte der Deutsche Akademie der Wissenschaften zu Berlin*: 207–220. Also published in *Mathematische Werke von Karl Weierstrass: Erster Band*, 233–246. (Berlin: Mayer & Müller, 1894).

Chapter 5
Observation and Explanation: A Guide to Philosophy of Science

Preface

Norwood Russell Hanson (1924–1967) was a man out of his time, a character from the Florentine Renaissance growing up in the contemporary United States. Hanson showed how much can still be achieved, even within the professionalized techno-cratic society of the mid-twentieth century, by the true *amateur:* the man who makes himself the master of an art or science out of curiosity, love or sheer cussedness, quite unconnected with the business of earning a living. And he showed how such an amateur can achieve a kind of richness and variety of experience in a whole range of activities which spills over the boundaries between them. In this way, he became a "jack of many trades" and, in his own very special way, a master of them all.

When Russ Hanson died in April 1967, he was piloting his own personal Grumman Bearcat, in which he had been planning to attack the world airspeed record for piston-engined planes. (He had learned to fly as a U.S. Navy[1] pilot during World War II and, during his years at Yale, he would give summer aerobatic displays under the soubriquet of "The Flying Professor.") But he was also a talented musi-cian, improvising at the organ or on the horn or trumpet, and equally a remarkable draftsman, with a special flair for grotesque and imaginative figure drawings remi-niscent of Fuseli or Blake. If his own house needed attention, again, he would do the work himself; manhandling steel girders into position which many builders would have blanched at. Even the theoretical physics which he wrote about so authorita-tively and confidently as a philosopher was largely self-taught; and, by the last years of his life, he could discuss the most technical problems of quantum mechanics with leading professionals in the field, in a way that won their respect—in strange con-trast to the frustrated exasperation with which working scientists regard the argu-ments of most professional philosophers of science.

---

[1] Hanson was actually a Marine, but he ended up serving on the USS Franklin, a Navy aircraft car-rier. For details on Hanson's military service, see Lund (2010, 21–25). –*MDL*

© Springer Nature B.V. 2020
N. R. Hanson, *What I Do Not Believe, and Other Essays*, Synthese Library 38,
https://doi.org/10.1007/978-94-024-1739-5_5

Hanson got his university education largely as a returned serviceman, at Chicago and Columbia Universities, and he went on as a graduate to Oxford. There he quickly added a mastery of the methods of post-war British philosophical analysis to his earlier skills, and was appointed to a University Lecturership in Philosophy of Science at Cambridge University. After the Suez affair of 1956, disillusioned with Britain, he moved back to his native U.S.A. and threw himself into the task of organizing the newly-created interdisciplinary department for the History and Philosophy of Science[2] at Indiana University, which owes its continuing impetus largely to his efforts.

Hanson's essays and polemical writings cover the whole spectrum from philosophical logic to theology—the theology, needless to say, of an unbeliever; for in religion, as in all things, he was strongly counter-suggestible. Dogmatism, even in defense of views he happened to support, would rouse his disputatiousness; worse still than believing "the right thing for the wrong reason," was believing anything for no particular reason at all. The two books published in his lifetime, on *Patterns of Discovery* and *The Concept of the Positron,* were both of them intellectual plumcakes; variable in texture, but stuffed with good things. The essay which follows gives us more characteristic snatches of that flavory, idiomatic style, which he made so much his own and which his friends came so much to appreciate.

Stephen Toulmin
January 1971

5.1   Observation

Pascal placed man midway between the angels and the brutes. It is from this positioning, he felt, that the 'human predicament' arises. Science, the glorious achievement of modern man, is itself analogously situated between pure mathematics and raw sense experience; it is from the conceptual tension generated between these polar coordinates that *philosophical* perplexities about science arise.

This is the format of all that follows. Our 'Guide to Philosophy of Science' will course through a conceptual terrain of standard issues—focusing first on the Scylla of formalism, and then sighting on the Charybdis of sensationalism. Most philosophical discussions of science move initially toward the bare, jagged rocks of symbology, and then back toward the other extreme—the turbulent, teeming maelstrom of phenomenology. Frightening formalism to starboard; formless empiricism to port! The most fruitful of these *engagements du voyage* resist toppling toward either

---

[2]Actually, Hanson was originally hired into Indiana University's Philosophy Department. The History and Philosophy of Science Department, an entirely new phenomenon on the American higher education scene, was later founded by Hanson. For details on Hanson's career at Indiana University and the founding of the History and Philosophy of Science Department, see Lund (2010, 31–34). –*MDL*

disaster, steering rather along the thin line of balanced reason and cautious moderation faintly discernible between.

Our strategy here will be to delineate these extremes (of 'sensationalistic thesis' and 'formalistic antithesis'), noting some of their attractions and their disadvantages. A balanced 'middle of the channel' resolution (a 'stable and realistic synthesis'), will be the objective sought at each stage in the winding journey which we now begin.

Natural science is concerned with the facts of this world. The results of that concern are articulated in factual statements. (No collection of *non*factual statements could ever constitute a natural science.) But factual statements have this property: they are 'synthetic' *vis-à-vis* their sign-design. That is, every factual statement is such that its denial does *not* generate any formal inconsistency. S is synthetic (within a language L), if and only if it is consistent *while* not-S is also consistent; neither S, nor not-S, generates (in L) anything of the form Q *and not-Q*. Such an S could be a factual statement in L just because the question "Is S a statement of fact?" cannot be answered through analysis alone.

How then can one determine of any given factual statement S whether or not it is true—i.e., a statement of fact? Semantical and symbolical scrutiny is not enough; nothing by way of syntax study, or of meaning analysis, can single out via inspection alone some one S as acceptable, while not-S is discarded. This, because both S and not-S are logically consistent—which is just to say again that S, since it is a factual statement, is synthetic.

Observational experience is required to screen those factual statements which 'obtain' from those which do not—the former alone being candidates for inclusion within systems of informative natural science. (No collection of factual statements known to be *false* could ever constitute a natural science.)

But this makes the process of observation sound somewhat 'Pavlovian', does it not? It suggests that factual statements come in pairs (S and not-S), then to be 'subjected' to observational testing ('ding-a-ling'), so that either S or not-S will emerge corroborated (salivation!). It suggests something like 'acid' or 'base' being indicated as the response of a litmus test-paper thrust into some liquid. What is it about scientific observations which corresponds to such a litmus paper reaction? Where, and when, does data-registration *simpliciter* dominate? (How passive can an observational determination of truth or falsity get?)

Granted, in experimental situations involving titrations, or litmus paper reactions, or salivation in response to a bell-ring, the observer's sensation-report *may* be a datum of significance. 'Red now' or 'ding-a-ling' may be observation-signals of primary importance in such contexts. The tastes of acids, the odors of gases, the textures of surfaces, the colors of fluids, the warmth of circuits, etc.—these all require normal observers, with standard sense-neuro-circuitry, in order to determine which factual claims are true, and which ones false. 'The observer' in these cases is no more than an animated detector; depersonalized, he is just a reticulum of signal receivers integrated with considerable mechanical efficiency and reliability. To this extent and on this account, *any* normal person could make scientifically valuable observations. The color-blind chemist needs help from someone with normal vision to complete his titration work—whether this someone be another chemist, or his 6-year-old son, does not matter. But, now, are there any observations that the latter, the child, could *not* make?

Consider the following passage written by Pierre Duhem (1914, 218):

> Enter a laboratory; approach the table crowded with an assortment of apparatus, an electric cell, silk-covered copper wire, small cups of Mercury, spools, a mirror mounted on an iron bar; the experimenter is inserting into small openings the metal ends of ebony-headed pins; the iron bar oscillates, and the mirror attached to it throws a luminous band upon a celluloid scale; the forward-backward motion of this spot enables the physicist to observe the minute oscillations of the iron bar. But ask him what he is doing. Will he answer 'I am studying the oscillations of an iron bar which carries a mirror'? No, he will say that he is measuring the electrical resistance of the spools. If you are astonished, if you ask him what his words mean, what relation they have with the phenomenon he has been observing and which you have noted at the same time as he, he will answer that your question requires a long explanation and that you should take a course in electricity.[3]

Thus, to observe what Duhem's physicist takes himself to be observing requires somewhat more than normal vision. Optical signal-receptors, however sensitive and acute, cannot provide everything needed for observing electrical resistance. *Knowledge* is also presupposed; scientific observation is thus a 'theory-laden' activity (to use an expression, from *Patterns of Discovery* (Hanson [1958] 2010) which seems now to be in vogue). Brainless, photosensitive computers— infants and squirrels too—do not make scientific observations, however remarkable their signal-reception and storage may be. This can be no surprise to any reader of this book. That the motion of Mars is retrograde, that a fluid's flow is laminar, that a plane's wing-skin friction increases rapidly with descent, that there is a calcium deficiency in Connecticut soil, that the North American water table has dropped—these all concern observations which by far exceed the order of sophistication possible through raw sense experience. Nor are these cases of simply requiring physicobiological 'extensions' to the senses we already have; for telescopes, microscopes, heat sensors, etc., are not sufficient to determine that Mars's motion is retrograde, that blood poisoning is setting in, that volcanic activity is immanent. Being able to make sense of the sensors requires knowledge and theory—not simply more sense signals. (Understanding the significance of the signal flags fluttering from the bridge of the *Queen Elizabeth* does not usually require still *more* flags to be flown!)

This recognition of a strong theoretical element within scientific observation sometimes drives philosophers to hint that the incoming signals from 'the subject matter' are less important than they really are. For a Descartes, a Poincaré, or an Eddington, observation can shrink to being little more than the provision of values for variables in a theoretical algorithm—in a framework of 'understanding'. Laboratory instruments, measurement, and experimental design, for such thinkers, may seem geared only to the supplying of 'initial conditions'—the barest localized starting points for scientific reflection. Such conditions only resemble 'the given' within mathematical computation, the *occasion* for problems, not their solutions—

---

[3] This appears to be Hanson's own translation. –*MDL*.

not 'knowledge' properly so-called. Even as such, they must often be 'corrected', reshaped and processed for further usefulness within a computational system. The theorist presses observers with challenges like "To what degree are departures from the 'ideal case' attributable merely to the crudeness of the experimental apparatus?", "How fundamental to our *understanding* of phenomena are your detected deflections, error-spreads, frictions, dislocations, deformations, etc.—all things inseparable from your instruments and techniques of measurement?" "In short, where can we ignore the 'side effects' of the insufficiently sensitive probes you use (and which, alas, hold our attention) and ponder instead the *essential* aspects of the events themselves?"

Here, it is as if the 'conceptual shape' of one's theories, the posture and stature of one's presuppositions, determine where observations have to be 'cleaned up'—where they should be realigned and reprocessed effectively to be plugged into a science's theoretical framework, its structure for intelligibility.

Doubtless, recognizing this central feature of scientific observation is very important. Understanding of actual phenomena is often advanced by studies of ideal fluids, frictionless surfaces, strictly rigid levers, perfectly elastic bodies, infinite wing spans, one-dimensional translations, point particles, and 'pure cases' generally. When our ideas of processes are structured by such crisp conceptions, the thousand natural shocks of actual observation can be smoothed and made tractable in terms of 'what is reasonable'. Inexpert plumbing, bad carpentry, and poor laboratory-bench technique need not shape our comprehension of a science's subject matter. This attitude was well conveyed by Laplace when he wrote:

> Astronomy is a great problem of mechanics in which the elements of the motions are the arbitrary constant quantities. The solution depends upon the accuracy of the observations, and upon the perfection of the analysis. It is very important to reject every empirical process, and to complete the analysis, *so that it shall not be necessary to derive from observation any but indispensable data.* ([1829–1839] 1966, 1.xxiii)[4]

Although Laplace recognized the indispensability of observation at *some* point (if ever scientific theory is to be harnessed to the natural world 'outside'), it was yet his wish to keep the observational-descriptive content of analytical science down to the barest minimum. Thus the major function of the scientific enterprise—to wit, the attainment of theoretical understanding, of knowledge—should be hampered as little as possible by laboratory 'busywork'. Refinements in conduitry and circuitry, in beam-focusing and detector-positioning, in spectrometry, thermometry, and hydrometry—these may lead to more decimal places as one reports the results of measurements, but they rarely determine a new *form* for an equation, or a new kind of *inference* concerning an old subject matter.

Periodically, however, theoreticians get caught up in a 'so-much-the-worse-for-the facts' attitude. Historically, such confidence almost appears to be understandable—especially in the wake of 'discoveries required by theory' such as those of the antiproton, the antineutron, the neutrino, Anderson's positron, the planet Neptune

---

[4] The italics were added by Hanson. *–MDL.*

(whose discoverer was Leverrier the theoretician; not d'Arrest, Galle's assistant, the first man whose retina distinguished that new light-point), etc. But still, the philosophical 'middle way' must always be the one which recognizes *significant* observations within a science as those which at once meet the criteria of relevance embodied within extant theory, while also being capable of *modifying* that theory by the hard, stubborn recognition of 'what is the case', of *the facts*. Science does not make the facts, however much it may shape, color, and sort them!

5.2   Facts

Observations are *of* such things as particle-pairs, perturbations, and pollination. Facts are *that* (or to the effect that), for example, pair production occurred in a cosmic ray shower at *x*, *y*, *z* and *t*. Another fact is that our moon perturbs, or deflects, artificial satellites from their 'pure' (Keplerian) circumterrestrial orbits. Still another fact is to the effect that our sun is located 30,000 light-years from the center of our galaxy.

'That'-clause linguistic constructions are always close to any articulation of what the facts are—necessarily so. This should suggest some conceptual intimacy between what we count as facts and the language we state them in—or at least between the facts and the types of logical entity we designate as 'statements'. Statement *S* states *that x, y,* and *z*. If *S* is true, then *the fact* must be that *x, y, z*. Such a verification may have proceeded via a scientist's observation *of* whatever was described truly by *S*. (The conceptual lines are not sharp here, of course; the scientist might be said to have observed *that x, y, z,* thereby rendering the truth of *S* perspicuous. But the trend seems clear.) Our observations of, for example, flowers and bees, and what they do (*S*) may establish it for us as a fact *that S,* in which case— should we choose to express ourselves linguistically to that effect—we shall state *that S*. We observe objects, processes, and events. But facts must be a different kind of denotatum, logically different. We do not observe facts (what would they look like?). Facts are not objects or collections of objects or constellations of objects. Facts are *to the effect that,* e.g., a bee, while sipping a flower's nectar, gathers pollen on its limbs, later deposits it on other plants, thereby fertilizing them. A statement to that effect would be true, or false, in virtue of facts of this type—and not because of the simple existence of bees and flowers, and certainly not because such facts *are* bees and flowers, or their geometrical interrelationships, or true statements about them. *Facts are what true statements state.*

Attempts to construe facts as objects, or constellations of objects, have been notorious during this century. The motivation is always the same; if color-terms are to be directly correlated with colors, and names directly correlated with objects, then statements (presumably) are directly to be related with facts—such as might be photographed, transported, or boxed—just as with any arrangement of objects and processes. 'Red' links with the poppy's color; 'Fido' links with that kennel's occupant; so, also, what is expressed 'Fido's tongue is red' links with Fido's tongue as it is colored—i.e., the fact that Fido's tongue is red. The statement describes the tongue, states a fact about it. *Ergo* the fact is the red tongue—so slides the slippery argument. But the notions of photographing facts (red tongues?), of transporting what true statements state, of fencing in, or wrapping, what experiment and observation reveals to be the case (can we box the fact that Fido's tongue is red?)—something goes awry with such notions and with these ways of expressing them. We state facts, we list them, we file them for further reference. Object-clusters do not accommodate to such locutions. (We cannot state Fido.) Still, it is the hard, stubborn, objective, and intractable feature of 'the facts'—which *are what they are* irrespective of anyone's pet theories to the contrary— it is this that has drawn some philosophical worthies toward the position that facts are but another kind of furniture within the warehouse of the world. (Bertrand Russell's *Logical Atomism* lectures are a case in point.) In principle they are no different, for such philosophers, from object-clusters, event-constellations, and configurations of situations—all these being photographable, enclosable, datable denota, which facts are not. To this way of thinking, then, the *direct* outcome of experimentation, observation, testing, and measurement is, and always has been, the facts—the objects, events, and situations exposed on the bench, at the observatory, in the field.

Noting the conceptual intimacy which obtains between 'the facts' and statements of the facts, however, suggests to other philosophers that there can be nothing logically less complicated about facts than about statements themselves. Since statements are conceptually more intricate than names, so also facts must be conceptually more intricate than objects; more intricate than object-clusters too. The theory-laden character of 'the facts' soon comes to impress such thinkers even more forcibly than is the case with observation. For whatever it is 'out there' that makes us say (truly) that the space immediately adjacent to our sun is non-Euclidean, or that the symmetry properties obtaining within our universe indicate the existence of an antiparticle corresponding to each kind of 'familiar' particle now known—these 'whatevers' must count as *facts*. Such 'whatevers' are accorded 'fact-hood' because they 'anchor' the least vulnerable statements within extant theoretical physics. The philosophical tendency here, then, will be to construe 'the facts' as those objective organizations of the objects, events, and states of affairs within a scientific subject matter which render true the theories we do hold. The view thus arises that 'the facts' are just those conditions a subject matter meets such that a given theory might be applied to it—the boundary conditions. In that sense the facts are 'theoretically determined'—somewhat as the rules of chess determine what layout the chessboard must have at the onset, and what moves will be permissible therefrom so that the subsequent interchange could be describable as 'chess'. Thus, in a Wittgensteinian view:

> ...the fact that it can be described by Newtonian mechanics asserts nothing about the world; but *this* asserts something, namely, that it can be described in that particular way in which as a matter of fact it is described. (Wittgenstein 1922, 6.342)

'Possible science' is thus a potential infinitude of possible theories—scientific idea-games—some of which will apply to *de facto* subject matters, but most of which will not. Applicable or not, such concept-networks are identifiable statement-structures, as were the phlogiston and caloric theories. *The facts,* on this breathtaking view, are just those objective, 3-D conditions a subject matter must meet in order to qualify as tractable and intelligible through the lenses of *this* theory, or *that* one, or *those;* and in some cases, of course, 'the phenomenal facts' meet the boundary conditions of no extant theory whatever (e.g., inversion layers in ancient Greece, lodestone-magnetism for Charlemagne, firefly luminescence in Galileo's day, ESP today). Then, the subject matter in question is (temporarily, it is always hoped) 'beyond science'. With respect to such complexes of phenomena, considerable confusion concerning what *are* the facts always abounds.

Here once again we see philosophical attitudes (now those concerning *facts)* ranging all the way from brute empiricism to an almost abstract theoretical idealism. But here, as everywhere in the philosophical firmament, there is a *via media.*

Note a conceptual feature of *scenes* and *landscapes.* As a skilled artist confronts the scene at dawn he may be moved to convey those colors, shapes, and textures to canvas. After his having done so, we may remark the likeness between the scene 'out there' (to the east) and the scene we view on his canvas. He is indeed a skilled *landscape* painter! As the landscape gardener works through the actual trees and shrubs on the eastern slopes, so our artist 'works through' the corresponding patches on his canvas. The landscape is 'out there'; but it is also captured forever in his painting. 'Scene' and 'landscape' are thus Janus-faced terms. The complex of 3-D objects to the east at dawn is such that it *can* be captured on canvas: it is that *kind* of designatum. The scenes continually before our eyes comprise the *possibilities* inherent in objects and events *to be* captured on canvas, or to be fixed in photographs. The painting (if it makes a successful 'capture') will be 'true to life'. *The same scene* is thus apprehendable in several ways—'out there' *in rerum natura,* and as on canvas, in home movies, in mirrors, etc.

The analogy with *facts* should be quite apparent. The possibilities of their being described in this way, or in that way, constitute objective features of events and processes in our world. To the extent that our descriptions (rendered more articulate and precise through algebra) instantiate such possibilities—to that same extent they are *true.*

The scientific description is true, then, when it states 'the facts'. And again, what are 'the facts'? Just those *structural* possibilities inherent within states of affairs such that some statements made about these states will be certifiably true, and some will be certifiably false. *What* the statements *state* when they are true (and, of course, what they deny when they are false)—these are the facts. The facts, then, constitute true-statement-possibilities ('describabilia') within the subject matter. 'Fact' is also a Janus-faced term, then. The facts are 'out there' in the subject matter—'there' and potentially describable even before anyone has articulated them. Yet, once embodied within a language, those same facts are *stated,* i.e., expressed explicitly. Facts are 'out there', yet statable. Facts, then, are the *describabilia* of this world. Before being captured by language they are 'natural *describabilia';* after language-capture they are 'expressed *describabilia'* (i.e., described). So, just as *landscapes* are the structural possibilities actual 3-D countrysides present to painters who aspire to set out what is 'true to life' (in painted landscapes!), *facts* are those structural possibilities within (infinitely) diverse varieties of subject matters such that scientists who aspire to do so may succeed in stating of those subject matters what is 'true of life' (in their systems of statements of fact!).

So, 'taking cognizance of the facts' is much more than simulating and emulating a hypersensitive data receptor. On the other hand, it also seems to be more than just the clamping of a scientific theory's rules and definitions upon the world, thereby selecting of study only those subject matters which are 'cooperative' with the extant theories. Rather, 'the facts' emerge here as the world's possibilities for *being* described in some available language—which possibilities will be every bit as 'theory-laden' as the descriptions themselves are disclosed to be. (Could '$E = mc^2$' have expressed a fact a million years ago? For whom?) And this will be so whether those descriptions concern only simple color-registrations, as in titrations, or intricate subtle number-assignments, as within most standard cases of measurement.

5.3   Measurement

Once again the inclination is to view science's subject matters as chunks of the world, as 'out there'—reposed, quiescent, and richly bedecked with properties— passively awaiting our theodolites, telescopes, microscopes, balances, centrifuges, galvanometers, accelerators, etc. As the camera records what is posed and exposed in front of the lens to *be* recorded, so (apparently) these instruments of measurement objectively register and record the degrees to which certain objects, processes, and events possess and manifest certain properties.

Without doubt, what is derisively designated as 'dust-bowl empiricism' derives some of its appeal from such an uncriticized view of the nature of measurement. Only during the scientific revolutions of this century—Relativity Theory and in Quantum Mechanics—have modifications of such a pervasive, powerful, and perennial view been lastingly effected. When practical operationalists (like Mach and Einstein and Bridgman) began searching for the 'cash value' of terms like *mass* and *simultaneity* and *time,* a certain 'involvement' of subject matter and observer became clear. 'Interaction' is now the watchword.

What is it to say, as all celestial mechanicians before 1900 would have felt it meaningful to say, that an explosion in Alpha Centauri took place 'at the same time' as did some event here on earth (e.g., the eruption of Vesuvius)? True or false, such a claim would have at least seemed significant to everyone at Maxwell's Cavendish Laboratory. Since a photon of light emitted from Alpha Centauri would take over four of our years to traverse the distance to us, the synchronization of timepieces, and the calibration of all associated optical instrumentation, would present a calculational-computational problem of the greatest magnitude. The techniques of measurement used in so (relatively) simple a case traverse acres of physical theory, much of which would be 'built into' the 'measurement' which resulted. Just as Archimedes' principle can never be refuted by measurements made with a beam balance, nor Hooke's law falsified *via* readings rendered by a spring balance—these laws being the *basis* of these balances—so also nothing involving *terrestrial* chronometry and optical theory is going to be upset during our measurements of explosive disturbances near Alpha Centauri. These traditional disciplines are 'enshrined' within our measurements of celestial events. Whatever information our instruments *do* convey to us is what it is because such disciplines are the vehicle for interpretation of needle-deflections, signal-strengths and counter-clicks. To that extent there is a pervasive interaction between such events and our theories of measuring technique. Whatever numbers emerge from measuring encounters may be the result not of simple, objective data-registration, but of a most intricate enravelment of subject matter, probe, and theory. (Somewhat like using porpoises to gather information about whales! or like using treacle droplet-probes to gather information about hot syrup!)

Little need be added here to the immense literature concerning the quantum theory of measurement. Only note this again: that information from the microphysical world reaches us in units no smaller than h (the quantum of action), and is always and necessarily the result of an interaction between some microphenomenon and a macroexperimental probe. Since the effect of the probe on the phenomenon is incalculable (in principle), our information must always be related to the *system* of phenomenon-*plus*-detector—information which, again, is restricted to event units greater than h.

It does not follow from this, of course, that knowledge which accrues to us from such measurements is no longer 'objective'. Rather, we must now recognize that 'objectivity' (in its classical meaning) may no longer be an appropriate conception for isolated (i.e., detector-independent) particles and processes. It is always a system, an ensemble, of processes about which we gain objective knowledge in microscience today. Perhaps the idea that once we were able to obtain more than just ensemble-knowledge (i.e., knowledge of microindividuals), was itself unsound? We have displaced the notion of measurement, nudged it from its unexamined pinnacle of classical objectivity, to a turbulent flux of detector-and-detected—and even to an occasional deep of inconstant subjectivity.

Again, the reasonable way courses midway between: objectivity is no less available to us today than it was to our predecessors. But it can no longer be construed as an objectivity of isolated particulars 'out there', a construction that was always unjustified. Just as sociologists *can* report objectively about groups of which they are members, so also laboratory detectors can report objectively about intricate situations within which they are inextricably entwined and intertwined. Nothing in our responsible conceptions of what inductive science is will require radical modification because of this 'realistic' appraisal of measurement. Scientific measuring instruments are not passive blotters; but neither are they so disturbing that they churn subject matters like eggbeaters. Rather, they record the properties of complex phenomena by disturbing them in a controlled and largely calculable way. The surgeon must cut to cure; the experimental scientist must dislodge and perturb in order to learn of a subject matter's properties when unperturbed and 'according to nature'.

5.4   Induction

I once remarked of a senior scientist that he had had 40 years of experience. A critic rejoined that the individual in question did *not* have 40 years' experience; rather, he had had the *same* experience over and over, 40 years in a row. Is induction simply a rote repetition of stimulus and response, of anterior events followed by later events? Or can one *learn* something from induction—learn something about the *nature* of, and interconnections between, the phenomena before us, and not simply how they are sequentially distributed! Is induction a superficial survey of event-pairs, or does it permit us to peer 'inside' processes—to see what *makes* them 'tick', and not merely that they *do* 'tick'!

If *this x* is *y* and *that x* is *y*, and those, and those, and those—indeed, if *all x's* ever encountered have also been *y*, will the claim *'all x are y'* be just a kind of actuarial shorthand for saying quickly what experience has revealed at length, *seriatim* and in detail? Or will *'all x are y'* reveal something 'deeper' than we could have learned just through repetitive experience—something to the effect that there is something '*y*-ish' about each *x?* Every possible position on this spectrum has been entertained by philosophers partisan to one polar extreme or to the other. Reichenbach congratulates Hume for having been the first to recognize that all induction, however intricate and 'theoretical', is ultimately dependent upon 'inductio per enumerationem simplicem' (Reichenbach 1938, 389). On the other hand, Aristotle, and a millennium of Aristotelians, urge that from noting *this x* to be a *y*, and *that* one, and *those* too—one can become conceptually positioned to make 'an inductive leap' to the (unrestricted) conclusion *'all x are y'*, which latter somehow discloses the *essence* of *x*. For Aristotle, induction reveals that 'it is in the nature' of an *x* to be *y;* this cannot be a matter of logical necessity, of course, but it is, nonetheless, an unexceptional feature of the constitution of the actual world in which we live (*Posterior Analytics* II.19).

On this issue, as on most others, philosophies of science divide into (1) 'philosophies of nothing but', (2) 'philosophies of something more', and (3) 'philosophies of what's what'. Inductive generalizations (even when stated as Laws of Nature), are *nothing but* empirical expressions which sum over enormous ranges of repeated observation-pairs (1). Or, induction may be represented as a process through which, by experience, we learn something of the fundamental structure of objects, events, and processes—where this 'fundamental something' is always qualitatively more than is disclosed *via* mere repetition (2). [Human beings *learn* about nature through induction—they come to understand it; animals and machines do not (although they may gather a great deal concerning how best to avoid mishaps and how to function efficiently.)]

The reasonable way (3) again seems to lie between mere instance-enumeration, on the one hand, and mysterious essence-divining, on the other. Perhaps just by noting that induction is rarely undertaken aimlessly, without some theoretically determined objective, it will become clear that generalizations are usually built on experience which is itself already highly selective. Scientists are not like manufacturers of ball-bearing toy skates. It is not 'quality and quantity control' but *understanding* which is (or should be) science's primary objective. This was long reflected in the reference to physics as 'natural philosophy'—a scientific discipline which has always required learning what is the case concerning classes of phenomena earmarked, by experience and by theory, for further reflection and study. We learn what obtains within phenomena—what 'makes them go'—by way of our perceptual linkages with the world through sense experience. We *understand* those experiences, even when they occur in profuse and diverse arrays, only when we can pattern them within conceptual frameworks; these provide structures to 'the scientific mind', idea-structures which are *sometimes* related to the structures of processes 'out there' in the actual subject matter. Induction is thus an epistemic tube; if phenomena come through it in pairs, or in trios, etc., often enough, then they may he recognized as not 'merely accidental' *vis-à-vis* their correlation. But the tube still has to be *aimed* in a given direction of inquiry, just as a telescope must be intentionally 'pointed' (for some purpose) at some restricted portion of the sky. It is such nonaccidental features of our inquiries into the world which, when understood, render whole classes of phenomena intelligible. Such *guided* uses of induction, however, have made modern scientific experimentation a virtual embodiment of theoretical understanding—for every datum encountered is detected along a line of inquiry, within a framework of interest bounded by the criteria of relevance and significance which aims our efforts *this* way rather than *that* way. Experiments may indeed be 'the senses extended'. But sharp eyes without a quick brain make Jack a dull idiot—a 'telescope flailer'. Ingenious experimentation, without the constant control of careful theory, could soon overstock laboratories with 'number-finders', but leave them somewhat short on new directions for the scientific understanding. "The discovery of new facts is open to any blockhead with patience, and manual dexterity and acute senses" (Sir William Hamilton).[5]

5.5   Experiment

For Galileo experimentation was important, but only as an *ex post facto* display and confirmation of what (for him) had already been disclosed by reason. Once the world, as created by God-the-Mathematician, had surrendered itself up for geometrical description, its miniscule properties and hidden details were epistemically foreordained—just as are all the consequences within Euclidean geometry for any student who accepts the Axioms and the Rules. Then, setting out a lively demonstration of those truths (with sloping boards, pulleys, and wires) was about as necessary in natural philosophy as it was within geometry—namely, not at all. Such recourse was mainly for those too slow-witted to follow the argument. Still, Galileo would

---

[5] Hanson likely got this quotation from (Beveridge 1957, 144). The unacknowledged quotation from Hamilton there reads, "In physical sciences the discovery of new facts is open to any blockhead with patience and manual dexterity and acute senses." –*MDL*.

have viewed it as a cardinal sin for anyone who was unable to follow the argument *also* to have ignored the 3-D 'experiment'. Some of his contemporaries did just that—and, in so doing, they sinned against reason. For although the structure of experience was construed as geometrically designed, that same design was clearly *in* the experiment, just as it was also in the *argument* which articulated the structure of that experiment. (This echoes our earlier remarks on the concept of *scene.*) Physical reality appeared as a geometrical creation for Galileo; physical facts were structured *à la* Euclid. Phenomena, experimentation, and argumentation could all share the same structure. Indeed, they *must* do so even to be related as subject matter—demonstration—and description. So the same insights seemed available to the natural philosopher by either one of two different routes of inquiry: geometrical argumentation or laboratory experimentation. The structure of physical facts could be delineated by either kind of inquiry. Even so, for Galileo (as for many contemporary scientific heroes) the 'rightness' of an experiment, of its design, was to some extent disclosed in the degree to which it embodied purely theoretical arguments. The failure of experimental results to support anterior theoretical reflections—this has always been, for some, an initial indication of something wrong in the experimental design itself. Herein lies the power of *gedankenexperiments,* such as Galileo's Pisa-cannon-balls, Newton's bucket, Einstein's elevator, Schrödinger's cat, etc.; the theoretical issues in such examples just overwhelm the virtues of pushing or pulling or cutting or heating chunks of matter in order to *show* 'what is the case' to the unconvinced.

Contrast this view of experiment with a diametrically different one. The position parodied as 'dust-bowl empiricism' construes experimentation and controlled observation as the very source, the development and the fulfillment, of everything worthwhile in science. All else is "mere speculation", or even "metaphysics"! In extreme form a scientist so oriented will 'let the facts speak for themselves'; he will tinker, roam, and ruminate at random, giving 'the world' (i.e., his chosen subject matter) every opportunity to 'express itself'. Scientific theories, on this account, will be like X-ray photographs of what given subject matters reveal of themselves during careful, precise, quantitatively circumscribed experimental inquiry. Experimentation provides its own direction, on this view. Preconceptions, hypotheses, hunches, intuitions, and errant speculations will, apparently, be pulped beneath the relentless advance of such an experimenter. Instrumental accuracy, control, mensurational detail—these will become the criteria and the very consummation of careful inquiry, alongside which all the elaborate, clever constructions of abstract theoreticians fade into the oblivion of history (and even mythology).

How is it possible to articulate either of the above positions without a modicum of caricature? Caricature or not, there *is* a contrast to be drawn between such extreme conceptions of the nature and function of controlled laboratory experience. The one view is that excellence in experimentation lies at the terminus of successful theorizing—as a final corroboration of what reason suggests to be the case. On this account the experimenter is directed by considerations of how the processes he is contriving to set into motion are relevant to some conceptual framework, the latter being central to whether or not we *understand* a given subject matter. Experiment here is theory-laden, theory-directed, and theory-oriented. It is simply the probe which ideas, concept-clusters, and arguments extend into actual, 3-D subject matters. On the other account, however, theory is the *product* of experimentation. It is just the terse, elegant, symbolic embodiment of what the theoretician has extracted from out of the writhing, multi-parametric subject matter itself. Here the theorist is subject to the judgments of the experimenter. The latter will always be 'letting the facts speak for themselves', and will be rendering them as perspicuous as possible. The theoretician, in straining to 'see the reality beneath the facts', might sometimes suppress what is all too obvious in the experiment—searching beyond for the 'something more'. But such a quest too often exceeds what experience can sanction. Thus this is a counsel of restraint against unbridled theorizing. The creative imagination must always 'knuckle under' to the data, the evidence, the facts. One way of ensuring this may be to stress the 'shorthand' function of theories; i.e., they are just systematically neat description-sets.

Again, history of science supports both positions. Hoyle's steady state theory has just given ground before a fusillade of facts from quasar astronomy. Eddington's second edition of *Fundamental Theory* records the 'fine structure constant' as related to the number *137*—in accord with certain observations made after the first edition was published. Without warning he thereby modified the first edition, where the constant was theoretically determined to be related to the number *136!*[6] Mesons turned out *not* to be 'electrons with queer properties at high energies', as some theorists had urged (Wilson and Blackett, 1936).[7] Electromagnetic radiation turned out *not* to be uniformly continuous and undulatory—extant theory to the contrary notwithstanding (Planck, 1901). On the other hand, there was in fact a trans-Uranic planet (Neptune), just as theory required (Leverrier and Adams, 1846). There have been multiform discovered things like neutrinos (Pauli, 1929), positrons (Dirac, 1931), antiprotons, and antineutrons (Segre et al., 1956), as well as the planet Pluto (Tombaugh, 1931)—all as theory required. Finding very often requires knowing where to look, the former being a function of the latter— experimental discovery being a function of theoretical strategy.

So it would appear that the verdict of history of science is impartial as between these two philosophical claimants. Examples of (1) theory leading experiment by the nose, and of (2) experiment correcting, 'and even generating, theory—such are ample enough within the ancestry of science. The *via media* is thus somewhat difficult to discern in this context. But clearly there can be no 'all or nothing' and *final* philosophical answer to the question 'What is experiment?' *Experimentation as demonstration of, or as corroboration of, theory* is surely different from *experiment as a generative source of theory.* When laboratory activities are this diverse, it is idle

---

[6] See Eddington (1944, 216) for a discussion of the reasons for the change to the value of the fine structure constant. The whole matter was complicated by Eddington's having died, in late 1944, with the work in an unfinished and unedited state. See Slater (1957) for a discussion of the difficulties attending the publication of the work. –*MDL.*

[7] In what follows, Hanson is referring to developments and discoveries by scientists and years, and not necessarily to publications. –*MDL.*

to seek a single philosophical formula to embrace everything called 'experiment'. Better to explore each case of inquiry on its own merits, learning thereby what epistemological or semantical or methodological role *this* individual experiment may have played relative to *this* particular theory. (A single given experiment may impinge upon different theories in quite distinct ways; it may bear on the same theory in different ways at different times.) Better also to ask how *this* theory may have been supported, defined and clarified by that particular experiment. (A given theory may relate to a host of independent experiments in a host of conceptually different ways.) What a monumental mistake it is, therefore, to seek some quasi-causal connection operative always between the design of an experiment and the creation of a theory. As if the idea of *cause* were sufficiently clear even at the level of billiard balls! It is not. *Eo ipso* it is not generally clear how experiments cause theories to possess certain properties, nor how theories cause experiments to have whatever design characteristics they may manifest.

5.6 Causality

A funny thing happened to 'cause' on its way from the Lyceum. Aristotle's word, αἰτία, as we still find traces of it in terms like 'aetiology', was beautifully articulated in The Philosopher's Doctrine of the Four Causes. Therein Aristotle was concerned with the reasons for, or the explanations of, distinguishable aspects of particular happenings. Of the massive Mayan earthwork structure which houses Yale's Tandem Van der Graff Accelerator one could be expected to ask 'why?' (Frank Lloyd Wright's constant question.)

Why what?

What *is* the question? Is one concerned to know what that great mound *is?* What is it meant to achieve? (John Dewey's constant question.) What is inside of it; how is it designed? What makes it all 'go'? (James Clerk Maxwell's constant question.) Of what material is it constituted? These are requests for an explanation of the genesis, the design, the *modus operandi* and the objectives of this imposing scientific addition to the Ivy. And the cloven hooves of (1) *material,* (2) *efficient,* (3) *formal,* and (4) *final* causation are clearly chiselled into such questioning.

1. "Explain to me what the 'Emperor' accelerator is made of—the metals, crystals, plastics, etc." Is there any wood in its construction? Or silk? Or 'animal fibers?
2. "What is the reason for the heavy earthen mound over the great instrument?" Why not a thinner, more appealing-to-the-eye metallic shell? Why not reinforced concrete?
3. "Make me understand how such an accelerator works, its design. "Is it like a synchro-cyclotron? How does it differ from Stanford's linear accelerator?

4. "What are the expectations, the intended accomplishments, of such a machine? What did Yale and the NSF hope to achieve?" What will we have learned by 1975 which, without Emperor, we should not have suspected?

Aristotle's *Metaphysics* journeyed to the Near East, from whence it was slowly percolated northwest to Latin-thinking lands—considerably the worse for its orientation, αίτία became rendered as *causa,* a term which in ancient Latin has much the same significance of the original Greek term; that is, the four *causae* also concern explanations of, or reasons for, things being as we find them to be—their aetiology. The Scientific Revolution, however, had a forceful effect upon our general understanding of the nature of causation. Efficient causation—the 'go' of things, the pushes and pulls, the drives and linkages, the perturbations and deflections—snared the attention of most philosophers, and became so energetically articulated within the emerging sciences of Descartes, Galileo, Barrow, Newton, and Leibniz that it by now seems extraordinarily difficult to think of causation in any sense other than that focused upon in the expression 'efficient causation'. Philosophers still become perplexed concerning what kind of *efficient* causation, final causation really is! What kind of 'nudges from astern' are material and formal causation? Due to such anxious confusions final causation (a pull from the future?), was banished peremptorily from natural science; material and formal causations were discussed, if at all, only *sotto voce.* Indeed, the whole *idea* of causation developed in remarkably Baconian terms, in the sense that *x* was construed as the cause of *y* if and only if the existence of *x* could 'bring about' the existence of *y,* or if the absence of *x* could prevent the presence of *y*. A cause came to be thought of as a trigger; you caused *y* when you contrived to bring about a chain of events which terminated in *y*. There is no dearth of this 'Rube Goldberg' notion of causal chain efficacy even in the recent history of experimental science. (See Hanson 1955) The challenge too often is that of seeking to fabricate in the laboratory complex conditions which seemed before to obtain only in nature. Wohler's laboratory synthesis of urea was a triumph of experimental science because, through its ingenious precision, its controlled and quantitative care, a substance was produced which had theretofore been construed as beyond human contrivance. Similarly with some large proteins, and a few short-lived microparticles. Ultimately, experimental science has come to seem to some an enormous 'Erector Set' challenge such that, by analysis and decomposition of natural events, men can conspire to construct corresponding events in the laboratory— finally to 'bring about' whatever 'natural' state of affairs one could adequately describe. Or, in experimental medicine, the challenge often appears to be to analyze a malady such that the appropriate *breaking* of a link in the 'causal chain' will prevent some bodily malfunction. Either way, the conceptual intimacy between all scientific experiment and 'causal chain' laboratory productions has suggested to some thinkers that science be construed as a sophisticated engineering operation, replete with levers, triggers, wires, pulleys, circuits—indeed all the paraphernalia that science in the Michelson-Millikan tradition does actually require! The creative scientific imagination may just be our scientists' ability to imagine laboratory conditions which will create naturally appearing phenomena (e.g., urea, lightning, chemo-luminescence).

Yet, there must be 'something more' to this kind of account. The request for the cause of an event is still a request for some *explanation* of that event. It is a plea for understanding—a plea that the event in question be rendered comprehensible in terms of other 'unsurprisabilia' known to obtain. (As Peirce intimates, a perplexity X is explained when it is shown to follow "as a matter of course" from the unperplexing y and z.) The 'dust-bowl' conception of causality blows away before a simple example like this: An airplane crashes; the pilot is killed; the FAA seeks for the *cause* of the accident.

Consider the possibilities:

1. The engine stopped, at night, over the Rockies.
2. Insufficient care had been given, during the last 100-hour inspection, to the fuel strainers, which *post mortem* examination proved to be clogged.
3. The pilot had probably not acquainted himself with meteorological conditions en route; at least the FAA Flight Service has no record of his having done so.
4. The weather data broadcast during the fatal flight was not current for the locale of the disaster.
5. Local thunderstorms hampered radio reception and transmission.
6. The pilot was not in practice *vis-à-vis* night flying and instrument procedures; his log book records his most recent night flight as having taken place 6 months previously.
7. Financial and personal anxieties affected the overall state of the aviator's psyche: so testifies his next of kin.

Now *all* these states of affairs could obtain simultaneously. Within suitable and specific frames of inquiry each one of references (1)–(7) cited above could be designated as *the cause* of the accident. To the aircraft designer stoppage of the engine was the cause. (It won't fly without power.) To the repair station supervisor shortcomings in the inspection procedure led to the accident. (The engine won't run without fuel.) To the psychologist pilot-anxieties were at fault. (A man can't 'think instruments' with a brain soaked in worry.) To the FAA flight examiner the lack of recent practice was responsible. (Rusty pilots, like rusty nails, don't drive well.) What counts as *the cause* of such an event will, in most cases, be that happening which (within a given framework of orientation—aeronautical, familial, legal, psychological) will render it intelligible that the accident took place at all. Such specialist-examiners as these will designate some anterior state (i.e., the clogged strainer, the lack of proficiency, the anxiety, etc.), concluding therefrom that the ensuing accident was 'all but inevitable', given such an antecedent preparation as that.

Thus the assignment of a cause is also a highly 'theory-laden' undertaking. Extensive networks of theoretical concern overlap upon the event in question. It is then a matter of the specialist's own theoretical posture, his interests, and his immediate professional concern which will load the very language used in his description of the accident with 'proto-explanations', with tacit semantical commitments, any one of which may make the event intelligible for someone—the natural terminus of some sequence of happenings.

Once granting all this, however, the pushing and pulling, the linkages and feedbacks—all so dear to the 'causal chain' view of experimentation—these fade before a more sophisticated conception of science, one which seems to concern only earlier and later states of affairs, theoretically construed.

Charge separation? Discharge!
Satellite deceleration? Fall to earth!
Tilt airfoil up? Increase drag! Etc.

This has become so apparent within theoretical physics that the very notion of *cause* has virtually been exorcised. Consider astrophysics and cosmology, within which disciplines the mathematical treatment of 'cause' and 'effect' is indistinguishable from calculations involving the time parameter t; state descriptions of events at $t - \Delta t$, and $t + \Delta t$ are construed as doing everything for science that 'traditional talk' of causes and effects used to do. Since a designating of the *causes* of a phenomenon is itself a theory-vectored performance, any careful explication of the full theory, plus detailed references to earlier and later states of the phenomenon in question, *must* be operationally equivalent to our more anthropomorphic ordinary language—that involving physical exertion in pursuit of physical goals. Much of our causal discourse derives from the recognition that events are often *effective*. *(Human beings* are effective, sometimes, in bringing about what they desire. But are other agencies within the cosmos similarly effective, for similar reasons? Do extragalactic processes cause specific things to happen as *desired?* As *they* desire?)

Within quantum mechanics the 'de-contentization' of causal talk has gone even further. It is not just that somewhat less anthropomorphic chat has been substituted for 'classical' causal discourse inside microphysics; rather, the very logic of microparticulate state descriptions is in many ways *incompatible* with the conceptual framework which structures our everyday ('classical') thinking about causes and effects. That is, *complete* state descriptions (i.e., 3 sharp spatial coordinates, plus a precise specification of energy) of individual particles at times t, $t - \Delta t$, and $t + \Delta t$—these are wholly ruled out of quantum mechanics by the 'formal rules' which structure that complex discipline. *Partial* state descriptions of microphenomena ('partial' as against the 'completeness' possible in a classical sense) constitute the maximal theoretical and epistemological possibilities within contemporary microphysics. As a matter of the *logic* of quantum mechanics, there is a theoretical limit to the joint precision obtainable for each of two conjugate parameters, such as *time* and *energy* (Heisenberg, 1927) or *position* (*P*) and *momentum* (*M*) (Bohr, von Neumann and Dirac, 1928).[8] The latter parameters are treated as operators within a noncommutative algebra such that $PM - MP = n$ (some number other than 0) (Graves, 1854). It should be stressed, again, that this is not merely technological limitation—something resulting only from the gross crudity of our present probes.

---

[8] Throughout this paragraph, Hanson is referring to the dates of discoveries, not necessarily specific publications. –MDL.

It is, rather, a feature of the rule-network— the formal concept-framework—of the mathematical algorithm of quantum mechanics. The complete state descriptions required in classical cause-and-effect relationships demand a thoroughgoing commutativity between all dynamical operators, such that it is theoretically irrelevant whether one determines first a bullet's velocity and then its position, or vice versa. This independence of dynamical operators constitutes a possibility totally excluded from the formalism of any workable version of quantum mechanics set out during the past 40 years, e.g., those of von Neumann and Dirac, 1930–1935. (Some 'systems' of microphysics have been speculated about, systems which abrogate these 'Uncertainty Relations'; but they turn out to be so badly attuned to the experimental facts that they should not be called 'quantum mechanics' at all. The author sees no reason for not including in this category the work of De Broglie, Bohm, Vigier, Bopp, Janossy, and Alexandrov. The parallel philosophical speculations of Popper, Feyerabend, Mehlberg, Toulmin, and other 'counter-Copenhagen' interpreters might also be noted here as being somewhat out of touch with the experimental realities of contemporary Elementary Particle physics.)

What is the philosophical upshot of such an oscillation between the chainlike conception of causality ("For want of a nail the shoe was lost; for want of a shoe the horse was lost... all for want of a nail.") as against an abstract representation of theoretical parameters such as typifies modern physics, within which a classical concept of causality is difficult even to detect? It is just this: what is even to *count* as a causal connection between phenomena within any context always depends upon one's special queries concerning the subject matter in question. That is, a single event-sequence 'viewed' via two different theories, might suggest quite different candidates for the status of the *cause* and the *effect*. (The "cause" of conflict in Vietnam has been assigned to notoriously many and diverse situations and individuals.) Moreover, some theories seem not to require the causal ideas at all!

Perhaps the 'middle way' here would be just to acknowledge that much of our everyday experience, our thinking, and our discourse, *does* depend upon classical conceptions of causality. (Magistrates, policemen, mechanics and plumbers cannot afford the luxuries of algebraic abstraction.) Laboratory experience 'links up' at many points with such everyday experience; to that extent it will always seem (to some degree) natural for scientists to discern and identify those linkages between experiences inside and outside the laboratory (where the causality concept is most applicable). Thus even in those theories in which *cause* and *effect* are very much modified concepts, or even dispensed with altogether, when such theories make contact with laboratory experimentation and observation (as ultimately they *must* do), there will be a human tendency to accommodate theoretical discourse to classical notions of causality—even when doing so can be somewhat misleading. The universal design *may* be that of a pulsing, organismic abstraction; but our representations of it, like our talk about it, will always 'click-click' off in single-file orderings of words, formulas, descriptions, and experiments. This is part of the price science pays for analysis. For analysis of complex wholes must be unit by unit. *We* are beings who are effective to the degree that we can cause things to come about, piece by piece, as we please. This is true also in scientific laboratories and for much the same reason. When our theories, however abstract, are linked with activities in the laboratory (as they must be, sooner or later, in order to be intelligible to us sentients), further associations with the causal nexus are inevitable.

But what is semantically inevitable in the course of explaining natural phenomena need not be pernicious, not so long as we remain aware that our explications of microtheories and macrotheories via causal talk could be misleading if construed as being literally true of the theory's conceptual fine-structure. Causal discourse seems to be most effective when explaining phenomena 'across' languages— when discussing quantum mechanics with engineers, or general relativity with amateur astronomers. It is sometimes dubious as an explanatory reference within a single language. Dirac and Heisenberg have no need of the causal hypothesis when discussing with each other the present state of their perplexing art.

5.7   Explanation

Causal explanations are important to us. There are, of course, many other ways of rendering phenomena understandable. A drawing of the heart is not a causal explanation. But philosophical controversy concerning explanation has often placed causal explanation at the forefront.

What is it to explain a perplexing natural phenomenon? Within a diversity of answers to this question, the thesis of Hempel and Oppenheim deserves special attention.

Many well-educated persons are still capable of being surprised when, after casual star-gazing for a few nights, they note a bright point of light to have come to a halt, and then to move in the direction opposite to its original course. Even today such individuals might request an explanation of so startling an observation. What can they be told? A modicum of heliostatic planetary theory will be gestured at, of course, with some inevitable Newtonian asides. Then some further, specific, references will be made concerning the joint-distribution of the planets, and stars, as they appeared on the celestial globe at some earlier moment in time. This 'state description' of the planetary array at $t - \Delta t$, plus some understanding of the dynamics of our local system, will quite often resolve the perplexity and allay the surprise of our sky gazer. For what had seemed problematic was then inferentially linked to prior conditions, none of which was in any way problematic. The retrogradation follows "as a matter of course" (Peirce). Psychologically it then appears that explaining some surprising x consists in decomposing it into smaller elements each one of which reflects some previous commitment totally lacking in surprise or novelty of any kind. Philosophically this may be put, as Hempel put it, by noting that an anomaly is explained by tracing it back, through laws, to initial conditions established through observation. (This is a kind of logician's analogue to 'causal chain' thinking. Thus, if the *cause* of the kingdom's collapse can be traced back to the want of a nail in a horseshoe, then the explanation of the kingdom's collapse consists in tracing back through a statement-series until one reaches a premise from which the entire series is generable—including the consequence that the kingdom will collapse.)

But if *that* is what explaining consists in, then one might have *predicted* the 'anomaly' earlier when, while confronted with initial conditions and Laws, it could have been deduced that, e.g., the planet would appear to halt and then move 'backwards'—just as we can now predict that a Piper Cub will appear to move backwards from the window of a Boeing-707!

Since what is predictable can hardly be anomalous, explaining x becomes tantamount to showing that x is predictable, i.e., could have been predicted! Premisses describing observations at $t - \Delta t$ entail conclusions describing events at $t + \Delta t$.

Explanation and *prediction* are thus conceptually linked within the Hempel-Oppenheim account. Explaining x is predicting x after it has actually happened. (Clearly, the predictable is not a matter for perplexity.) Predicting x is explaining it before it has actually happened. (What could be more predictable than a recurrent phenomenon which is nonproblematic?) Moreover, this relationship between the concepts of *prediction* and *explanation* must be 'managed' within a deductive framework—a theory. The latter allows one to infer from initial conditions (through laws) to predictions of future states. It also permits one to reason from observed anomalies, 'back' through laws, to initial conditions whose lack of novelty leaves nothing to be perplexed about, at least not within the original context of inquiry. (The 'arrow of inference' has a different 'sense' within these two undertakings. Inferring to conclusions from known premisses is radically different from 'inferring to', or 'reasoning towards' premisses from known conclusions. Of this more later.)

This analysis reduces questions about explanation and prediction to questions concerning whether or not there are *deductive connections* between anomalies and initial conditions—whether, that is, there exists a theory within whose systematic capillaries one's surprise can be deployed, diverted, scattered, and diluted. Big question marks disappear when one attends to the sharp inkdots of which they are constituted.

Several critics have argued that this makes of theories little more than inferential connecting-rods, or connecting-reticula. *Any* calculus which allows one to 'predict' future states of affairs (however strange the theory and incomprehensible the future state) would thereby also be the instrument through which explanations of those future states must also proceed. The way is open, apparently, for all manner of 'nutty' correlation-schemes such that whatever (e.g., increased sunspot activity) inclines us to predict some future state (e.g., a wheat failure in Kansas) would thereby also have provided the conduitry for explaining the latter. Dissatisfaction with this has been expressed by many philosophers; Schoolmen anxious over the fallacy of 'post hoc ergo propter hoc' would have been predictably anxious over Hempel's account. Many philosophers are not content to construe theories merely as 'predicting calculi '. Still less will they grant that an anomaly has been explained when one merely designates other conditions from the obtaining of which that event could have been predicted (e.g., from the distant storm warning we may predict the high winds; but are the high winds 'explained' by the storm warning? Are they explained by the cumulus mammatus and nimboid clouds everywhere above?)

Since we have found some point in contrasting radically empirical attitudes toward a scientific conception (e.g., observation) with alternative abstract treatments, it might be worth attempting that again. The Hempel-Oppenheim account of *explanation* and *prediction* is surely a theoretician's delight. It suggests that an explaining of *x* is not a rubbing of one's nose into *x*, or an attempt to empathize with the 'pure essence' of *x*. Rather it is an inferential-linking of *x* with a variety of other nonproblematic data, or data-claims. This delineates an important feature of theories themselves; linking the unfamiliar with the familiar has always been a glory of theoretical science.

Now, what empirical counterposture is to be adopted in contrast to the relatively formalistic and abstract analysis of Hempel? Perhaps it is this: there is no substitute for old-fashioned familiarity when one seeks to understand a subject matter. Truly, there is little to be explained (at least about fish) to the old fisherman who, like his father and grandfather before him, has lived all his life with net and hook, gaff and oar. What questions will perplex *him?* What will the 20-year-old ichthyologist explain to *him?* There is something the old sea captain has which the young fluid mechanician lacks. There is something the experienced electrical repairman surely has which the junior electrodynamicist may lack. Deep and abiding familiarity with a subject matter can render it totally understood, unproblematic and comprehensible—sometimes in the face of a total lack of theoretical or inferential sophistication. Midwives do not have records remarkably inferior to that of M.D.s. Or will we say that the ancient mariner, since he lacks calculational skill in hydrodynamics, therefore does not understand the sea around him, and could not explain its properties to others? That would be too absurd. Must we pronounce the midwife too ignorant of the process of childbirth fully to comprehend the vital drama being enacted before her eyes? Doesn't the senior electrician even know what he is doing? Is the algebra of electron theory *that* critical to his work?

Thus the suspicion of some philosophers that explanation (in Hempel's sense) may be possible without understanding. They find the equation 'Premiss for $X = $ Explanation of X' repugnant. And in our counterpoised empirical view, understanding may be possible in the absence of any 'Hempelian' explanation. Anatomists are not notorious as keen arguers. Perhaps this latter point need not be taken too seriously, however; a pancreatic sympathy should never be confused with articulate and detailed understanding. Nijinsky understood the dance. But, apparently, he could not explain it to others. He was inarticulate about it; others could not understand *from what he said* what the dance was! Similarly the Wright brothers understood flight; but they were largely inarticulate with respect to it. Others could hardly gather, from their words, what flight was. The brothers were powerless to make them see. Our midwife, electrician, and fisherman *could not explain* (to others) childbirth, circuitry, and seasonal spawning. The *sentiment of comprehension* should therefore never be confused with the *structure of explanation. Feeling* and *logic* are as different as *brain* and *mind. Knowing how* and *knowing that* are as unlike as *retinal reaction* and *observing*. The distinction between understanding in the sense of intuitive familiarity and understanding in the sense of rationally comprehending the 'go' of things must never be collapsed.

Still, the middle way might again be the one which seeks to gain strength from both positions. What can be wrong with our seeking examples of scientific theory which are capable both of explaining *à la* Hempel and of providing understanding and illumination of the nature of the phenomena in question! Even if distinguishable, the two are genuinely worthwhile objectives for scientific enquiry; they are wholly compatible. And, it may be noted, the second is unattainable without the first. So although Hempel's account of scientific explanation may not be sufficient, it seems to be necessary. Ontological insight, unstructured by quantitatively precise argument and analysis, is mere speculation at best, and navel-contemplatory twaddle at worst.

5.8   Theories

In his *Syntaxis Mathematica,* Claudius Ptolemy put together a detailed calculational scheme of prediction. It was quantitatively accurate to a degree unsurpassed until late in the sixteenth century. Predictions of the future positions of the planets were thus genuine inferential possibilities within Ptolemy's astronomy. But, by his own account, and by way of subsequent criticisms advanced by timid heliocentrists and Schoolmen, no explanation, no understanding, no comprehension of the planets, and their interrelationships, was forthcoming from the hand or mind of Ptolemy. His positional astronomy was restricted to studying the kinematics of otherwise inscrutable lights in the sky. Understanding that they moved, and that scholars were able roughly to predict where they would move to, was totally different from understanding *why* and *how* they moved as they did. Copernicus' heliocentric alternative was, at first, not as successful a predicting device as was Ptolemy's *Almagest.* But it did offer a theory, a conceptual framework, an idea-structure within which one seemed able to relate the actual behavior and appearances of the planets (i.e., their observed kinematics) with a physical account of what sort of things such objects really were. What has percolated through to us is the Copernican recognition that our understanding of what planets *are* is intimately connected with our ability to predict where they will be at future times, and to describe precisely where they are now.

So, in this historical example, we have an instance of two different theories (Ptolemy's and Copernicus') which were *not* equal in explanatory power, despite the fact that they were (for a short time) equally efficient as predicting machines. Put in another way: the inferential connections proposed by Ptolemy, despite their success in prediction, did not foster an understanding of the heavens to the same degree, nor in the same way, as did those inferential connections offered by Copernicus. Yet these latter were no more 'successful' in predictions than those within the Ptolemaic alternative (at least this obtained during the late sixteenth century).

At work here are different notions of the nature of scientific theories. What are theories? What are they supposed to do?

Hempel's account of explanation may be but a specialized reflection of an overall view of theories. If the major responsibility of a scientist is to supply statement-systems for deducing future state descriptions from earlier ones, then to speak of *x* as 'explained' is to say only that it has been located within an acceptable inference network. Prom *any* description over which the theory and its sundry laws range one should be able to infer to any other intratheoretic description whatsoever. (The inference may be *deductive*—toward the bottom of the page; or retroductive—toward the top.) But although this may be a necessary feature of any system of propositions 'properly' to be called "a theory", it is surely not a sufficient condition. It must always be logically possible for two theories, I and II, to be equally powerful in prediction yet wholly dissimilar *vis-à-vis* the degree to which they are felt to give an 'understanding' of their single subject matter. The astronomies of Ptolemy, Copernicus, and Brahe, as widely understood in A.D. 1600, were indistinguishable from the point of view of their forecasting capabilities. But that they constituted different idea-frameworks about the planets was clear enough within the intellectual revolutions of the seventeenth century. (Again, wave mechanics and matrix mechanics were shown to be predictively equivalent in 1926 (by Eckardt and Schrödinger), and equivalent in an even stronger sense in 1930 (by Dirac, within his operator calculus). But Dirac himself has recently been lecturing to the effect that 'the conceptual pictures' provided by wave mechanics and matrix mechanics are so different as to make one of them far preferable to the other. The *understanding* of microphenomena provided by one is different from that afforded by the other—despite their indistinguishability at the level of observational number-production).

In practice, therefore, distinctions *are* made between (1) theories that interlink descriptions within arbitrary inferential networks, and (2) different theories, the inferential linkages within which are patterned in terms of idea-clusters, analogies, and models such that to have succeeded *both* in inferring 'anomalies' from initial conditions (via the standard principles of deductive inference) and *also* to have placed that 'anomaly' within an intelligible framework of ideas (wherein further principles are now construed as 'laws of nature'), is to have *explained* the phenomenon in question in the fullest sense modern science can provide.

'How to succeed in prediction without ever explaining anything' is thus something more than a parody of the Hempel position. Rather, it indicates that view as not having said enough about explanation (and about scientific theory) to exclude *'mere* predicting devices' as being serious candidates for such titles. There *are* (and have always been) nonexplaining predicting devices in the history of science; philosophers will insist on being shown how, on the Hempel theory—all of whose criteria are met, e.g., by Ptolemaic astronomy and by mere correlation studies in several disciplines—such nonexplaining predictors are yet to be excluded from the circle of genuine scientific explanations and scientific theories, in the fullest sense of those expressions. Hempel's view needs supplementation, not revision.

5.9 Laws

The framework that structures an explanatory scientific theory derives its shape from the Laws of Nature set high up in the deductively fertile realms of the algorithm. The understanding of a subject matter conveyed *via* a theory is connected with the idea-Gestalt packed into each law. Since laws such as $F = G(m_1m_2)/r^2$ are replete with variable terms, they are not, like propositions, directly true or false. They are not propositions at all, but rather *propositional schemata*. Of course, if the numerical values for each of the variables is specified, or if the entire expression is universally quantified, the result *will* be a proposition. It will then be true if it expresses a 'law of nature'. However, philosophers have learned that law-statements are not like ordinary statements in any significant way.

'Ordinary' empirical statements can be located in logical space by delineating their (1) Syntax, (2) Semantics, and (3) their Epistemological status.

Syntax (1) concerns what might be called 'sign-design'—such that to designate a proposition as 'synthetic' is to characterize the symbol-structure of that assertion. Thus S is *synthetic* if and only if its negation, not-S, entails no inconsistencies (i.e., nothing of the form Q *and not-Q*). Knowing this much about a proposition is, of course, not yet to know anything of its contingent truth or falsity; either S or not-S constitute logical possibilities: both are consistent. Reflection alone is thus insufficient to determine the truth value of S, or of not-S. Which brings us to the second, Semantical, point above.

Besides being synthetic, empirical claims are usually vulnerable; this is the locus of Semantics (2). There is nothing about any uninterpreted cluster of signs, whatever their structure, which relates to vulnerability or invulnerability in any sense. Sign-design *simpliciter* is sense-neutral. Some consideration of the *meaning* of symbols, such that on one interpretation a claim may be defeasible, while on another the possibility of counterevidence may be inconceivable—this is the thrust of the vulnerable-invulnerable dichotomy intended within (2) above. Invulnerable claims (whatever the genesis of that invulnerability) are often designated as 'necessary' or 'necessarily true.' Vulnerable claims, on the other hand, are said to be 'contingent', e.g., on the way the world is, or on the rules of the game, or on the conditions of inquiry within a given context.

Now, besides being constituted of a synthetic sign-design (1) and of a contingent (vulnerable) semantical status (2), empirical claims are such that the information they carry can be gained only through experience of one kind or another. Since reflection alone is insufficient to decide the truth or falsity of a synthetic/contingent claim, something else must be involved. *Experience* is that 'something else', which is clear from the attempts we make to justify such claims against all challenges to the contrary. Such a justification would be logically unlike that appropriate to demonstrating that, e.g., all equiangular triangles are equilateral.

Empirical propositions are, therefore, (1) synthetic, (2) contingent, and (3) *a posteriori*.

Other propositions, however, true or false in a genuine sense are different from empirical propositions in matters of syntax, semantics, and epistemological status. Claims like 'All fathers are parents', 'Bicycles have two wheels', and 'All equiangular triangles are equilateral' are (re: their sign-design) the opposite of synthetic. The denials of such claims generate inconsistencies—by which is meant no more, at this stage, than sign-designs of the form Q *and not-Q*. If such a sign-design as this is diametrically opposed to that set out earlier under the title 'synthetic', we might designate such a propositional state as 'analytic'. A proposition will then be analytic if and only if its negation *does* generate (via rules of the 'language' within which it figures) some symbol-cluster of the form Q *and not-Q*. (Nothing has been said yet about the semantical status (2) of S, or of Q *and not-Q*.)

Moreover, if synthetic claims are vulnerable, their opposites (i.e., 'analytic claims') might well be termed 'invulnerable'. This is the semantic force of 'necessary' or 'necessarily true' in most ordinary contexts. A claim such as 'all Euclidean equiangular triangles are equilateral', since nothing intelligible could count against it, will be felt to be necessarily true within the language (L) of which it is a part. Now this much goes beyond mere sign-design. For we are here noting that a necessary claim will, within L, always be *true*. *Truth* and *consistency* differ typically. So now it is not just the *form* of an assertion, but also the meaning of its terms (i.e., the semantical values inserted into the symbolic variables) which is at issue.

Tautologies (claims which cannot be false, and whose negations are inconsistent) are therefore (1) analytic, and (2) necessary—i.e., true by legislation.

Furthermore, such claims—if they *are* 'claims'— since reflection is sufficient to reveal their necessity, are what they are independently of experience. The 'knowledge' they convey is not drawn from experience. Their justification requires no appeal to experience. If 'a posteriori' indicated of empirical claims that they were epistemologically dependent upon experience, 'a priori' may indicate of tautologies that they are epistemologically independent of experience.

This much logic-chopping demarcates two kinds of propositions. On the one hand, we have empirical propositions that are synthetic, contingent, and *a posteriori*. On the other, we have tautologies that are analytic, necessary, and *a priori*. If laws of nature are expressed in propositions (and it would be hard to deny this; one feature of a law is that its linguistic articulation is invariably said to be of what is 'true'), then what *kind* of proposition is a law-statement? Is it just one more empirical proposition (synthetic in its design, contingent in its meaning, and *a posteriori* in its relation to experience)? Or is a law of nature expressed by way of a tautology (analytic in its design, necessary in its meaning, and *a priori* in its relation to experience)? The history of discussions of laws consists either in attempts to analyze them as if they were no more than empirical generalizations, or as if they were but definitions. Some philosophers, dissatisfied with either of these accounts, have undertaken to find some third, more realistic, analysis of laws of nature.

'Dust-bowl empiricists' seem unanimous in viewing laws as being nothing more than generalizations. In this frame of mind $F = ma$ emerges as synthetic, and *a posteriori*. Even empiricists of a sophistication somewhat beyond the dust bowl may press for a similar analysis. Thus Ernst Mach, Bertrand Russell, and C. D. Broad—

all construed $F = ma$ as a generalized and highly abstract descriptive account of experiences we have while pushing Steinways and lifting leaden weights. The inclination to treat other classical laws of motion similarly has been manifest in many thinkers, not all of whom have been philosophers. Statisticians, sociologists, and subtle laboratory men often join hands around *laws* and *generalizations*, pronouncing them to be essentially the same.

There are difficulties with such an interpretation. A law like 'All unsupported bodies in terrestrial space move towards center of the earth' is *not* such that we can easily entertain its negation; the conception of a *genuinely* 'levitating' terrestrial body makes the mind boggle. Whereas an exception to a mere generalization like 'All white, blue-eyed tomcats are deaf' does not have such conceptual consequences. If curled up before you now there purrs a white, male, blue-eyed feline— one which possessed a perfect sense of hearing—you would not suddenly have doubts concerning the meaning of 'cat', or of the other words in the generalization. But if this book lifts up off your hands and 'floats', there might be some question in your mind concerning whether you had heretofore understood what *bodies* were and what their 'normal' behavior was really like. In short, genuine laws of nature, although they have a logical form identical to that of generalizations, i.e., $(x) f_x \rightarrow g_x$, nonetheless exert a 'conceptual grip' on the elements of experience—a grip often absent in an actuarial regularity. The mere fact that x and y have always occurred together provides no reason for thinking of x as a 'y-ish'; x and y as related inside a theoretical framework, however, 'hooked together' within what we know, may have just this effect on us. In terrestrial space, unsupported bodies *are* 'freely falling' kinds of things. Classical mechanics provides a conceptual structure in terms of which that relationship is articulated. The theory 'makes' the relationship intelligible, or at least captures whatever it is in the relationship which can be made intelligible. *In fact* all unsupported bodies do (as a matter of actuarial generalization) move toward the center of the earth! This fact makes *sense* only when appreciated as an instantiation of that theory of classical mechanics.

Granted, in a different world our laws of nature might be quite different. But it remains that treating such universal claims as *mere* empirical propositions gives no glimmering of the function of law-statements—to wit, that they interrelate conceptions in a semantically most intimate way.

Other thinkers, less impressed with the context-dependent, synthetic aspects of law-statements, and more taken with their relative indefeasibility, view them as being closer to tautology than do the philosophers mentioned above. Poincaré, Vaihinger, Kolin, and Bullard are philosopher-scientists, who have characterized Newton's laws of motion as being 'mere definitions'. Nothing can count against them because they *define* the relationship between a theory and all its possible subject matters. Thus the laws of ideal fluids generate a theory of fluid mechanics which (since there are *no* ideal fluids) can perhaps be applied to greater or lesser degree to actual fluids in specific states. What a given real fluid does or does not do is thus irrelevant to the design and conception of such a theory. Of particularly recalcitrant fluids, all one can say is that the definitions within such a theory do not make its 'application' to such a subject matter a genuine possibility. Whatever the ultimate account of such a fluid may be, *that* theory cannot supply it. Its boundary conditions just do not provide the links and hooks for grappling liquescent phenomena of that type. All laws of nature may be said to be like this, in principle; that is the position being espoused here. Since it is at the level of its laws that a subject matter *identifies* itself as being suitable for description *via* a given theory, nothing that ever happens within that subject matter could possibly refute such laws. Just as nothing weighed in the pans of a beam-balance can ever upset Archimedes' law of the lever (since the latter is built into the very construction of a beam-balance), and just as nothing weighed by a spring-balance can ever refute Hooke's law (since that law is 'enshrined' in the construction of the spring-balance), so with all laws of nature. They are 'built into' the instruments that do the measuring, into the theories that interpret the measurements, and thus determine which subject matters will, or will not, be managed in accordance with a particular theory. Hence, nothing actually observed or experienced could possibly count against such a law of nature. (Could the actual gait of the Bishop Berkeley ever have refuted the chess rule: "All Bishops move diagonally to the edge of the board"?)

There is something illuminating about this characterization of laws of nature as being somewhat like definitions, or even like tautologies. It underscores their 'invulnerability', at least so far as the theory of which they are integral parts is applicable to a subject matter at all. It italicizes how laws shape our conceptions of given phenomena by dovetailing particular theories to those phenomena, more or less. Still, it is hard to go 'all the way' with this conception. Statements of laws of nature are synthetic. Their negations entail nothing of the form *Q-and- not-Q*. Even a law like that concerning the *impossibility of a perpetual motion machine (first kind)* is such that its negation, although conceptually untenable to an advanced degree, is not inconsistent in its sign-design. It *must* be factually false for a scientist to claim that he has built a perpetual motion machine. But his claim will not itself contain a contradictory symbol-structure, nor entail such. This being so—i.e., it being clear that law-statements are synthetic—their functions as definitions, stipulations, and 'subject-shapers' is not at all clear, philosophically. Better perhaps to allow that law- statements are formally synthetic, that their negations are not logically inconsistent, and even to grant that there is something descriptively important in their function. All this can be conceded while yet insisting that, within a given scientific theory, a law-statement may be so much built into the 'rules of the game' as to be virtually invulnerable so long as one continues to use that theory (i.e., 'play that game').

Recognizing this latter, that laws of nature are 'invulnerable within the theory they serve to structure', inclines some philosophers to dub them *'necessary* within special scientific languages'. Thus $F = G(m_1 m_2)/r^2$ is 'necessary' (i.e., invulnerable) within every part of classical celestial mechanics. It is inconceivably difficult to think of phenomena such that the general understanding of them proceeds by way of classical mechanics, but with respect to which the law of universal gravitation does not obtain. This, then, is 'provisional necessity', 'relative necessity'— sometimes misleadingly referred to as the 'functionally *a priori*'. $F = G(m_1m_2)/r^2$ thus functions within classical celestial mechanics *as if* it were a necessary claim.

Concerning the *a posteriority* of laws, this again is beyond serious dispute. In a different universe, our laws of nature could be other than they are. (Kant once entertained the possibility that our world might have been constituted according to an inverse threefold ratio [$F = G(m_1m_2)/r^3$], rather than the inverse square—a possibility in every way self-consistent and meaningful. The most one can say of such suggestion is that it is empirically false of our world as it *is* constituted, which clearly marks the *a posteriority* of the law in question.)

Thus laws are such that their statements are (1) synthetic in sign-design, (2) 'necessary' within the theory which they help to constitute, and (3) *a posteriori* with respect to their epistemological status. Are they then both 'synthetic and necessary' [*à la* (1) and (2)]? Yes, if it is *relative* necessity, and not absolute necessity, which is understood. Some synthetic claims (*L*) are virtually invulnerable within the descriptive statement-systems of which they are part.

The quest for a synthetic *a priori*—a quest central to much of the history of modern philosophy— strikes this author as being confused in several ways. Granted, it was Hume's unacceptable analysis of Newton's laws of motion (as 'statistical regularities plus psychological expectations') that awakened Kant from his "dogmatic slumbers". The specifically Kantian resolution, however, seems somehow conceptually unsound today. That law-statements are synthetic (they have consistent negations)—and are yet invulnerable within the languages they help to shape—provides no occasion for any illicit philosophical 'mixing' of logical alternatives which are sharply distinct. Both Broad and Peierls conclude that, since law-statements have both *synthetic* and *necessary* elements, they must therefore be 'mixtures' of these constituents. But the 'unmixability' of oil and water is as nothing when compared to the 'unmixability' of the *synthetic* and the absolutely *necessary*.

A further feature of law-statements lends to this confusing picture of 'mixing the unmixable'. Law-statements are expressed in sentences. *Which* statement a given law-sentence does express will itself be a context-dependent matter.

Thus *'Le ciel est bleu'* and 'The sky is blue' are two sentences that make the same assertion: the sky is blue. Two sentences, one claim. On the other hand, the sentence 'The sun rises in the east' may express either an empirical truth, or a tautology. If 'east' is the name of that place where the sun rises (wherever that may be), then, even should tomorrow's dawn be in the direction of Antarctica—the sun must still be rising in the east. But if 'east' is determined by an appeal to celestial coordinates rather than by definition, then it will be a contingent matter that the sun, on any given day, does rise in the east. One sentence, two claims. Law-sentences derive their maddening versatility from this same context dependence. The uses to which the sentence 'F = ma' can be put are wide ranging enough to determine law-statements of almost any location on the analytic-synthetic spectrum (See Hanson [1958] 2010, ch. V).

5.10   Hypothetico-Deduction

So much for the 'overview' of theories; we have spoken of what they are, what they do, and how they relate to some wider philosophical issues. This still leaves the conceptual analysis of the 'fine structure' of theories largely undiscussed; how are they constituted, what distinguishes their logical structure, what is their 'grammar'? One significant response here consists in characterizing theories as 'hypothetico-deductive systems'. On this view, scientific theories are first and foremost systems of inference, within which every component proposition is located either at the 'bottom of the page' of inquiry (where the propositions resemble theorems in purely deductive systems), or else it falls 'mid-page'—'beneath' claims of greater generality, and 'above' claims of less generality. Or, finally, propositions within a theory may be of the very highest level, the 'from which' everything else in the system inferentially follows. Still, these highest-order claims are not just posited, or assumed, or presupposed simply for the purpose of deducing everything below—as in a deductive theory. Rather, on the hypothetico- deductive view, even such highest order claims are themselves ultimately empirical in nature. They are *a posteriori,* factually true, or factually false, even though determination of this may require subtle techniques of analysis.

There may be some question concerning whether scientific theories are systems of 'propositions' at all! An intratheoretic expression like '$F = G(m_1 m_2)/r^2$' is a propositional schema—and as such it is neither true nor false. Such a symbol-cluster can express a proposition only when observational values are inserted for the variable terms, or when the entire expression is universally quantified. Such quantification is often tacitly assumed, within the boundaries set by particular languages. Thus the law of universal gravitation usually begins "For any bodies whatever..." But note that the business of rendering a symbol-network into being a 'system of propositions' requires explicit attention somewhere (1) in the correspondence rules, or (2) in the 'coordinating definitions', or (3) in the abstract interpretations which transform the 'symbology' of a 'theory into something semantically intelligible.

Within this thought-framework scientific theories are much like isolated symbolic games, whose properties are determined by formal 'algorithmic' considerations. Such a theory can effect contact with its subject matter only 'at the bottom of the page', where one finds expressions which, when suitably linked to the 'outside world' via coordinating definitions and correspondence rules, will generate observation statements. These latter will be certifiably true or false (certifiable by observation). The measure of a theory's empirical utility, then, will be the degree to which these observation statements turn out to be true, rather than false. Should the verdict go toward recognizing a given theory's predictions as true far more often than as false, this will tend, indirectly, to confirm the theory as a whole. To the extent that this happens, all the laws within the theory—i.e., all the high-order hypotheses—will, insofar, also be confirmed.

This position has a great deal going for it. Worthies like Hempel, Braithwaite, Popper, Carnap, and J. S. Mill, by articulating variations of this hypothetico-deductive analysis, have succeeded well in accounting for some aspects of scientific theories. The 'all-or-nothing' feature of our acceptance of most theories at once becomes clear. One cannot decompose an inferential system piecemeal, preserving just those components in it which have been favored in experience. A deductive system (and a hypothetico-deductive system also) stands or falls *en bloc*. When experience fails to support its consequences, to that extent the entire theory has been revealed as vulnerable—marked as possibly untrustworthy within all further inquiries. One solid body observed floating in mid-air, supported by *nothing*, will send all of classical mechanics to the bottom. The hypothetico-deductive account of theories illuminates this point as well as, or better than, alternative analyses.

That account also demarcates what we understand by an *anomaly*. Within an established theory our 'expectations' of a given subject matter can be exfoliated with deductive precision. 'Normal expectations' are identical with a hypothetico-deductive exposition of our knowledge. Anomalies, then, are those happenings whose descriptions express the *negation* of observation statements entailed within a hypothetico-deductive 'unpacking' of some well-established theory. Thus, it was expected that the emission of radiation would always be a continuous process; we expected Sirius to move rectilinearly in its translation across the stellar background—not to 'wiggle'; we anticipated that Uranus would 'keep time' in its orbit like all other then-known (pre-1846) Newtonian objects. When such 'theoretical expectations' are not fulfilled, the resulting situation is said to be 'anomalous', as were the undulations of Sirius and the decelerations and accelerations of Uranus. Indeed, most features of the anatomy of theories receive a clear elucidation within hypothetico-deductive accounts.

5.11 Retroduction

To what extent, however, does such a position enable us to understand the dynamics of 'theory construction'! How are the rational strategies of scientific problem-solving illuminated by works such as those of Braithwaite and Reichenbach? The latter duo will quickly remark that they are not concerned with the 'process of discovery'. Matters of sociology, psychology, and inspired intuition are of no interest to the thinker for whom 'rational reconstruction' and 'axiomatization' are primary objectives. It thus appears to 'hypothetico-deductive' philosophers that any analysis, such as that of Peirce, which passes under the name of 'retroduction', must be irrelevant to conceptual analysis. Retroduction *must* concern itself, apparently, with mere matters of fact, with sundry issues of psychology, with sociological and historical considerations having much to do with the *process*, the 'psycho-dynamics', of problem-solving, but little, if anything, to do with comprehending the conceptual structure thereof.

There are reasons for resisting such a final and uncompromising appraisal of retroduction. Aristotle *(Posterior Analytics* II.19), and Peirce himself *(Collected Papers,* I), certainly knew the differences between matters of fact and matters of analysis. It misrepresents the positions of these thinkers to suggest, that, when writing about science, they concerned themselves only with the former. They took themselves to be doing philosophy. They *were* doing philosophy!

Granted, they were not *(à la* Braithwaite and Reichenbach) undertaking *ex post facto* logical reconstructions of the 'argument-anatomy' of Finished Research Reports. But it does not follow that they were hence exclusively concerned with psychology. The exhaustive and exclusive dichotomy 'Psychology or Logic?' may win debates occasionally, but it cannot win the guerdon of truth. Many features of the actual problem solving of ordinary people, and of ordinary scientists, require understanding the *criteria* in virtue of which one can distinguish *good* reasons from *bad* reasons. Long before an investigator has finished his inquiry, has solved his problem, and has finally written up his research report, *there* must have been many occasions when he found himself forced to use his head, to invoke his reason, and to decide between those speculations which seemed potentially fruitful, and those which did not. There are such things as 'proto-hypotheses'; these test our capacity to delineate *ranges* of plausible conjectures within which we would be prepared to argue that our final solution is most likely to lie. Determining these ranges of possibility and plausibility will often be based on reasoning of a fairly exacting variety. Thus, while still an undergraduate (and long before he succeeded in fashioning the final form of the law of universal gravitation), Newton reasoned that the law, whatever its ultimately divined form, would certainly be of inverse square structure. His reasoning was trenchant, resting upon the deductive linkages between Kepler's third law $[T^2 \propto r^3]$ and Huygens's law of centrifugal and centripetal force $[F \propto r/T^2]$. From these it follows that *if* the sun exerts a centripetal force upon the planets, then that F will be proportional to r/r^3, or $1/r^2$. Newton had good reasons for anticipating that the Law would be of that certain *kind.* His reasons then (1661–1665) appear to us even today to be good reasons. Yet such cerebrations obtained 20 years before any final formulation of the law in question.

Regarding the functioning of theories within technical science, the hypothetico-deductive account seems illuminating *vis-à-vis* our ideas of hypothesis-testing, and terse expositions of the results of that testing. The retroductive emphasis, however, is more centered upon the conceptual aspects of problem-solving. The primary datum within the latter is the *anomaly* itself—the perplexing occasion for further inquiry. The leading consideration within hypothetico-deductive thinking is the well-formed exposition of the problem's *solution.* This is in answer to the question "What follows from these premises (i.e., hypotheses, laws and initial conditions)?" The leading question in retroductive thinking is "From what premises can this anomaly be shown to follow?"

Can one solve problems *reasonably* within scientific inquiry? Of course. To that extent there are canons of reason, criteria of rationality, which distinguish good technique from bad, promising conjectures from dubious ones, likely directions of inquiry from unpromising courses of research. Such criteria, or strategies, might well be examined by philosophers of science in terms which do not 'reduce' to being mere psychological speculation. Aristotle, Mill, Whewell, Peirce, Toulmin, and Hanson may have made some faltering starts along this path of inquiry. But, faltering or not, such a philosophical interest is to be distinguished, on the one hand, from the 'formalizing' tendencies of Axiomatizers within the hypothetico-deductive 'school', and, on the other hand, from the psychological patter of scientific biographers concerned with the thought processes and psychological conditioning of discoverers. Examining the rational strategies of scientific problem-solving, therefore, does not collapse into being logical reconstruction or psycho-factual recitation. It is, in principle, philosophical inquiry, of a different kind. There are too many important events within history of science that are philosophically deformed by treatments guided solely via hypothetico-deductive formal structure of finished research reports. The discovery of Sirius' companion, of Neptune, of the neutrino, the positron, etc.—these were disclosures responding to perplexing, anomalous situations, deviation from the expected. Such discoveries often result from positing (or 'divining') theoretical entities in the course of 'putting anomalies to sleep.' This epistemological point is lost while philosophers shuffle to provide elegant, formally economical presentations of the support for claims that such entities as the neutrino and Neptune actually *exist;* this in addition to the earlier reflections that *if* they did exist, then some specific anomaly would be thereby resolved. What is called "The Astronomy of the Invisible" is a continuing instantiation of this point (See "Sirius" in the *Encyclopaedia Britannica.*) Perceiving that *if* some such things as Sirius' companion and Neptune did exist our observational perplexities would evaporate —*this* is neither a formal argument in support of the *de facto* existence of trans-*observabilia,* nor is it a subtle sampling of the intracranial processes of discoverers. Many significant solutions to scientific problems have been generated via rationally directed appeals to 'as if' entities, theoretical entities, the intellectual need for which has provided practical occasion for experimentalists to seek after such *denotata* within the subject matter.

5.12   Theoretical Entities

'One cannot give a mechanical explanation of the very things that make mechanical explanation possible. ' This sentiment is natural for the thinker who puzzles over the epistemological and semantical status of *denotata* such as force-free bodies, rigid levers, frictionless surfaces, ideal fluids, etc. One can provide mechanical explanations of actual, observable phenomena—the trajectories of hockey pucks, the steering linkage in automobiles, billiard balls on felt-covered slate tables, the properties of cold air and hot oil—by anchoring descriptions of such familiar objects and processes to inferential networks which hook up with what is calculated to happen in 'the pure case'. Knowing how objects would move through an ideal fluid is what allows us to calculate the actual behavior of non-ideal objects (e.g., aircraft) through non-ideal fluids (e.g., the air above us). So the very comprehension of everyday processes and laboratory events depends on conceptual extrapolations to what would obtain with 'pure designata' released-in-thought from the 'imperfections' of their empirical embodiment (another droplet of Aristotelian metaphor which generates seas of philosophical punch).

Full scientific understanding then, may require such arguments from 'the pure case', these latter resting in references to theoretical entities. Some philosophers, of course, will deny the essential irreducibility of such theoretical references. They will decompose the latter into their 'observational' components, and other components the precise appraisal of which will depend upon the philosophical objectives of the appraiser.

5.13 Craig's Theorem

Consider any scientific statement whatever—e.g., "The transverse magnetic field deflected the beta-particles." It might be argued that every such claim could be restated, without any loss of 'operational' meaning, in strictly nontheoretical, observational form. The only way we can even ascertain the existence of a magnet field is to shoot a 'test particle' through it, recording the resultant deflections to the particle's original trajectory. The only way we can ascertain the existence of a beta-particle is to note the molar-observable 'footprints' it leaves (e.g., in a cloud or bubble chamber, or in an emulsion). Hence a strictly operational equivalent of the statement above would concern the observable deflection-curves traced by water droplets formed on ionized gas particles, or traced by bubbles in a super cooled fluid; it would record thus the 'electron' path's sense of curvature and the degree to which that track is to be distinguished from others noted on various occasions in the past; and it will cite the conditions of experimentation which were operative on all relevant occasions, past and present. The resulting paraphrase would, of course, be very long indeed—possibly interminable. But it would have the philosophical 'virtue' of excluding all theoretical entities, whose properties surpass the information gained via such observational encounters in the laboratory.

The message?: every operationally sound scientific sentence will, sooner or later, be statable in observationally responsible terms.

What if some component of a scientific sentence is not completely capturable within such a corresponding operational translation! Then that term may have had no business being within a proper scientific language at all; it may not be 'operationally meaningful'. At the very least, such a term must be segregated from all those other terms in a science which do pass the 'operational' test. The explanation of such a term may turn on 'metaphysical' or 'psychological' considerations.

But what of science itself! The Craig theorem is to the effect that every scientific theory can be cleft atwain—into two sets of sentences. In the one set will be all those sentences which contain no theoretical terms at all. They will be operationally respectable. The other set will contain all other terms, whose appraisal will hinge on multiform considerations. And these will be all the sentences in the theory. This harmless representation seems to be all that Craig had in mind—that the observationally tractable part of a theory could be represented as a conjunction of all its

constituent observation-sentences. Some philosophers of science have carried
Craig's message much further. They have construed the theorem as claiming, *à la*
our earlier remarks about ' operational translations', that *only* the conjunction of a
theory's observation-sentences has a rightful place in the scientists' attention.
Everything respectable within science itself can therefore be conveyed in operationally significant terms (is this a tautology!), and the appropriate semantical vehicle
for this is the string of observation-statements generable therefrom—the possibility
of generating which for the entire theory, without remainder, is the main test of a
theory's 'operational respectability' and the main burden of this extreme construal
of the Craig theorem, a construal which surely exceeds the original intentions of
Professor Craig himself.

The inspiration, and motivation, behind this extremism is clear. The 'formal'
lineage is obvious. (So is the dead horse of radical empiricism.) Any continuous
function F is representable as the complete series (usually infinite) of all those
n-tuple terms which comprise membership of the expanded series over which F
ranges. Analytically there is no more to F than what can be represented thus. So,
also, it may seem to some philosophers that there is no more to any theoretically
impregnated scientific statement than the observation-sentence 'unpacking' which
sets out in a linear semantical sequence all the meanings 'confusedly' compressed
within the statement itself.

On then to the next level of abstraction: Every scientific theory (usually an algebraic network of terms functionally related) is representable as the infinite conjunction of all its constituent observation- sentences. This is the radical and extreme
extension of Craig's theorem, which was originally a pronouncement of much more
modest scope.

By this extended account there is no function for a theoretical term (like 'electron', 'force', 'psi', etc.) beyond what can be unpacked from it at the most uncom-
plicated level of observation-statement. Anything in excess of this is 'metaphysical
embroidery' or perhaps a 'heuristic carrot' to be held before the scientist's nose. In
a sophisticated and uncompromising form, this position has obvious similarities to
the older and newer positivisms, empiricisms, and operationalisms. Is there any way
of questioning such an analysis without at the same time collapsing spinelessly into
a soft idealism, or becoming a Rococo Kantian—or (even worse) a navel-
contemplatory Aristotelian?

One *could* consider the intratheoretic roles of theoretical terms, bypassing the
ontological and operational issues thereby. It just isn't true that after having
remarked all the extrasystematic correlations of a term like *e* with *observabilia* (e.g.,
pointer-readings, titration-colors, deflection-measures), one has said everything
about the intrasystematic functioning of such terms. Indeed with this much only at
hand, one has said *nothing* about the latter. Reflect upon the largely uninterpreted
term 'straight line' within pure Euclidean geometry; its functions there are in no
way illuminated by our considerable conversations concerning correspondence
rules and coordinating definitions, in virtue of which 'straight line' is observationally strapped to things like rays of light, taut strings, or the surfaces of optically flat
glass (when viewed on edge). No attempt to set out the reduction-sentence equivalent

of geometrical optics can indicate how 'straight line' helps the inference-nets which structure optical theories to operate as they do.

There may be many intrasystematic functions of theoretical terms which are not collapsible into conjoint references to observational experiences. Nor, on the other hand, are they representable as being but symbolic shorthand (i.e., substitution rules making some short expression like '4' or 'Σ' equivalent to several longer, operationally transparent expressions). Indeed, to the extent that a *systematic* exposition of a subject matter is involved in 'understanding' it, to that same degree theoretical terms help us to comprehend phenomena; they function so as to systematize and structure observation-sentence clusters within a discipline.

How then does one determine whether a statement, one which irreducibly includes a theoretical term, is *true* or *false?* How does the thesis of verification bear upon our appreciation of these diversities of discourse within science?

5.14   Verification

The 'verification-meaning controversy' of the 1930s was at once a cause of, and an effect of, philosophical concern about the nature of science. Once one has made observability a virtual criterion of the legitimacy of scientific discourse, it is but a small step to argue that any unit of discourse will count as descriptively significant only to the extent that it 'deals with' observables—only to the degree to which one can have a clear idea of what such discourse describes, what state of affairs it seeks to delineate. A scientific statement is meaningful to us only insofar as we can specify what kind of observations would disclose it as certifiably true. It is at the heart of reduction-sentence translation that terms which are in no way correctable with observables, and statements which (because they include such terms) are not decisively testable by way of experience— that such terms, and such statements, are just not semantically well formed. They are deficient in descriptive meaning. Psychology, and tradition to the contrary notwithstanding, such discourse is not 'scientifically meaningful'. Thus, the verification principle.

The principle generated conversation, consternation, and considerable confusion. It encouraged an 'antitheoretical' development within the sciences, and within philosophical and historical commentaries about science. 'Dust-bowl empiricism', as found amongst Questionnaire-designing Sociologists, Rat-running Psychologists, Species-counting Biologists, Substance-analyzing Chemists and Data-gathering Physicists—is not too far removed from the radical philosophy of verification. Intense concern with techniques of corroboration, with success in prediction, and maximizing accuracy within the statistical gathering of data—all this has often led to de-emphasis on explaining and understanding perplexities. It might be urged that success in prediction, and the understanding of a subject matter, are one and the same; but that is a question *for* philosophical discussion, not a sergeant-major's command preparatory to all discussion.

The issue here centers on a perhaps too-restricted conception of what verifying a claim, or a theory, actually consists in. [A philosophical position, extended beyond a concern with single propositions to a study of *systems* of propositions (i.e., theories), can sometimes be broadening and comprehensive.] Might it not be the case then that, in addition to observational felicity, predictive power, and success in confirmation, one should also consider how far *understanding* is conveyed via a proposition-system? Ptolemaic astronomy and Copernican astronomy were, during the latter half of the sixteenth century, on a par *vis-à-vis* success and confirmation. At that moment in history observation could not decide between these two statement-networks. This alone, however, was insufficient to prevent some thinkers from regarding the latter as 'verified' and the former as not so—it being clear that such thinkers packed the meaning "provides us with an instrument for *understanding* phenomena" into the expression 'is verified'. Too often "is successful in observation and prediction" is all that is allowed to serve as the semantical content of 'is verified'.

Such a broadening being granted *pro tem,* we can perceive again our contrast between Scylloid and Charybdoid philosophical postures; where verification is restricted to corroboration-in-terms-of-sense-experience, one result is a distrust of all theoretical science and, indeed, of all statements of fact which transcend the lowest order of our perceptual encounters. 'Dust-bowl empiricists' thus seem indifferently hostile toward astrologists, cosmologists, and sociologists. But, surely, to the degree that sociologists and cosmologists, by their special theories and techniques of inquiry, can explain perplexing aspects of their intricate subject matters, and can render intelligible what might otherwise have been a chaotic confusion of conceptual concerns—to that degree such disciplines will justifiably be said to be verified. So the *via media* here lies this side of the dust bowl; phenomenalistic sense-encounters are not the answer to the question "How can scientific inquiry both explain, and yet remain responsible to observation and experience?"

5.15 Falsification

There is no reason for letting our verdict oscillate to the other extreme, however. Another reason why it can be misleading to rivet attention exclusively to matters of verification is that this criterion allows disciplines like astrology, graphology, and phrenology, all to pass as responsible and respectable. These 'sciences' experience no difficulty in indicating *ex post facto* events which were vaguely forecast by the practitioners thereof. (Compare the contemporary best-seller.) How many astrologers did, after the fact, proudly refer back to their horoscopes of 1930, mysterious scribblings which foretold great changes being brought about by a dark man of erratic temperament? Hitler was seen by them all as the fulfillment of their prophecy. But such unfocused hunches were not predictions. They were never respectable components within genuine scientific theories. Because it would not have been possible, at the time of their first enunciation, to specify in detail *then* what would count as evidence *against* such claims. Even knowing what is meant by a prediction requires being able to articulate in advance what events, were they to take place, would falsify that prediction. All too often the forecasts of astrologers, graphologists, and phrenologists seem compatible with anything whatever. Occasionally this same verdict has been applied to theories inside ostensibly 'proper science', e.g., those of Yelikowsky, Lysenko, Miller, Ehrenhaft, Spencer, Priestley, Stahl, and Hooke. But insofar as the meaning of scientific terms is to connect with operations and observations, *falsification* is no less essential than *verification* as a criterion of acceptability. The great theories of yesterday have indeed been verified; but we know what it would have been like to falsify them. Therein lies their semantical strength. The former, verification, in the absence of the latter, falsifiability, is not a reliable guide to the achievements of science.

5.16 Models

Akin to these considerations is the scientific model, about which so much has been written during the recent past. The conceptual structure displayed via the articulation of a model—such as the Saturn-model of the hydrogen atom, the shell-model, of the atomic nucleus, the telephone switchboard-model of the human brain, the hydraulic conduit-model of the neural fibers—that structure suggests a possible idea-framework for otherwise unstructured ensembles of descriptions. Such frameworks of ideas hook the descriptions together with inferential links. The model which suggests these inferential linkages between statements fosters intelligibility; it aids in our understanding of a subject matter; it provides channels of interconnexity between states of affairs which (except for these links) might remain conceptually isolated and independent of each other. Explaining perplexities requires linking them to the normal cases—the unperplexing. The unusual becomes unsurprising only when inferentially hooked to the usual. Models suggest to us ranges of possible explanations—routes to the unsurprising. Knowledge may begin in astonishment, as Aristotle observed, but it surely doesn't end there. Full knowledge of anything consists in expecting every feature of that thing "as a matter of course."

But saying what models *do* does not indicate how differently different models may function. Nor does it suggest how with every model there may be disadvantages as well as virtues. Since it is an objective of every model to provide an inference-structure for propositions descriptive of a subject matter—a structure which is neither simple nor perspicuous within the descriptions themselves (else why would a model be needed in explaining them?)—it follows that the structure must be presented in a *different* way through the model from what obtains within the subject matter itself. To the extent that one appreciates the model's structure as superimposed on the data, without also swallowing the differences (without, that is, investing the subject matter with features unique to the model only, and foreign to the former), to that extent the model is serving well. Still, the scientist who uses models in his reflections must always remain alert to the possibility that his questions are inspired only by properties of the model, having nothing directly to do with the subject matter itself. Thus the water-closet model of animal instinct, as invoked by Niko Tinbergen, helped us understand the 'all-or-nothing' response of instinctive impulse once the 'cistern chain handle' had been pulled. Instinctive behavior 'flows' all at once when the pipes are opened (to use another model). But Tinbergen was at once on the hunt for 'leaks in the instinctual conduitry', 'blockages', 'displacements', etc. Some of these were suggestive and helpful. Others not at all. Again, is the area of electronic influence around the nucleus of the hydrogen atom in any way like the rings of Saturn? To a Nagoaka, or a Rutherford, the suggestion was attention consuming. But the question is now seldom asked.

Thus a model, persuasively to present an idea-structure as a possible linkage-format for descriptions of a given subject matter, *must* differ from the subject matter. If it were not different, the original structure would itself be observationally obvious to everyone who confronted the descriptions, or at least as obvious as in the model. Either that or it would be obvious to no one, not even the would-be model-builder. Models are thus a way of presenting structures that might *possibly* inforce subject matters. They do so in ways psychologically more compelling (i.e., simpler and more focused) than would just another confrontation with the subject matter itself.

Suppose one undertook to minimize, indeed eliminate, the differences between the model and the original phenomena. Scientists have felt that whenever it is necessary to articulate a structure (the model's) in terms different from those directly applicable to the subject matter itself, this constitutes an imperfection in the ' state of the art ' at that time. The Saturnian model, the Shell model, the two-fluid model— all were advanced as expositional ploys, always with the embarrassed reassurance that more knowledge of the subject matter would render the model unnecessary. A hypothetical objective of science then, might be, systematically and surely to minimize the area of divergence and disparity between the original phenomena and the theoretical model. Ultimately science will articulate 'what's what' of phenomena, *sans* models and all other toys.

Thus the 5-inch balsa wood model of a Spitfire airplane is 'less faithful' to the original than a 15-inch metal-covered, flying model, one with movable controls. Both of these are 'less faithful' to the real thing than would be a construction half the size of the original machine, a model possessed of *every* structural component within the actual Spitfire. Even this last disparity (i.e., being half-size) might be eliminated in a model faithful to the original in *every* way. The result, however, would not be a *model* of a Spitfire; it would *be* the production of another operational Spitfire! Whatever might have been one's motivation to have a model of the original, it would still remain unfulfilled if the result was 'only' another original! (Harvey didn't puzzle out the circulation of the blood just by becoming a father—thereby helping to bring another circulatory system into existence. Reproducing perplexities exactly is not the same as highlighting their structures.)

Therefore, by completely eliminating *all* differences between the model and the original state of affairs one ends up destroying the very thing the model was meant to achieve—namely, the provision of an 'awareness of structure' absent from the original confrontation with a complex of phenomena. [Of course, if the full-size Spitfire reproduction was amplified with display circuitry which illuminated the fuel system in green lights when one pressed a certain button, or the ignition system in red lights, or the hydraulic system in blue lights—this would again have become a model. It would be reinstated as a model because, by such differences from the original, aspects of the latter would now stand out in a way unrealized in any mere replication, however faithful it may be to the facts. This is itself a clue to one of the main differences between laws of nature and statistical generalizations. Although both have the form $(x)(f_x \rightarrow g_x)$, the generalization is just a descriptive replication of *observabilia;* the law is never just that. And a single sentence of the form $(x)(f_x \rightarrow g_x)$ may *now* express a law, and at another time express a mere generalization—depending on whether the context of its employment is one within which our attention is directed to a recitation of data, or to the form of those data, Kepler's third law, $T^2 \propto r^3$, has had both kinds of uses.]

What models must do to *be* models is related to what theories must do to *be* theories, and related also to what sciences must do in order to *be* sciences. Understanding perplexing phenomena requires attending to what "makes them go." Within the staggering variety of ways of directing attention to special features of complex subject matters, one thing is common to them all; there must always be *some* differences between (1) the mode of presentation, or the representation (i.e., the model, the theory, the science), and (2) the big, blooming, confusing, phenomenal perplexities which drove men to try to understand them all in the first place. Detailed photographs of jumbled jigsaw puzzles are just as puzzling as what was photographed; they are not different *enough*.

At one end of this philosophical world there will always be those for whom differences between the mode of representation and the properties of the original phenomena (assuming such differences to be articulable at all) will always constitute a blemish, an imperfection, an unwarranted heuristic prop —or even a metaphysical excrescence—upon the primary business of 'telling the truth' in science. Such thinkers will recoil from 'larger pictures' of phenomena, toward ever-refined descriptive techniques, toward more precise laboratory equipment, toward descriptions which are increasingly close to the reports of sense experience. Depending upon how far the philosopher of science is prepared to move along this spectral world line, he can be located somewhere with the empiricists, or the positivists, or the experimentalists, or the observationalists, or the operationalists—all of them possible, valuable philosophies of science. All such philosophers, like Raphael's depiction of Aristotle in *The School of Athens*, 'point downward', to the foundation of experience. To the extent that the scholar stresses 'the bigger picture'—even at the possible risk of doing fleeting injustice to experience—to that extent he will, like Raphael's Plato, 'point upwards'; he may even end up saying, with Hegel, "so much the worse for the facts." This is an extreme position rarely to be found within responsible philosophy of science. But there are surely occasions when, while reading Whewell, Meyerson, Poincare, Cassirer, Natorp, Cohen, Blanshard, and Capek, one is almost prepared to encounter just such extremities. But, somewhere short of that 'far-out' pole, one can locate conceptual coordinates occupied by philosophers of all persuasions who are unready to allow theory, thought, hypothesis, and experiment to be reduced to 'nothing more than' a congeries of wide-eyed encounters with the phenomena of this world. The scientific encounter, say such philosophers, is more than a scratching amongst the data of experience, yet never so much more as to become indistinguishable from artistic or even mystical experience.

Our objective in everything that has gone before has been to locate these polar-types of philosopher as they wander through ranges of connected subjects within philosophy of science. In each foray *our* search has been for some nonextreme 'middle way' —some resolution sensitive both to the existence of our Scylla and of our Charybdis—but sensitive also to the practices of respected scientists, and to the best analyses of responsible philosophers. It was hoped that a kind of conceptual chart might have been set out, a map representing many of the problems perennial within philosophy of science. Perhaps this 'Baedeker tour' which we have sketched in dotted lines—a journey which moves (conservatively) between the prominences so dear to idea-cartographers—may seem mild and unimaginative to the reader. (This verdict is usually passed on tours as conceived by Baedeker, AAA, and American Express.) But even so, the route might be sufficiently well marked and annotated to provide occasions for adventurous digressions off the beaten path, into untrod and unrowed regions of philosophical inquiry, the traversal of which may be made more exciting by the proximity of Scylla and Charybdis.

References

Beveridge, W.I.B. 1957. *The art of scientific investigation*. New York: Norton.
Duhem, Pierre Maurice Marie. 1914. *La théorie physique: son objet, sa structure*. Paris: M. Rivière.
Eddington, Arthur. 1946. *Fundamental theory*. Cambridge: Cambridge University Press.
Hanson, Norwood Russell. 1955. Causal chains. *Mind* 64 (205): 4–7.
———. [1958] 2010. *Patterns of discovery: An inquiry into the conceptual foundations of science*. Cambridge: Cambridge University Press.
Laplace, Pierre-Simon. [1829–1839] 1966. *Celestial Mechanics*. Trans. N. Bowditch. Bronx: Chelsea Publishing Company.
Lund, Matthew D. 2010. *N. R. Hanson: Observation, discovery, and scientific change*. Amherst: Humanity Books.
Reichenbach, Hans. 1938. *Experience and prediction: An analysis of the foundations and the structure of knowledge*. Chicago: The University of Chicago Press.
Slater, Noel B. 1957. *The development & meaning of Eddington's 'Fundamental Theory'; including a compilation from Eddington's unpublished manuscripts*. Cambridge: Cambridge University Press.
Wittgenstein, Ludwig. 1922. *Tractatus logico-philosophicus*. Trans. C.K. Ogden, with German original (*Logisch-philosophische Abhandlung*). London: Routledge and Kegan Paul Ltd.

Part II
History of Science

Chapter 6
Leverrier: The Zenith and Nadir of Newtonian Mechanics

U. J. J. Leverrier was a colossus of nineteenth-century science. But for philosophers and historians his work has lain largely undiscovered; a suboceanic mountain beneath the scientific sea. We have surrendered 1859 to another giant. This is indeed the year of the watershed; it divides history into everything which went before, and everything which has flowed to us since. But 1859 is not only the year of *The Origin of Species*. It is also when Leverrier announced his ill-starred "hidden planet hypothesis" to account for the precession of the perihelion of Mercury. The very discovery of this aberration is itself due to Leverrier. Its negative consequences for history of science are at least as great as was the positive work of Darwin.

Who else can be said to have raised a scientific theory to its pinnacle of achievement – and then, shortly later, to have discovered those discrepancies which dashed the theory to defeat? By pressing Newton's mechanics to the limit of its capacities to explain and predict, Leverrier revealed Uranus' aberrations as intelligible; he also predicted the existence of the then-unseen planet Neptune, which has just those properties required dynamically to explain Uranus' misbehavior. In history few have approached Leverrier's achievement as a human resolution of an intricate natural problem. When he detected a somewhat analogous misbehavior in Mercury, Leverrier naturally pressed the same pattern of explanation into service. He calculated, via the law of gravitation, the elements of some as-yet-unseen planet which would do for Mercury just what Neptune had done for Uranus. In this Leverrier failed. In a sense, his failure was one for Newtonian mechanics itself.

This is the skeleton of this article. Now I shall place an anatomy of historical detail on these bones, and bend a few logical limbs and joints to delineate how key concepts of this period both fitted with each other and mesh with distinctions familiar to historians and philosophers today.

© Springer Nature B.V. 2020
N. R. Hanson, *What I Do Not Believe, and Other Essays*, Synthese Library 38,
https://doi.org/10.1007/978-94-024-1739-5_6

I

Leverrier is usually coupled with John Couch Adams – these two being the independent co-discoverers of the cause of Uranus' aberrations. It is sometimes said that their work was equivalent. This is not so. Although Adams knew (in 1843) that no undisturbed orbit would fit the observations of Uranus he *assumed* that the disturbance was due to an unknown body. Leverrier's first move was to *establish* this by careful analysis of all available observations. Furthermore, the actual orbit plotted for Neptune by Adams is different from that plotted by Leverrier; indeed, both plots depart from the actual path of the planet, Adams' rather more widely. Fortunately, Adams and Leverrier began to worry about Uranus just when its deviation from the predicted path had reached its maximum: it was beginning to decrease.[1] For a short time, and through a small segment of arc, "The Leverrier orbit", "the Adams orbit", and Neptune's actual orbit were in close proximity. At any other time and place in the sky all three would have diverged.

Incidentally, although finding Neptune was a triumph for Newtonian theory, it was not a triumph for astronomy *per se*. Everything needed for this discovery was well-known 15 years before Leverrier's coup. Professional astronomy was initially somewhat discredited. For astronomers so long to have tolerated a defective Uranian theory when all the remedial data and algebra were available – this was felt to constitute a measure of scientific irresponsibility. The international storm which raged over the priority of this discovery, and over Airy's unfortunate role, reflects this general reaction – this plus the fact that the exact determination of the orbital elements of both Uranus and Neptune had to wait until late in the nineteenth century for its confident publication.

Uranus was discovered by Herschel in 1781. It had been observed 19 times between 1690 and 1771. When Alexis Bouvard, fortified with Laplace's *Mécanique Céleste*, sought to embrace those observations, it transpired that no one general calculation represented both the old observations and the numerous modern ones. He remarked:

---

[1] Compare: "...Mr. Adams told Somerville that the following sentence in the sixth edition of the "Connexion of the Physical Sciences," published in the year 1842, put it into his head to calculate the orbit of Neptune. 'If after the lapse of years the tables formed from a combination of numerous observations should be still inadequate to represent the motions of Uranus, the discrepancies may reveal the existence, nay, even the mass and orbit of a body placed for ever beyond the sphere of vision.'" (Somerville 1873, 290) At an even earlier date Adams recorded this memorandum: "1841, July 3. Formed a design, in the beginning of this week, of investigating... the irregularities in the motion of Uranus which are yet unaccounted for; in order to find whether they may be attributed to the action of an undiscovered planet beyond it; and if possible thence to determine the elements of its orbit, etc. approximately, which would probably lead to its discovery." (Preserved in the Library of St. John's College, Cambridge.) [While Hanson seems to have encountered this memorandum in the library, it is reproduced in Adams (1847, 428). Adams's paper was read to the Royal Astronomical Society on November 13, 1846. –*MDL*]

I had to construct tables with all the required accuracy *vis-à-vis* the modern observations; but these did not harmonize with the old observations.... I leave it to the future to discover if the difficulty of reconciling the two systems (the old observations and the new ones) is due to the inexactitude of the old observations, or if it depends on some foreign action which would have influenced the course of the planet. (Bouvard 1821, xiv–xv)[2]

That was in 1821. But even Bouvard's Tables, which abandoned the old observations, began to deviate from still newer observations. In 1829 Hansen wrote to Bouvard that

... in order to explain the discrepancy between theory and observation, it was necessary to take into account the perturbations of two unknown planets. [Hansen later denied that he referred to *two* such planets.][3]

In 1837 Bouvard's nephew wrote (to Airy)

... one sees that the differences between the observations and the calculated longitudes are very great, and that they are becoming greater. Is this due to an unknown perturbation introduced into the movements of this planet by a body situated beyond?[4]

In 1840 Bessel wrote:

I think the time will come when the solution of Uranus' mystery may be furnished by a new planet whose elements would be known according to its action on Uranus, and verified according to that which it exercises on Saturn.[5]

In 1845 Arago encouraged Leverrier to attend to this problem. This was already 2 years after Adams had begun work and some months after he had arrived at his first provisional results. But within 1 year Leverrier presented three *Memoirs*. The first (November 10, 1845) established precisely the disturbing actions of Jupiter and Saturn on coordinates of Uranus. By this alone Leverrier showed that Uranus' longitude, as calculated for 1845 *via* Bouvard's adjusted Tables, had diminished by 40 sexigesimal seconds.

The second *Memoir* (June, 1846) compares the elliptical orbit of Uranus – as required by Newtonian theory and exaggerated by Jupiter and Saturn – with all earlier observations (as well as 262 contemporary ones). The observations and the theory are irreconcilable. Leverrier wrote:

---

[2] Throughout this essay, Hanson seems to have produced his own translations of French and German originals. –*MDL*

[3] Hanson provided no source for this citation. Hanson seems to have based his conclusions about Hansen's theorizing on Smart (1947, 8). That Bouvard was told of Hansen's two planet hypothesis was related to Airy through the Reverend T.J. Hussey's letter of November 17, 1834. The pertinent extract of that letter is to found in Airy (1846, 123). I am indebted to William Sheehan for providing the reference to Hussey's letter. –*MDL*

[4] Airy published an extract of the letter from Eugene Bouvard (dated October 6, 1837) in (1846, 125) containing the text Hanson quotes. Airy published the letter in French, and it is not clear where Hanson obtained his translation of it. Airy's memoir was read to the Royal Astronomical Society on November 13, 1846. –*MDL*

[5] This statement comes from a letter to Alexander von Humboldt dated May 8, 1840. The letter can be found in Humboldt and Bessel (1994, 130). –*MDL*

I have demonstrated… a formal incompatibility between the observations of Uranus and the hypothesis that this planet is subject only to the actions of the sun and of other planets acting in accordance with the principle of universal gravitation. (912).

Leverrier does not doubt the exactitude of the latter. He even notes many earlier occasions when this law had been suspect, and had always emerged victorious after further examination by worthies like Laplace and Lagrange.[6]

Now the hypothesis of an unseen planet acting on Uranus is publicly entertained. (Note again that this is where Adams began his inquiry.) Reasoning back from Uranus' aberrations, Leverrier proved that the "planet" responsible must be transuranic, and that its mass must be great. He, like Adams, allows himself the guidance of Bode's "law" or rule – a merely empirical relation between planetary distances. From this he argues that the major axis of the unknown planet's orbit must be roughly double that of Uranus', a false conclusion (derived from a false premise) which nonetheless squared with the planet's 1846 positions.[7]

Leverrier assumes that the unseen planet is coplanar with the ecliptic, which is reasonable considering Uranus' minute disturbance in latitude. He asks:

It is possible that Uranus' inequalities may be due to a planet located in the ecliptic, at a mean distance double that of Uranus? And were this so, where is the planet now? What is its mass? What are the elements of its orbit?

Notice the difficulties here. If one knows a planet's mass and its orbital elements, the disturbance it produces in another body is easily determined. This is the classical problem of perturbations. Leverrier's problem, like Adams', consists in describing the disturbances in Uranus, from which he then *infers* the mass and orbital elements

---

[6] Huygens, Leibniz, John Bernoulli, Cassini and Miraldi were immediately hostile to Newton's *Principia* and to the initial formulation of the law of gravitation. The theoretical motion of the lunar perigee was but half that observed; the anti-Newtonians seized energetically on this flaw. Clairaut suggested that the inverse-square law required another term involving the inverse-fourth power of the distance. Later, Clairaut and Euler overcame this lunar difficulty completely: residual flaws in Clairaut's first analyses had been at fault. Later, Euler declared that the Moon's secular acceleration could not be produced by the Newtonian estimate of gravitational force. Lagrange reached the same conclusion. But in 1787 Laplace resolved the entire problem in a triumphant display of the power of the *Principia*. After 1826 the Uranian discrepancies again raised doubts about the inverse-square law, and minor modifications were regularly entertained by astronomers. Even Airy, in 1846, wrote "If the law of force differed slightly from that of the inverse-square of distance…," a conjecture later entertained with full force by Hall in 1895. It is against this persistent push and pull *vis-à-vis* the status of '$F \propto \gamma^{(Mn)}/r^2$' that Leverrier's achievement may best be appreciated.

[7] Bode's "law" consists in adding the number 4 to each member of the series 0, 3, 6, 12, 24, 48, etc. – the result being the units of planetary distance. Out to Uranus the law works fairly well, predicting in this latter case 196 units against 192 actual. For "Neptune" it predicts 388 [i.e., roughly "double that of Uranus"], but the correct figure is 301. Bode gives 772 for "Pluto," which is actually observed at 396; this latter figure is close to that predicted for Neptune itself. Moreover, nothing genuinely planetary is to be found at Bode 28, although a few thousand planetoids *are* at that distance. So the "law" is false. Leverrier's conclusion about Neptune's major axis is thus also false. But part of Leverrier's plot, drawn in accordance with this conclusion, overlaps with Neptune's actual orbit for the period 1845–1846.

of the disturbing planet.[8] This is the problem of perturbations in reverse, sometimes called "the inverse perturbation problem"; it is considerably more intricate to resolve than the classical problem. The difference is worth stressing since even so towering a figure as Airy failed to get the logical point of this kind of argument – with the result that he misunderstood Adams' work until the very day Leverrier's success was assured.[9] The contrast resembles that between having a set of premises and being asked to reach a conclusion from them – as against being given a conclusion along with instructions to determine a set of premises from which the former can be shown to follow. Logicians know the latter undertaking to be much more difficult. Indeed, the "conclusion" Arago gave to Leverrier was not even well formulated; the true elements of Uranus' orbit were not then precisely known. So he had to give some finality to the determination of the properties of Uranus as well as to the properties of Uranus' tormentor. This is like having first to design a conclusion which exactly *fits* the facts, and then constructing an argument "from the bottom of the page upwards" – terminating in a set of premises, no one of which is clearly false (and some of which are clearly true), from which the conclusion can be generated. Leverrier's problem included eight unknowns, requiring him to solve a host of complex transcendental equations. He concludes that one could account for Uranus' irregularities via the action of this new planet, whose very position he fixes for January 1, 1847, claiming an accuracy for this prediction of within 10°.

In his third *Memoir* (August 31, 1846) Leverrier announces the new planet's mass and orbital elements. He observes that every extant observation of Uranus is at last representable within the limits of observational error. He even delimits a small celestial zone within which observers should seek the new planet. Finally, having estimated the latter's mass, and granting it a density comparable to Uranus, Leverrier predicts an apparent diameter for the new body of 3″.

On September 23, 1846, Dr. Galle (a Berlin astronomer) received Leverrier's letter asking him to seek the planet. Within an hour Galle and his assistant d'Arrest noticed a star (of the eighth magnitude) unrecorded on Bremiker's celestial map. Next day he and d'Arrest observed the star in another place. Here was Leverrier's planet! It was within 1° from the assigned position; its apparent diameter was just under 3″. Galle wrote to Leverrier:

---

[8] Compare again Martha Somerville: "The mass of Neptune, the size and position of his orbit in space, and his periodic time, were determined from his disturbing action on Uranus before the planet itself had been seen." (Somerville 1873, 290)

[9] Thus, in reply to Adams' "short statement" of his results, written in late 1845, Airy wrote: "I am very much obliged by the paper of results which you left here a few days since, shewing the perturbations on the place of Uranus produced by a planet with *certain assumed elements*..." (quoted in Smart 1947, 23). As W. M. Smart observes of these italics, they make "it uncertain whether Airy realized that Adams' calculations... were associated with the inverse problem of perturbations...." (23). [The italics were Smart's. –*MDL*]

> La planète dont vous avez signalé la position existe réellement. *Le jour même* où j'ai reçu votre Lettre, je trouvai une étoile de huitième grandeur, qui n'était pas inscrite dans l'excellente carte *Hora* XXI (dessinée par M. le docteur Bremiker), de la collection de Cartes célestes publiée par l'Académie royale de Berlin. L'observation du jour suivant décida que c'était la planète cherchée. (Académie des Sciences 1846, 659)[10]

Encke wrote Leverrier: "Your name will be linked forever with the clearest proof of universal attraction one can imagine." (660) Airy wrote: "[Leverrier] told the astronomical observers: 'Look in the place which I have indicated to you, and you will see the planet very well....' Nothing... so legitimately bold, [since Copernicus] has been enunciated by way of astronomical prediction." (1846, 142) When the full story of Adams' work came to light Airy had to eat some of these words – but they represent the standard scientific reaction in late 1846.

Leverrier had carried Newtonian mechanics into the brightest heaven of scientific achievement. What achievement? It consisted of an argument of this form:

(1) Uranus' aberrations are formally incompatible with the Newtonian predictions,
(2) But Newtonian mechanics is unquestionably true,
(3) And the observations of Uranus' orbit are unquestionably accurate.
(4) This tension would be resolved were there some mass having just those dynamical properties required by theory to generate Uranus' observed positions.

The pattern of this argument is somewhat reminiscent of what Peirce called "retroduction". This is not a case of taking Newton's laws as one long conjunctive premise, and the described positions of Uranus as another premise – and then deducing Neptune's existence. All that follows from these is, as Leverrier observes, "a formal incompatibility". But neither does the argument proceed in any easily recognized *inductive* manner. Leverrier is not summarizing Uranus' observed positions and then "generalizing" from these that Neptune exists. He argues: From what hypothesis could Uranus' positions be shown to follow *in accordance* with Newton's laws? Aristotle notes (in the *Prior Analytics*) that one can answer this question only by formulating a new, comprehensive idea.

A further comparison with logic is apposite. Leverrier's question is not "What follows from these premises (Newton's laws and Uranus' positions)?" Nor is it "How can I summarize and generalize these data?" The question is rather, "Given Uranus' aberrant positions as a *conclusion,* from what further premise (besides Newton's laws) could this conclusion be generated?" The logical structure of retroduction is easily appreciated when viewed against Hempel's analysis of explanation and prediction. On that analysis explanation and prediction are "logically symmetrical". Thus if one explains an event (e) by tracing it back through a theory's laws and principles to a statement of known initial conditions (or data) – then one might as well have begun with the initial conditions, "process" these via the laws and then predict (e). E.g., Mars' present retrograde arc is thus explained by tracing it through the laws of Newton's *Principia* back to initial conditions concerning Mars' mean

---

[10] Hanson here refers to the transcript of the meeting of October 5, 1846 of the Académie des Sciences, in which Arago read excerpts from the letters of Galle and Encke. *–MDL*

period, its mass, its mean radial vector, its perihelial point, its position (heliocentric and geocentric) last Christmas, etc. But from these data – *all well-known* – one could have predicted (via Newton's laws) that Mars would be in retrograde motion right now.

Suppose, however, that not all of the initial conditions are known. Suppose that one element is hypothetical, and that one's very reason for including it amongst the premises is that it can serve with the other initial conditions to generate the anomalous event (e) via some established theory. The result is a "hypothetical explanation" – and the inquiry which turns up such hypothetical conditions may be called "retroduction". And (of course) retroductions are verified or falsified by hypothetico-deductive methods. But we are concerned here not with the logical form of the argument which *tested* the Neptune hypothesis, but with the historical development of the argument which *generated* that hypothesis.

I am not here advocating some unusual way of *justifying* arguments. My concern is not with justification at all. I am, rather, delineating how Leverrier *argued* – how indeed most scientists seem to argue when they seek resolutions of theoretical and observational perplexities. Determining an argument's validity, justifying it, constitutes inquiry into the soundness of deductions. But this does not mean that everyone who argues is deducing. A statistician, e.g., building up generalizations by systematically accumulating data, is not deducing. But when he argues *that* certain conclusions are warranted by data so accumulated – this turns on deductive criteria. The *ex post facto* inquiry concerning the soundness of Leverrier's argument can be designed in deductive terms. But the way Leverrier *actually argued* is perhaps best characterized as "retroductive". Every line of his attack on the inverse problem of perturbations would sustain this designation. The expression "Leverrier's argument" conflates these two issues. It runs together the actual order of Leverrier's moves as he engages his problem, with the logical order which formally structures his final argument. The expression "how Leverrier argues", since it can have a different force depending on whether "Leverrier" or "argues" is stressed, has confused many philosophers and historians of science, including myself. But although "premise-unpacking" and "premise-hunting" may be subject ultimately to the same criteria for valid argumentation, they constitute different intellectual activities. Analogously, the traveler's question "Where do I go from here?" is different from his question "How will I be able to get back here from there?" Each reflects a different conceptual context; but the criteria for appraising possible answers are the same – purely cartographical: viz., "Is there a geographical route connecting these two positions?" Similarly, the logical criteria for good scientific argument are purely deductive: "Is there a logical route connecting these two propositions?" But the queries "What follows?" and "From what does this follow?" remain conceptually distinguishable. It is a scholar's duty to understand the conceptual anatomy of each, because Leverrier's premise-hunt reveals at once the Excalibur and the sword of Damocles of Newton's mechanics: the sharpest intellectual tool of modern science. The same pattern of retroduction which had fortified mechanics in the case of Neptune, slowly decomposed it in the case of Mercury.

We have "excalibrated" Leverrier's prize. Let us now "damoclize" Newton's enterprise.

II

> No planet has exacted more pain and trouble of astronomers than Mercury, and has awarded them with so much anxiety and so many obstacles.

Leverrier thus indicated that he had not escaped the perplexities this planet had caused his precursors.[11] Before he had solved the problem of Uranus, Leverrier had already engaged with Mercury (1842).

In 1843 he presented to the Academy a 'Détermination nouvelle de l'orbite de Mercure et de ses perturbations' (1843a). Leverrier remarks the great precision with which the planet's position can be fixed when Mercury begins a transit across the sun. By 1631 it was calculated that the instant Mercury's disc and that of the sun are in visuo-geometrical contact, the angular distance (θ) between the centers of each body equals the sum (or the difference) of the apparent semi-diameters of the two discs. A formula then permits calculation of (θ) for the duration of the transit. After rejecting clearly faulty observations, Leverrier retained nine November transits (dating from 1677 to 1848) and five May transits (dating from 1753 to 1845).[12] To these Leverrier joined 397 meridian observations (made in Paris from 1801 to 1842). His observational basis was therefore sound and finely drawn.

Almost immediately, however, Leverrier was in trouble:

> ... the observations of passages by the ascending node give rise to but slight errors; whereas passages by the descending node generate an error of 12″.05 in 1753, this diminishing regularly until by 1845 it is reduced to −1″.03. These 13″ of variation in 92 years must be seriously considered because of the exactitude of the observations from which they result. They cannot derive from unreliability in the observations; this would make it necessary to suppose that all astronomers have made great mistakes in measuring the time of the contacts. These mistakes, moreover, would have to vary progressively in time, and differ by several minutes at the end of the period of 92 years. This is utterly absurd! Hence, one will eliminate the errors of the May transit, without introducing new ones into the November passages, only by modifying the values in the time-dependent elements of the orbit. The two corrections required must cancel out in the November passages while also accounting for the divergencies observed in the May passages. (1859b, 76)[13]

For Leverrier the theory of Mercury was like a frayed garden hose conveying high-pressure steam. When he adjusted it here, it sprang a leak there; he hadn't enough hands, or ideas, to plug every hole at once.

On July 2, 1849, Leverrier said to the Academy:

> ... That the interactions of planets do not alter their mean movements... is a condition of order in our planetary system. I was surprised then, when theorizing on Mercury, I saw that the mean movement of this planet over the last forty years was notably weaker than that determined by... the ancient observations... this result is not due to insufficient observations; but my efforts to arrive at a theory... have not been fruitful.... (1849, 3)

---

[11] Leverrier specifically refers to Moestlin, Riccioli, Brahe, Copernicus, Schöner, Kepler, Gassendi, Skakerloeus, Cassini, Hevelius, Roemer, Halley, Lalande, Lindenau, and Arago. (Leverrier 1843a, 1054 ff.) Indeed, Leverrier (1843b) describes in detail Biot's account of 13 observations of Mercury made in ancient China between 118 A.D. and 1098 A.D.

[12] Mercury passes its ascending node in November and its descending node in May.

[13] Hanson seems to be presenting the development of Leverrier's thinking in the 1840s here, but takes this quote from 1859. –MDL

He concludes:

> ... it is impossible to determine Venus' mass so as to account both for Mercury's transits and the observed obliquity of the ecliptic. If one determines this value so as to resolve the discordance in Mercury, it only reappears in the theory of the earth – and conversely. Moreover, adjustments in this value force anomalies into other conditional equations basic to the theories of Mercury and earth... One can shape these theories to fit the observations, provided that in a century Mercury's perihelion turns not merely 527″ as a result of the combined actions of the other planets, as [Newtonian theory] requires, but rather 527″ + 38″. There is, then, with the perihelion of Mercury, a progressive displacement reaching 38″ per century, and this is not explained.[14]

Again, this 38″ could not be denied without supposing that skilled observers, like LaCaille, DeLisle, Bougour and Cassini, had erred by several *minutes* in determining the visual contact times of Mercury and the sun. Given the precision possible, this seemed implausible. Moreover, these errors would have regularly to recur in a progressive way. "It remains to discover the cause of this anomaly in the perihelial movement of Mercury."

Just as the decline in strength of the Roman Empire had begun in the golden age of the Antonines, so here a foundation-crack in the structure of Newtonian mechanics appears just as its most spectacular spire is built up to its greatest height. Mercury's anomalies were known before Neptune's discovery, and they remained unresolved. By 1859 these constituted such a fissure at the base of mechanics that Leverrier tried again to cement it. Now the contrast between Newton's achievements and the nagging anomalies of Mercury, made resolving the latter astronomy's prime problem. So while Darwin swept his vast collection of biological data into a loose scientific system, Leverrier was struggling to save the very archetype of exact science.

In the *Annales de l'Observatoire Impérial de Paris* (1859b) Leverrier yet again calculates Mercury's perturbations, now by two independent methods: (1) by further and deeper analysis of Laplace's theoretical determination of this effect, and (2) by the 411 observations cited. Each of these was simultaneously made by several astronomers. The result was a precise plot in every case. Re-examination of these forced Leverrier again to conclude that the orthodox theory fitted the observations quite well, provided only that the secular perihelial movement of Mercury be augmented by 38″.

The whole astronomical fraternity being convinced of the need for this correction, Leverrier now wonders how to bring it about. It seemed that this problem was clearly one of gravitational perturbations. He again tinkers with the values of the mass variables in the law of gravitation. By increasing Venus' estimated mass by $\frac{1}{10}$ its usually accepted value this kind of adjustment can be just made to work (1859a, 381). But myriad meridian observations of the sun reveal the accepted mass of Venus' as exact; only a slight correction at most could enter here. Indeed, Venus'

---

[14] Cf. Leverrier (1859a, 381)

mass plays *the* dominant role in the oblique "creep" of the ecliptic. Were this mass increased by $\frac{1}{10}$, errors of very great magnitude would infect this part of the planetary theory.

Impelled by the very pattern of explanation which disclosed Neptune, Leverrier entertains the possibility of yet another new planet moving between Mercury and the sun – perturbing Mercury just as Uranus had been perturbed.[15]

Conjectures about an intra-Mercurial planet had some vogue even before 1850. Uranus' aberrations had been held up to Mercury's somewhat similar transgressions. It was supposed, qualitatively, that an unseen mass was upsetting Mercury just as Neptune had upset Uranus. But such a body ought to have been brilliant during an eclipse. Why was it never seen? It was once suggested the planet was hidden *behind* the sun's 30′ disc, hence never visible. (Shades of Philolaus' anticthon!) But three obvious objections destroyed this conjecture: (1) Mid-nineteenth century astronomers were operationalist enough to be suspicious of any explanation which depended on a phenomenon's being permanently unobservable. (2) For the perturbing planet thus to remain obscured, it must have the same period of revolution as the earth. This would rupture Kepler's Third Law ($T^2 \propto r^3$), which states a proportionality between the square of a planet's revolutionary period and the cube of its mean distance from the sun. Here, both earth and the new planet would have the same period, although at very different distances from the sun. (3) The formal deathblow to any such hypothesis was dealt by Liouville in 1842 – a vital moment. Addressing himself to the classical three-body problem, Liouville demonstrated that a straight-line solution is unstable. Grant that the arrangement *earth-sun-intra-Mercurial planet* forms a straight line for a moment. It is physically impossible that they should persevere in this disposition.[16]

Naturally, Leverrier's intra-Mercurial hypothesis was not so crude as these early conjectures – although he probably entertained the idea of a sun-obscured body in the mid-1840s. After recognizing an augmentation of Mercury's perihelial movement as an astronomical necessity, and after rejecting facile explanations of this (e.g., Venus' gain of weight, or a sun-obscured planet), Leverrier writes:

> A planet, or if you will, a group of planetoids orbiting inside Mercury, could produce the abnormal perturbation to which this star is subject... let us examine the effect of a single perturbing mass....

Here the saviour of Newton's celestial mechanics is doing more good works. Now in 1859 (as earlier in 1846), Leverrier is about to argue "backwards" from a discrepancy detected between observations and theory to a fully described "planet". Such a planet, as-yet-unseen, would once again reconcile the data with the system. His task is to determine each dynamical property of such a hypothetical mass. All of

---

[15] "Considérons, pour fixer nos idées, une planète qui serait située entre Mercure et le Soleil, et, comme nous n'avons point remarqué dans le mouvement de noeud de l'orbite de Mercure une variation pareille à celle du périhélie, imaginons que la planète supposée se meuve dans une orbite peu inclinée à celle de Mercure...." (1859a, 382).

[16] See Appendix.

these when combined, would require (via Newtonian theory) that Mercury's perihelion *must* advance 38″ per century. This, so far from being anomalous, as before, now becomes a physical necessity. Thus:

> The disturbing mass, if it exists, has no perceptible effect on the movement of Earth. We do not know whether it would affect Venus; until this point has been clarified, we shall say that this action may be imperceptible or at least weaker than on Mercury. On this hypothesis, the sought-for mass should be found within Mercury's orbit. If, in addition, its orbit is not to tangle with Mercury's, its aphelion distance must never exceed 8/10 of Mercury's mean distance – i.e., 3/10 the mean distance from Earth to sun. Our observations of Mercury have... shown no variation in the inclination of its orbit... [so] the orbit of the disturbing mass is [not] much inclined to Mercury's.

His argument intensifies algebraically, churning up further properties of the hypothetical intra-Mercurial mass. Thus:

> ... the orbit... has only a tiny eccentricity... the perturbing mass must be reckoned the more considerable the nearer the sun we place it... [there] it varies inversely as the square of its distance from the sun. *Thus, merely from the mechanical viewpoint, by the hypothesis of a perturbing mass – whose location remains indetermined – we can account for all observed phenomena.* It is nonetheless necessary to consider whether all solutions are equally admissible physically. [My italics][17]
>
> At the mean distance of 0.17 the perturbing mass would be equal to that of Mercury. Its greatest deviation would be just under 10°. This planet would shine more brilliantly than Mercury; would it not necessarily have been perceived grazing the horizon after sunset or before sunrise? Or could the intense, diffuse sunlight have permitted such a star to escape our glance?
>
> Considered as farther from the sun, the perturbing mass is weaker; the same is true of its volume; but the deviation is much greater. Nearer the sun, the inverse obtains; the glow of the body increases as its dimensions increase – and increases also by its proximity to the sun. But the deviation becomes so slight that a star of unknown position might not be perceived under ordinary circumstances.
>
> But even in this case, how could a planet, extremely bright and always near the sun, fail to have been glimpsed during a total eclipse? And would not such a planet pass between the sun and Earth, thereby making its presence known? (1859b, 103–105)

This is the march of Leverrier's argument; it traverses familiar logical ground. As before, his conclusion is given: that Mercury's perihelion advances 38″ per century. He hunts a premise, a hypothetical initial condition which, in accordance with the laws of celestial mechanics, will generate this conclusion. In an intra-Mercurial planet Leverrier has found just such an initial condition. He delineates its properties with all the care he exercised on Neptune 13 years earlier.

But to have designed an hypothesis from which the observational conclusions merely follow is not enough. Leverrier knew this. Thus, after seeing that the Neptune hypothesis *could* generate Uranus' observed aberrations, Leverrier still had to

---

[17] Contrast this with Leverrier's first paper on Uranus: "... qui me paraît très-propre à porter dans les esprits la conviction que la théorie que je viens d'exposer est l'expression de la vérité." (1846b, 438). This confidence he can never seem to muster when setting out his conjectures about Mercury. [This was not Leverrier's first paper on Uranus. Hanson presumably meant to say that this was Leverrier's first paper on the planet that later came to be known as Neptune. –*MDL*]

inquire of Galle whether this was true – whether Neptune did in fact exist. False premises can vividly generate true conclusions; hence just the discovery of an hypothesis from which a true conclusion follows leaves it unsettled whether that hypothesis is factually true.

Leverrier now felt no need for an elaborate telescopic search. The properties he fed back into his hypothetical planet *could* generate a perihelial advance for Mercury. But these same properties would also make such an intra-Mercurial body spectacularly observable – which it certainly is not.

> Such are the objections to the hypothesis of a single planet, comparable to Mercury in size and circling within its orbit. Those to whom these objections are serious will replace this single planet with a series of asteroids whose [combined] actions will produce the same total effect on the perihelion of Mercury. Apart from these asteroids' not being ordinarily visible, their circumsolar distribution may account for their not producing important periodic variations in the movement of Mercury.[18]

> … [further] observations of Mercury will disclose whether we should admit that such asteroids exist near the sun…. Among these there might exist some larger than others…. We cannot establish their existence other than by observing their motion across the sun's disc; this discussion should make astronomers the more zealous to study each day the sun's surface. It is important that every regular spot appearing on the sun's disc, however tiny, be carefully followed for a few months to determine its nature through familiarity with its motion.[19]

Newton had given astronomy a pattern of argument. Leverrier had pressed Uranus into that pattern and Neptune tumbled out. Now he presses Mercury into that same pattern. But nothing tumbles out – nothing observable. To save Newtonian theory from the 38 seconds of celestial arc ticking away its life, Leverrier needs intra-Mercurial matter. A single planet would be theoretically observable but *de facto* unseen. A ring of asteroids would be theoretically unobservable at our distance – save only in its effect on Mercury, which is what posed the problem! And *de facto* telescopic scrutiny cannot count against the hypothesis, but only for it.[20]

---

[18] Note that even here the unobservability of the asteroids constitutes a detraction. Compare: "Toutes les difficultés disparaîtraient en admettant, au lieu d'une seule planète, l'existence d'une série de corposcules circulant entre Mercure et le Soleil." (Leverrier 1859a, 382).

[19] Compare Faye (1859, 383 ff).

[20] Comparison with the early neutrino hypothesis is irresistible. That otherwise identical β particles are ejected from a radioactive source (RaE) at different energies placed several conservation principles into jeopardy – (Cf. Chadwick 1914, 383) – just as Mercury's perihelion-advance placed mechanics into jeopardy. Pauli "invented" the neutrino to resolve the tension. (This was stated in an open letter to Geiger and Meitner in 1930.) The neutrino's only function was to accompany each β particle emitted from the source, giving each neutrino-electron pair an aggregate energy identical to that of every other neutrino-electron pair. The principles of conservation were saved, the β spectrum notwithstanding; the neutrino's only function was to save it. Analogously, mechanics is saved by Leverrier's asteroid ring, Mercury's aberrations notwithstanding; and the asteroid ring's only function is to save it.

III

It may not surprise historians to learn that, immediately after Leverrier's announcement, and his exhortation that astronomers study the sun anew, many observations of "intra-Mercurial planets" were reported. An analysis of all these would fill a small book; indeed, Tisserand wrote 50 pages on the subject in 1882.[21] Most of these reports embody "wishful seeing" at best, and flat fabrication at worst. The one most deserving of attention was that of a Dr. Lescarbault, who (on 22 December 1860) wrote to Leverrier recounting an observation made on 26 March 1859.

After a detailed account of his telescope's construction, and its manner of employment, Lescarbault describes his observation of 26 March 1859:

> The planet appears as a black dot with a well-defined circular perimeter. Its angular diameter seen from the earth is very small... much less than $\frac{1}{4}$ that of Mercury, whose transit I observed with the same magnification on 8th May, 1845. (1860, 43)

Lescarbault located the encroachment of the planet on the solar disc at $57°22'30''$ west of "12 o'clock." Its exit was at $85°45'0''$ west of "6 o'clock" on the solar disc. Lescarbault fixes the times of each event, and also gives the duration of the entire transit (4 h 29 m 9 s). He determines the instant when the planet was closest to the apparent center of the solar disc ($0°15'22.3''$). He then gives the angle at which the planet's path departed from horizontal, as seen from Orgères [in France ($9'13.6''$)].

> ... someday someone will again observe the transit of a perfectly round, tiny, black dot, traversing a plane line inclined to the point of disappearance at angles between $5\frac{1}{3}°$ and $7\frac{1}{3}°$ the orbit described by this plane line will cut the earth's orbital plane at about $183°$ – from north to south; unless there is enormous eccentricity in the black dot's orbit it should be visible on the sun's disc for $4\frac{1}{2}$ hours... [the planet's] distance from the sun is less than Mercury's... this body is the planet, or one of the planets, whose existence you, Monsieur Directeur, recently revealed near the sun through your marvelously persuasive calculations; these are the same which in 1846 enabled you to recognize Neptune's existence, to fix its position within our planetary world and to trace its course through the depths of space. (44–45).

The celebrated Leverrier journeyed to Orgères to interrogate the village doctor. Picture the haughty, opinionated Leverrier almost crushing, by his manner and tone, the shy, obsequious Lescarbault. But Monsieur le Directeur was sufficiently con-

---

[21] Cf., Tisserand (1882, 729 ff). E.g., M. Herrick writes to Leverrier: "... sur la probabilité de l'existence d'une ou plusieurs planètes entre Mercure et le Soleil, je prends la liberté d'appeler votre attention sur certaines observations qui semblent démontrer qu'une semblable planète, accompagnée d'un gros satellite, a été plusieurs fois observée et toujours perdue." Herrick then cites observations reported by Pastorff, Gruithuisen, and Bradley in support of his contention (1859, 811 ff). The letter immediately following in *Comptes Rendus.*, by M. Buys-Ballot, speaks to the same matter.

vinced of the doctor's integrity and skill to pronounce before the Académie des Sciences (2 January 1860):

> M. Lescarbault's accounts, and the lack of fanfare with which he communicated them, have convinced me that his detailed observation should be admitted into the field of science. The long delay in publishing his findings springs only from modest reserve and from the quietude sustained far from the activity and agitation of cities. An article in the journal *Cosmos* concerning the work on Mercury was what led M. Lescarbault to break his silence. (46)

Le Directeur secured for the doctor the award of the Legion of Honor. So clearly, Leverrier's initial reaction to "Vulcan" – as the Abbé Moigno called the Leverrier-Lescarbault planet[22] – was optimistic enough, although not as unqualified as that which followed Galle's report 13 years before. From Lescarbault's numbers, Leverrier built up the elements of Vulcan.

(1) Its greatest apparent distance from the sun would be 8°.
(2) The largest opening of Vulcan's apparent orbit (on each July 3 and January 5) would be 3.5°.
(3) Transits would occur when Vulcan is in inferior conjunction (within 18 days before, and after, April 3 and October 6).
(4) One transit will occur when Vulcan is near its ascending node; another when near its descending node. Each year there would be at least two transits, and usually four. Each would be visible from more than half the Earth's surface.
(5) Vulcan's sidereal period is 19 days 7 h; hence its synodical period will be less than 20.5 days.
(6) Near elongation Vulcan should be brighter than Mercury – the former's close proximity to the sun compensating for its smaller disc. Since Mercury can be seen telescopically within 8° from the sun, Vulcan at the same distance would be visible to the naked eye. Vulcan attains that distance at 10-day intervals, remaining there for 2 days. Hence it should be clearly visible in different parts of the world 30 times each year.
(7) Fifty percent of all solar eclipses should disclose Vulcan shining like a star of the first magnitude at 20° from the darkened sun.
(8) With a volume of but $\frac{1}{17}$ Mercury's (and assuming corresponding masses), Vulcan would be unequal to the task of disturbing Mercury enough to advance its perihelion as the law of gravitation requires.

---

[22] French scientists of the nineteenth century were precipitate in naming "discoveries." Thus, after Galle and d'Arrest found Neptune, it is written in *Comptes Rendus* (Académie des Sciences 1846): "M. Arago: il s'est décide à la désigner par le nom de celui qui l'a si savamment découverte, a l'appeler *Le Verrier*." (662) Again, "M. Le Verrier dépose sur le bureau les feuilles 5, 6 et 7 du travail qu'il publie, sur les recherches qui l'ont conduit à la découverte de la planète qui porte son nom." (863) Compare Arago writing after Herschel's disclosure of Adams' independent theoretical discovery: "Avec son [Adams'] consentement, je mentionne *Oceanus* comme un nom qui pourrait probablement recevoir l'assentiment des astronomes." (751)

So even granting Lescarbault's discovery, Vulcan alone could not do what Leverrier, and celestial mechanics, needed to cope with Mercury's aberrations. Hence it constituted no real improvement over the asteroid-ring theory.

However, immediately after the Academy approval of Lescarbault's results, M. Emmanuel Liais, astronomer to the Emperor of Brazil, who was simultaneously studying the same area of the sun which Lescarbault describes, and with a better telescope, saw nothing of "Vulcan". Thenceforth Vulcan, the planet of romance (as the Abbé Moigno called it), became the planet of fiction. As late as (April 4) 1875 the German astronomer, Weber, observing the sun from China, reported a small round spot which moved much as Lescarbault had described in his observation of 1859. Moigno, Lescarbault, *and Leverrier* were at first over-joyed. But this turned to gloom when proof was given that Weber's dot was an uneven sunspot, penumbral fringe, et al.

Professor Tisserand's *Notice sur les Planètes Intra-Mercurielles* begins:

Durant ces vingt dernières années, la question des planètes intra-mercurielles a vive ment attiré l'attention des astronomes et du public scientifique... Nous rappellerons d'abord les raisons théoriques qui ont conduit à admettre l'existence de ces planètes.... (1882, 729)

This section of Tisserand's paper concludes:

Leverrier s'est trouvé conduit à admettre comme possible l'existence d'une planète inconnue, circulant entre Mercure et le Soleil.... (740)

Thence to the observations. The names of Faye, Wolf, Haase, Carrington, Peckelok, Stark, Decuppis, Sidebotham, Littrow, Steinhübel, Lofft, Fritsch, Scheuten, Scott, Wray, Hind, Lummis, Coumbary, Ventosa, Swift, Todd and Watson bulk large in the lists of those who made observations – both long before and long after Leverrier's announcement – of what appeared to them to be intra-Mercurial bodies. The last-named was as worthy of attention as Lescarbault himself. For Watson was already an astronomer of international repute, having personally discovered 20 planetoids, before he declared that on 29 July 1887 he had observed, from the observatory at Ann Arbor, Michigan, *two* intra-mercurial planets. It was later supposed by all observers that what he had really seen (during the total solar eclipse occurring on that day) were the stars Zeta Cancri and Theta Cancri. Watson himself remained convinced until his death that he had observed intra-mercurial planetoids; he even developed an elaborate plan, in conjunction with the University of Wisconsin, to dig an immensely long earth-shaft through which the sun might be viewed during its next total eclipse visible in America. Thus he hoped to establish conclusively the planetary character of his original observation.

Vulcan's forge cooled. An occasional spark was rekindled, as when a Mr. Tice of Kentucky claimed to have seen Lescarbault's planet; but he described it in a way that exposed him as a liar. Scores of intra-Mercurial observations are seriously recorded even on into the twentieth century. But none was ever really validated. Moreover, none of these even pretended to be of a body which had what it took to save mechanics from Mercury's relentless advance.

The Vulcan hypothesis is false. But the *way* it is false holds lessons for historians. By noting what Leverrier did with Mercury, even when he did *not* succeed, we can understand better what he did with Neptune when he *did* succeed. The pattern of argument in both cases was fundamental to the conceptual framework of Newton's mechanics. Each time the credentials of that great theory were placed at stake. With Neptune the theory, and its modes of explanation, triumphed; its credentials never looked better. With Mercury the theory and its patterns of argument failed; its credentials have rarely looked worse. The man holding the stakes in both cases was U. J. J. Leverrier. The triumph was his; he drew from the theory what few suspected it possessed. But the failure was not his – it was the theory's. He pressed it onto the problem of Mercury, just as Neptune had taught him to do. But it could not bring forth results it did not possess.

IV

All this reopened the issue anew. Some astronomers still favored the asteroid-ring hypothesis. Thus Proctor, writing in 1895, says:

> Whatsoever matter *must be assumed* to travel within the orbit of Mercury, to account for the motion of the planet's perihelion, is evidently neither gathered into a single planet nor distributed among several bodies which, though small, could be regarded nevertheless as planets. In the former case we could not fail to recognize an orb so important and so brilliantly illuminated during eclipse, or by telescopic aid without eclipse, or when crossing the sun's face (which it must do frequently). In the latter case, powerful telescopes could not fail to show each year many of the small planets in transit. The only supposition which remains available is, then, that the matter within the orbit of Mercury consists of multitudinous small bodies individually invisible. Many among these may be several tons, or hundreds of tons, in mass; but (when considered with reference to the enormous region they occupy, and compared with the masses of even the smallest planets) they must be regarded collectively as mere planetary dust. (Proctor and Ranyard 1895, par. 1023)[23]

However, in this same year Newcomb surveyed all the perplexities occasioned by Mercury's perihelial advance, examining a range of possible explanations. *Le roi Leverrier, l' était mort*; his system of celestial government lay moribund; a host of rival hypotheses quickened the dissension initiated by Mercury, finding flaws in other parts of Newton's plan; even confidence in the lawgiver, Sir Isaac, declined somewhat.

One mercurial proposal characterized the precession as resulting from a nonspherical distribution of matter within the solar body; this would give an excess of polar (as against equatorial) moments of inertia. This hypothesis is certainly compatible with our current conception of the sun being a molten-liquescent, or a gaseous body. A tiny inequality of this kind *could* account for the perihelial advance.

---

[23] Hanson commits a bit of an error here in attributing these words to Richard A. Proctor in 1895. Proctor died in 1888, and his book was completed by Arthur Cowper Ranyard before being published in 1895. Hanson himself never refers to Ranyard in this context. –*MDL*

However, were this hypothesis true, the equipotential solar surfaces would have an ellipticity greater than $\frac{1}{2}''$ of arc. Now the visible photosphere is an equipotential surface. Auwers' heliometer measures in 1874 and 1882 prove there is no such photosphericellipticity. Conclusion? The sun's interior is *not* distinguished by any such non-symmetrical distribution of matter. Indeed, even an equatorial ring of planetoids – or gaseous substances, for that matter – if near to the photosphere, would render the latter's equipotential surfaces elliptical to a degree ruled out by Auwers' measures.

An intra-Mercurial planetoid ring, to do its job for Mercury, would have to be tilted to the orbit of Mercury at a mechanically impossible angle. This inclination would be necessary since the orbits of Mercury and Venus are inclined to each other, making it geometrically impossible for the planetoids to lie in the plane of both orbits, and physically impossible to lie in the plane of either. To generate the required effect without disturbing Venus, this inclination is required. But planetoids so disposed would scatter themselves uniformly between Mercury's orbital plane and that of the solar system itself. It *is* dynamically possible for a single planet to have this inclination. But, as already noted, it would long since have been detected.[24]

Another hypothesis concerned an attenuated mass of diffused matter, like that which reflects zodiacal light. This is again the "interplanetary dust" conjecture. But again, to have any effect on Mercurial dynamics, this dust must circle near the sun. Again, it must be steeply inclined to the ecliptic plane. Observation reveals that such diffused matter as there is near the sun is not at all inclined to the ecliptic. The effect of this observed hypothetical dust on the Venusian and Mercurial nodes should be just opposite to that required by the unadjusted Newtonian theory. Another flaw in the hypothesis is this: the perihelial advance of Mercury *might* simply indicate an increase of its gravitation toward the sun *en route* from aphelion to perihelion – an increase slightly greater than that given by the unadjusted inverse-square law (involving the major planets and the sun alone). Such an increase could be produced by a ring of dust, lying wholly outside, or wholly inside, Mercury's orbit. But should Mercury's orbit ever pass *through* such a ring, as happens with other planets, the noted effect should be the reverse; Mercury's gravitation toward the sun should diminish from aphelion to perihelion. The perihelion would thus not advance, but retreat.

The hypothesis of planetoids between Mercury and Venus is likewise objectionable. Such planetoids would retard the orbital planes of Mercury and Venus. We should then detect a *retrograde* motion in the nodes, contrary to Leverrier's discovery.

Besides, that such planetoids should be individually invisible, yet collectively produce a band of light no brighter than the zodiacal arch – and at the same time have an aggregate mass great enough to advance Mercury's perihelion – all this seemed a highly improbable combination at the turn of the century.

The most arresting hypothesis of the late nineteenth century was that of Asaph Hall, who returned to the conjecture that solar gravitation may not function exactly as the inverse square of the distance. He argued that in $F \propto \gamma(Mn)/r^n$, n takes a value

---

[24] Its mass (at the solar surface) would be 1/1650 that of the sun itself – affording a remarkable reflecting surface.

not of 2, but rather of 2 + δ, δ being a small fraction, just great enough to effect Mercury's perihelial advance and just too weak to affect Venus' period in any detectable way. Newcomb himself observes:

> This hypothesis seems to me much more simple and unobjectionable than those which suppose the force to be a more or less complicated function of the relative velocity of the bodies. On this hypothesis the perihelion of each planet will have a direct motion found by multiplying its mean motion by one-half the excess of the exponent of gravitation. (1895, 118)

By very subtle computations. Hall's hypothesis is shown consistent with the general results of the law of gravitation in most astronomical applications. In fact, the effect of Hall's adjustment diminishes with distance. The earth is observationally unaffected. Venus' displacements are negligible. The hypothesis effects secular variations only in the planetary perihelia – nothing else. Hall's idea is designed to have its maximum effect on the perihelion of Mercury; this falls off sharply with the perihelion motion of all other planets.

Hall's theory is logically interesting. Leverrier had become accustomed to holding three "premises": (1) a single conjunction of Newton's laws of motion and gravitation, (2) the observed dispositions of all the major planets, and (3) the hypothesis of intra-Mercurial matter. From these Mercury's position at any time was inferable. Hall's appeal is different. The "hidden parameter" hypothesis of Leverrier, so compelling with Uranus, had now lost its attractions. So Hall chose the radical course of Leibniz, Poleni, Bernoulli, Euler and Laplace, the course of tinkering with Newton's laws themselves. He sought to manage all his computations on the basis of premises (1) [the "tinkered with" law of gravitation] and (2) where Leverrier undertook to reconcile recalcitrant data with his theoretical boundary conditions (i.e., Newton's laws) by adding further initial conditions (i.e., hidden matter), Hall reconciles the data with the laws directly – by effecting minor adjustments in the latter.

This conceptual shift is of historical significance. Once one is prepared to modify the foundations of Newtonian theory in order to accommodate the facts, the possibility of rejecting the *whole* theory – replacing it with new foundations – becomes genuinely a live option. Every line of Leverrier's work reveals that he could never have done this. Everything else would have been challenged, anything else would have been hunted, to preserve the Newtonian theory intact. But the theory's failure with Mercury quickened the march of ideas and the pace of conceptual adjustments. Hall was prepared to chip away at a cornerstone of the theory Leverrier had erected into a scientific fortress beneath the sign of Neptune. Einstein removed the cornerstone altogether and restructured a new celestial mechanics, of which the greatest triumph was the theory of Mercury itself.[25]

---

[25] According to the general theory of relativity, the elliptical orbit of a planet rotates in its own plane in the same direction as the planet moves, with a speed given by

$$\frac{\delta\tilde{\omega}}{\phi} = \frac{12\pi^2 a^2}{c^2 T^2 \left(1-e^2\right)}$$

[$\delta\tilde{\omega}/\phi$ is the motion of the perihelion per circumsolar revolution, a is half the ellipse's major axis, c is the velocity of light, T is the time of one planetary revolution, and e is the elliptical eccentricity.]

Appendix: On the Impossibility of a Straight Line Solution to the Three-Body Problem

Three bodies, M (sun), m (earth), and m' (intra-Mercurial planet) describe circles around their common center of gravity and are in a straight line. M is at rest and is the origin of the coordinates. Let r, θ be the coordinates of m, and r', θ' those of m'. Now m is acted on by $(M+n)\big/r^k$ along the straight line mM, and $m'\big/r'^k$ in a direction parallel to $m'M$.

The polar equations of m's motion are:

$$d^2r\big/dt^2 - r\left(d\theta\big/dt\right)^2 = -(M+m)\big/r^k - m'\big/r'^k \cos\omega - m'\big/R^k \cos\phi \qquad (6.1)$$

$$\frac{1}{r}\left[d\big/dt\left(r^2\left(d\theta\big/dt\right)\right)\right] = \left(m'\big/r'^k\right)\sin\omega + \left(m'\big/R^k\right)\left[(r'\sin\omega)/R\right].$$

[where ω, ϕ are angles at M, m of the triangle formed by the bodies, and R is the side mm'.]

Similarly, the polar equations of m' are:

$$d^2r'\big/dt^2 - r'\left(d\theta'\big/dt\right)^2 = -(M+m')\big/r'^k - \left(m\big/r^k\right)\cos\omega + \left(m\big/R^k\right)\cos\phi' \qquad (6.2)$$

$$\frac{1}{r'}\left[d\big/dt\left(r'^2\left(d\theta'\big/dt\right)\right)\right] = m\big/r^k \sin\omega - m\big/R^k \sin\phi'$$

[where ϕ' is the external $\angle$ of the $\triangle$ at m'. Here, the standard case is where $\theta' > \theta$ and $r' > r$.]

Now substitute $r = a(1 + x)$; $\theta = nt + y$; $r' = b(1 + y)$; $\theta' = nt + \eta$. Reject all powers beyond the first of small quantities x, y, ξ, η. (Remember, $\sin\theta\big/r' = \sin\omega\big/R = \sin\phi'\big/r$.) After reduction:

$$\left(\delta^2 - n^2 - kE\right)x - 2n\delta y + m'kB\xi + 0\cdot\eta = 0$$

$$2n\delta x + \left(\delta^2 + m'B\right)y + 0\cdot\xi - m'B\eta = 0$$

$$mkAx + 0\cdot y + \left(\delta^2 - n^2 - kF\right)\xi - 2n\delta\eta = 0$$

$$0\cdot x - mAy + 2n\delta\xi + \left(\delta^2 + mA\right)\eta = 0$$

In the case of Mercury, Einstein's theory (cf. Einstein 1916, 769), predicts an advance in the perihelial motion of $43''.03 \pm 0''.03$. The "observational" value, obtained by subtracting from the total perihelial advance all those gravitational contributions of the major planets (i.e. $= 5599''.74 \pm 0''.41) - (5557'' \pm 8.85)$), is exactly $42''.56 \pm 0''.94$. This close agreement constituted the first major argument in favor of the non-Newtonian planetary theory.

[where $\delta = d/dt$; and $c = a - b$]

$$A = \frac{a}{b}\left(\frac{1}{c^{k+1}} - \frac{1}{a^{k+1}}\right), \quad B = \frac{b}{a}\left(\frac{1}{c^{k+1}} - \frac{1}{b^{k+1}}\right)$$

$$E = M + \frac{m}{a^{k+1}} + \frac{m'}{c^{k+1}}, \quad F = M + \frac{m'}{b^{k+1}} + \frac{m}{c^{k+1}}$$

By equating the constants in Eqs. (1) and (2) above, the steady motion of the three bodies may be deduced. Thus $n^2 = E - m'B$, $n^2 = F - mA$.

Note that E and F are positive. When $K + 1$ is positive, $a > b > c$. Thus A, B and $E + F - 2n^2$ are also positive. And whatever k may be, $E + F - n^2$ is positive.

To solve the four equations, put $x = Ge^{\lambda t}$, $y = He^{\lambda t}$, $\xi = ke^{\lambda t}$, $\eta = Le^{\lambda t}$. Substituting, and eliminating ratios G, H, K, L, we get a determinantal equation constituted of coefficients of x, y, ξ, η (with λ written for δ). [This determinant is of the 8th degree in λ]

To find its factors we must simplify: add the ξ column to the x column, and the η column to the y column. Divide the 2nd column by λ and multiply by $2n$. Subtracting now from the first, we get $\lambda^2 - (K - 3)n^2$ as another factor to be divided out. Subtracting the first row from the third, and the second from the fourth, the 1st column acquires 3 zeros and the 2nd column 2.

The determinant is now expanded:

$$\lambda^2\left[\lambda^2 - (K-3)n^2\right]\left[\lambda^2 + C\right]\left[\lambda^2 - CK - (K+1)n^2 + 4n^2\lambda^2\right] = 0,$$

Where $C = E + F - 2n^2$.

If K is greater than 3, this gives a real positive value of λ, and the motion is thus unstable. If K is positive at all, C is too, and the 3rd factor will have the product of its roots negative. One value of λ^2 is real and positive; the other real and negative. *The motion is therefore unstable for all positive values of K.*

References

Académie des Sciences. 1846. Planète de M. Le Verrier. *Comptes rendus hebdomadaires des séances de l'Académie des sciences* 23: 659–662.

Académie des sciences (France), and Centre national de la recherche scientifique (France). 1835. *Comptes rendus hebdomadaires des séances de l'Académie des sciences.* Paris: Publiés avec le concours du Centre national de la recherche scientifique par MM. les secrétaires perpétuels.

Adams, John Couch. 1847. An explanation of the observed irregularities in the motion of Uranus, on the hypothesis of disturbances caused by a more distant planet; with a determination of the mass, orbit, and position of the disturbing body. *Memoirs of the Royal Astronomical Society* 16: 427–459.

Airy, George. 1846. Account of some circumstances historically connected with the discovery of the planet exterior to Uranus. *Monthly Notices of the Royal Astronomical Society* 7 (9): 121–144.

Bouvard, Alexis. 1821. *Tables astronomiques publiées par le Bureau des longitudes de France, contenant les tables de Jupiter, de Saturne et d'Uranus, construites d'après la théorie de la mécanique céleste, par M. A. Bouvard.* Paris: Bachelier et Huzard.

Chadwick, James. 1914. Intensitätsverteilung im magnetischen Spektrum der ß-Strahlen von Radium B+C. *Verhandlungen Deutsche Physikalische Gesellschaft* 16: 383–391.

Einstein, Albert. 1916. Die Grundlage der allgemeinen Relativitätstheorie. *Annalen Der Physik* 49 (7): 769–822.

Faye, Hervé. 1859. Remarques de M. Faye à l'occasion de la Lettre de M. Le Verrier. *Comptes rendus hebdomadaires des séances de l'Académie des sciences* 49: 383–385.

Humboldt, Alexander von, and Friedrich Wilhelm Bessel. 1994. *Briefwechsel zwischen Alexander von Humboldt und Friedrich Wilhelm Bessel*, ed. Hans-Joachim Felber. Berlin: Akademie Verlag.

Herrick, Edward. 1859. Lettre de M. Herrick à M. Le Verrier. *Comptes rendus hebdomadaires des séances de l'Académie des sciences* 49: 810–812.

Lescarbault, Edmond Modeste. 1860. Passage d'une planète sur le disque du soleil, observé le 26 mars 1859; Lettre de M. Lescarbault à M. Le Verrier. *Comptes rendus hebdomadaires des séances de l'Académie des sciences* 50: 40–45.

Leverrier, Urbain Jean Joseph. 1843a. Détermination nouvelle de l'orbite de Mercure et de ses perturbations. *Comptes rendus hebdomadaires des séances de l'Académie des sciences* 16: 1054–1065.

———. 1843b. Discussion d'anciennes observations de Mercure, extraites par M. Édouard Biot de la Collection des vingt-quatre historiens de la Chine. *Comptes rendus hebdomadaires des séances de l'Académie des sciences* 17: 732–735.

———. 1845. Premier Mémoire sur la théorie d'Uranus. *Comptes rendus hebdomadaires des séances de l'Académie des sciences* 21: 1050–1055.

———. 1846a. Recherches sur les mouvements d'Uranus. *Comptes rendus hebdomadaires des séances de l'Académie des sciences* 22: 907–918.

———. 1846b. Sur la planète qui produit les anomalies observées dans le mouvement d'Uranus. Détermination de sa masse, de son orbite et de sa position actuelle. *Comptes rendus hebdomadaires des séances de l'Académie des sciences* 23: 428–438.

———. 1849. Nouvelles recherches sur les mouvements des planètes. *Comptes rendus hebdomadaires des séances de l'Académie des sciences* 29: 1–5.

———. 1859a. Lettre de M. Le Verrier à M Faye sur la théorie de Mercure et sur le mouvement du périhélie de cette planète. *Comptes rendus hebdomadaires des séances de l'Académie des sciences* 49: 379–383.

———. 1859b. Théorie du mouvement de Mercure. *Annales de l'Observatoire Impérial de Paris* (Mémoirs) V: 1–195.

———. 1860. Remarques de M. Le Verrier au sujet de cette communication. *Comptes rendus hebdomadaires des séances de l'Académie des sciences* 50: 45–46.

Newcomb, Simon. 1895. *The elements of the four inner planets and the fundamental constants of astronomy: Supplement to the American ephemeris and nautical almanac for 1897.* Washington, DC: Government Printing Office.

Proctor, Richard A., and Arthur Cowper Ranyard. 1895. *Old and new astronomy.* London: Longmans, Green.

Smart, W.M. 1947. *John Couch Adams and the discovery of Neptune.* London: Royal Astronomical Society.

Somerville, Martha. 1873. *Personal recollections of Mary Somerville: from early life to old age; with selections of her correspondence.* London: John Murray.

Tisserand, Félix. 1882. Notice sur les planètes intra-mercurielles. *Annuaire du Bureau des Longitudes*: 729–772.

Chapter 7
The Contributions of Other Disciplines to Nineteenth Century Physics

Abstract My objective here is vulnerable to some spurious semantics. I aim to delineate how the development of physical theory in the later nineteenth century was *essentially* dependent on the reflections and discoveries of Chemists, Biologists and Mathematicians. But when one alludes to some professional Chemist or Biologist as having affected the history of physics, it can always be countered "to that extent he was *doing* physics, he was being a physicist". Thus when Urey helped to separate U^{235} from U^{238}, and when he invoked theoretical thermodynamics to determine the age of the solar system, he might be said then to have been *doing* physics – despite being a Nobel laureate in chemistry. And when Pauling applied quantum theory to studies of complex substances, and when he considered the rotation of molecules in crystals – he too was apparently *being* a physicist, although, again, a Nobel laureate in chemistry. So also of great moments in the work of Nernst, Boltzmann, Helmholtz, Faraday, Young… and so on. That is, one can always say that whatever helps physics *is* physics. But this is 20–20 hindsight focused to the point of tautology. For it suggests that a Biologist might one day awaken with the pronouncement "I think I'll do some physics today"; a Chemist or Mathematician might muse "this problem will put demands on the physicist in me". The history of science can be thus chopped up only by destroying the organic interplay between disciplines, an interplay which constitutes the very pulse of scientific research. The picture of disciplines rigidly fixed as to content, and of scientists as compartmentalized thinking machines – both pictures are unreliable reflections of the ways in which problems and their solutions have actually shaped the history of science. Just as being a 'natural philosopher' in the seventeenth century was not identical with being a theoretical physicist in the twentieth, so also the divisions between biology, chemistry and physics in the nineteenth century may not always have been drawn along the same lines as we should sketch them today. In short, one must be an *historian* when tracing the lines of development through nineteenth century science. Not everything that is embraced today in physics texts began in physics labs, or in the minds of professional physicists. Therefore, I propose to correlate the contributions of scientists now known to us as important in the histories of chemistry and biology-with moments in the development of nineteenth century physics. Should one then choose to dub all such individuals as *really* physicists, *à la bonne heure*.

© Springer Nature B.V. 2020
N. R. Hanson, *What I Do Not Believe, and Other Essays*, Synthese Library 38,
https://doi.org/10.1007/978-94-024-1739-5_7

I

In 1802 Joseph Louis Gay-Lussac extended the impressive work of Charles. He demonstrated the relationship between the volume and temperature of a gas (held at a constant pressure); he established the coefficient for the expansion of gases (as $\frac{1}{273}$ per °C); and he proposed the famous 'law' that gases combine in numerical ratio by volume – the volume of the resultant compound bearing some simple ratio to that of the constituents. This latter claim provided the conceptual foundation for Avogadro's Hypothesis. (I take it as beyond question that Gay-Lussac was primarily a Chemist.) In 1808 he generated potassium from fused potash, using K then to obtain boron from boric acid. In 1809 he analyzed chlorine, and then fermentation, and then hydrocyanic acid, iodine, cyanogen and its compounds – and in 1815 he investigated the nature of compound radicals and acids. He proved that acids need not contain oxygen. Gay-Lussac then undertook to improve processes for manufacturing sulphuric acid and oxalic acid (1829). All this must be recognized as 'typical' chemistry.

Amedeo Avogadro asserted (in 1811) that "equal volumes of all gases at the same temperature and pressure contain the same number of molecules". Thus he sought to explain Gay-Lussac's observation that gases which react chemically (at the same temperature and pressure) have volume ratios which are simple integers of each other. It thus appeared (contra Dalton) that substances are, normally, molecular and not atomic; reactions take place between molecules. One curious fact about the history of Avogadro's discovery was that, although he was himself certifiably a physicist, general acceptance of his hypothesis did not come about until 1858 when S. Cannizzaro laid the cornerstone of modern experimental chemistry; Cannizzaro used Avogadro's hypothesis to structure our understanding of molecular weights, gas behavior and the interpretation of reactions between gases. 'Avogadro's Number' is then readily cooked out of the hypothesis. This is just the number of molecules in 1 gram-molecular weight (g-m. wt.) of any substance. The oxygen molecule is arbitrarily assigned a m. wt. of 32. Thus 1 g-m. wt. is the number of grams equal to the m. wt., so 1 g-m. wt. weight of oxygen is 32 grams. At 0 °C, and at 1 atm, 32 grams of O has a volume of 22.4 liters. But, since the Hypothesis says that equal volumes of gases (at identical temperatures and pressures) contain the same number of molecules, it follows that the m. wt. of any gas can be determined by weighing 22.4 liters (at 0° and 1 atm). Thus 22.4 liters of N weighs 28 grams; 28 is thus the m. wt. of N. And from the m. wt. the atomic weight of elements in any gaseous compound can be readily determined.

These insights, of immediate use in experimental chemistry, had profound effects for the development of kinetic theory. A general rationalization of the gas laws, first steps towards which were undertaken by Avogadro, came slowly to form a single theory; Boyle's Law, Charles' Law and Gay-Lussac's Law formed integral parts of that theory. The ideas of "the average velocity of a molecule", the kinetic[1] energy of "randomly moving particles", the "Total energy" of a constant volume, absolute temperature, specific heat, latent heat, change of state... etc. all these were given a new significance within rigorous presentations of kinetic theory – which began with Daniel Bernoulli, later revived by Joule, and ultimately developed by Clausius, Clerk Maxwell, Gibbs and Boltzmann. Indeed, the entire discipline known as Statistical Mechanics was developed with an eye to its experimental applications in the kinetic theory – for here the techniques of probability were applied to accounts of the velocities of molecules; examining the effects of their motion in a quantitative fashion was the result.

One need hardly stress how much of our contemporary work in microphysics is the direct consequence of nineteenth century explorations in Statistical Mechanics, a discipline which itself has its roots in chemical inquiry of more than a hundred years ago. So *there* is a first bridge between theoretical physics and experimental chemistry in the nineteenth century; from the work of Boyle, Charles, Gay-Lussac and Avogadro on gases, through to the Kinetic Theory and Statistical Mechanics,[2] and then on to what we now know as Quantum Statistical Mechanics – as typified in the great exposition of von Neumann in 1929–1930.

II

These considerations lead at once to reflections on the concepts of 'molecule' and 'atom'. For Dalton atoms were certainly impenetrable and rigid. But was the atom elastic or inelastic? If elastic, the atom would have to have *parts* in relative motion. This would be incompatible with the 'elementary' nature of the 'uncuttables'. But if inelastic, collisions between atoms would lead to a loss of energy in motion; the kinetic energy of an aggregation would gradually diminish through internal collisions (contrary to what we actually experience). In 1811 Avogadro distinguished "molecules intégrantes" from "molecules élémentaires". The latter corresponded to our atom; the former was like our modern molecule – a constellation of atoms forming an elastic group. Regnault, Liebig and Roscoe seemed to have been quite clear about the distinction. But Clerk Maxwell in 1873 uses "molecule" just as a chemist would use the word "atom". He addressed those small chunks of matter "any further subdivision of which will deprive them of the properties belonging to the substance in question".[3] But later he spoke ([1891] 1902, 318–319) of the "internal motion of

---

[1] Cf. Bernstein (1963, 207–208).

[2] "... in the second half of the nineteenth century... the thermodynamics of C. Clapeyron and R. Clausius metamorphosed into a kinetic theory of matter which in its turn was linked to a novel kind of mechanics, namely the so-called statistical mechanics" (Bochner 1963, 189.)

[3] The exact quotation Hanson gives could not be located. However, Maxwell says essentially the same thing in (1873, 437): "A molecule of a substance is a small body such that if, on the one hand, a number of similar molecules were assembled together they would form a mass of that substance, while on the other hand, if any portion of this molecule were removed, it would no longer be able,

each molecule consisting partly of rotation and partly of vibrations among the component parts of the molecule" – the latter being made up of "atoms". Clausius also took the molecules to be elastic and their constituent atoms to be inelastic. Kelvin opts for "perfect elasticity in ultimate molecules". The chemists' molecule, then, became transformed by the requirements of theoretical physics into perfectly elastic spheres. These came to function (axiomatically) within the Kinetic Theory of gases – and molar physics generally – analogously to the ways the perfectly inelastic atoms were axiomatically determined to function within microphysics. Again, there is no question but that the modern concepts of chemical molecules and of physical atoms were developed and refined through the works on gases of Gay-Lussac and Avogadro. The molar-physical properties of the gas are determined by the number of molecules present per unit volume, the mass of the molecules, the average kinetic energy, and the magnitude of any forces of attraction existing between them. The chemical properties are determined by the number and arrangement of atoms within each molecule. Thus the inert gases are monatomic; easily vaporized chemical elements form poly-atomic molecules in the gaseous state. The forces that unite the atoms in a molecule (the chemical bonds) are much stronger than the forces of attraction between molecules. Now the term "molecule" is usually reserved for an electrically neutral assemblage of atoms having more than a momentary existence. Atomic assemblages having a charge are the ions. Doubtless, the experimental challenges to chemists – challenges deriving from gas theory and general theories of molecular combination – these forced increasing precision into the chemists' thinking about the nature of molecules. A derivative consequence of this precision was clearer and sharper thought within those more purely physical studies concerned with the nature of the structureless, inelastic atom. In general, the quantitative demands built into the experimental chemistry of the nineteenth century 'put the question' to much of the speculative natural philosophy that was being lavished on 'the fundamental properties of matter'. Chemists concerned with quantities and measurement, had to reach agreement on what a molecule was going *to be* for them. This set the corresponding problem for physics: "If molecules are such-and-such for chemists, what must we then say atoms are for us?" 'Naturphilosophie', spiced with sundry neo-Kantian condiments, was tending to eradicate distinctions between the precise, quantitatively-responsible reflections of exact science on the one hand, and the 'navel contemplation' which constituted *too* much academic philosophy in the latter nineteenth century. It might be conjectured that the rigor respected by experimental chemists is what bridled the otherwise free speculation of much natural philosophy; especially is this true of developments on the Continent.

---

along with an assemblage of other molecules similarly treated, to make up a mass of the original substance." –*MDL*

III

Consider the concept of 'the closed experiment' as it developed within chemistry. It was the notion of the perfect conservation of mass, and the ideas of *weight* and *ponderosity* which led by slow degrees to the later physical principle which we know as 'The Conservation of Energy'. In the form of the First Law of Thermodynamics this became the greatest generalization in all nineteenth century physics. The refined work of Rankine, Kelvin, Tait, Helmholtz, Thomson, Joule, Carnot – all this (don't forget) was anticipated in the insights of the *physician*, Robert Mayer. Helmholtz was a *physiologist*. Jacobi was a professional *mathematician*. The 'trade journal' of the physics in the mid-nineteenth century, Pogendorff's *Annalen*, was hostile to the work of Mayer, Joule and Helmholtz, insofar as it carried the *chemical* insight of the closed experiment up to the level of abstract principle. Here the picture is of academic physics being bombarded from outside; by the bold thinking of non-physicists and by conceptual innovations within sciences other than physics.

Julius Robert von Mayer, as a boy, had been deeply interested in perpetual motion; this led ultimately to his announcement of the general principle of the conservation of energy (1842). Later, as a physician, he became absorbed in questions about animal heat and in the work output of any organism. Mayer made very clear distinctions between potential energy and kinetic energy – as well as between heat, electro-magnetic and chemical energy. Indeed, he established a reasonable numerical value for the mechanical equivalent of heat. All this from a practicing physician looking 'from the outside in' on the professional physics of his day.

Then consider also Hermann von Helmholtz. He is known to us largely through his work in physiological optics and acoustics. He first measured the velocity of nerve impulses and invented the ophthalmoscope. He developed Young's theory of color vision and generated a completely original idea of auditory sensation and perception. Yet it was he who helped to establish the Faraday-Maxwell conception of electricity. He aided in the development of non-Euclidean geometries. And in 1847 he formulated a broad and definitive version of the principle of conservation of energy – a formulation which served as the foundation for most subsequent developments in thermodynamics and the study of energy transformations. Of Helmholtz in particular, a professional physician, physiologist and psychologist, it is *so* easy to say "but of course he was a physicist too, and a very good one". But again, his intellectual center of gravity lay well outside the academic physics of his time. By his insight and energy, Helmholtz's very existence caused research in physics to coagulate *around him*.[4]

---

[4] "It is a remarkable fact that, in the nineteenth century, chemistry and biology were more attracted to thermodynamics than to any other large theory of physics" (Bochner 1963, 190)

IV

Until the writings of W. K. Clifford and Karl Pearson had their considerable effect, the biological scientists seemed to be more articulate about objectives, presuppositions and methodology than were their physicist-colleagues. The work of Thomas Young began in physiological optics; but by his precision, clarity and forceful expression Young was soon a central figure within physical optics despite a considerable early opposition from 'the old school'. Claude Bernard and Thomas Huxley again distinguished themselves by energetic expository styles which had the effect of clamping the reader's attention onto precisely those things that these writers were concerned with. Clerk Maxwell had the same talent – however, his problems were so clearly within the higher echelons of physical theory that his success consisted largely in unifying his own discipline, rather than in spreading its boundaries into other fields. And yet even the young Maxwell turned astrophysicist, giving an account of the stability of Saturn's rings which was explicitly adapted in 1903 by the microphysicist Nagaoka in his theory of the 'Saturnian Atom'. This latter was explicitly acknowledged as the source of Rutherford's theory of atomic constitution which, through Bohr and Sommerfeld, has become built into our entire conception of atomic phenomena.

V

Consider more fully the flow of ideas from studies in physiological optics to works in physical optics later in the century. The name 'Thomas Young' again springs to mind. Here was a trained physician; he became interested in the nature of animal vision. The discovery of astigmatism is due entirely to him (he discovered it in himself). He explained optical accommodation, and initiated a study of color vision which later was formulated in a theory and then passed along to Helmholtz with remarkable success. It was in connection with his studies of the human eye that Young first addressed the problems of interference. He made measurements of the wave lengths of the various colors. From the standpoint of physiology, he moved on to problems involved in understanding sound and elasticity (remember Young's modulus!). Indeed, from considerations analogous to what we now identify as homeostasis, Young was moved to formulate the concept of energy in a form most useful for later nineteenth century physics. One could do as most of the Dictionaries of Science do – namely, refer to Young as "a physician *and* physicist". This only reflects the degree to which his problems, posed initially while examining living organisms, did later affect the course of physical optics. Young was a medical doctor who changed physics, not a physician who became a doctor of physics.

Much the same story applies with Helmholtz. His development of Young's theories of color vision, as well as his own theory of auditory sensation and perception, constituted a springboard for most of what is really important in the later works of Stokes and Kelvin.

In passing, it might be noted that the so-called 'personal equation' – the discovery of which affected all subsequent astronomy – was fundamentally a disclosure about the human physiology. To the degree that biologists can tell physical scientists more about the perceptual apparatus which contributes to and informs observations in astronomy and physics – to that degree such inquiry is important to an understanding of precisely what observations in physical science are observations *of.*

VI

Friedrich Wöhler is known to us for his many chemical achievements. Most significant was his synthesis of urea in 1828. From wholly inorganic material Wöhler created, in his laboratory, a substance until then known only as a product of kidney activity in animals. The conceptual impact can hardly be exaggerated. It suggested to physicists not only that there were no substances in principle beyond laboratory inquiry – it also stressed the much more general point that there were no non-physical types of causation, or even explanation. Now, this latter conclusion is not so strongly held today as it was 80 years ago, when Helmholtz wrote:

> the task of physical sciences is to reduce all phenomena of nature to forces of attraction and repulsion, the intensity of which is dependent only upon the mutual distance of bodies. Only if this problem is solved are we sure that nature is conceivable. (1847, 6)[5]

Indeed, the glorious possibility of there being a completely general and *Fundamental* Physics, to which all other disciplines might one day be 'reduced' – this was an insight due largely to pioneer work taking place *outside* of 'physics proper'. The quantitative accuracy of measurement within chemistry, and the analyses of biological processes in terms of Conservative principles; these forced physics ever more into *the* fundamental position. But this was not in any great measure due to the programmatic energies of professional physicists.

VII

As a penultimate example, consider the magnificent work of Euler and Bernoulli in theoretical hydrodynamics. These thinkers worked out a theoretical discipline which, although it is consonant with, and genetically related to, Book Two of Newton's *Principia*, was nonetheless a new departure within analytical physics. Euler and Bernoulli succeeded in making the properties of turbulent fluids conceptually tractable by appealing to the *idea* of an 'ideal fluid'. Such a fluid was inviscid, irrotational and a perfect physical continuum. Granted, had Nature ever been good enough to have provided such a liquescent substance, the Euler-Bernoulli theory

---

[5] Hanson refers to Helmholtz's original German publication. The translation Hanson gives is unknown. *–MDL*

would have embraced its complete understanding. But no fluids are ever *ideal*; they all show marked viscosity to some degree, a high degree of internal rotation(s), and are surprisingly granular in certain circumstances. The atmosphere of air within which we live is a thin, low viscosity, gaseous-fluid medium, the exact description of which eludes the Euler-Bernoulli approach. With respect to the theoretical possibility of mechanical *flight,* therefore, the best work of the greatest theoreticians in physics had provided little understanding of the medium through which the proposed ascent was to take place. It is significant that one of the foremost practical contributions in the understanding of flight was entitled "Bird Flight as the Basis of Aviation" (1889). The author, Otto Lilienthal, had profited enormously from studies of natural historians and zoologists concerned with the structural detail of birds' wings, and general bird-physiology; he had also learned much from ornithologists' careful descriptive research on the actual process of bird flight. Indeed, some of these accounts were so beautifully detailed that even Lord Rayleigh was encouraged to undertake studies of the planing motion of a swan's wing: these studies led to results which are significant even today. In short, here is another field – that concerned the physical properties of very thin viscous media – which, had its study been left entirely to professional physicists in the nineteenth century, would not have progressed very far; theoretical hydro-dynamics had little to do with the practical achievement of mechanical flight, and with aerodynamics as we know it today. But the interest which nineteenth century zoologists showed in birds, (fossil and contemporary), as well as in the flight of bats, *this* is what made the entire subject come alive as a problem for physical theory. We now hear so much about 'Mach numbers' and 'Reynolds numbers' that it might appear that the theoretical history of aerodynamics lies squarely within the history of physics. I submit that the achievements of the Wrights, Langley, and Chanute would not have come about without the epoch-making research of Lilienthal which, in its turn, would have been impossible without the ground-breaking work of nineteenth century biologists. No present-day study of the physics of mechanical flight would reflect these origins; all the more important, then, for historians to note them.

VIII

My final allusion concerns the vast and intricate connections between the formal sciences and work in late nineteenth century physics. I will do no more here than indicate how the work of mathematicians on the Theory of Functions, and in the non-Euclidean Geometries, provided material that determined the shape of much nineteenth century physics. We have noted how significant experimental chemistry was in the business of attaching numbers to observed phenomena. It was quickly realized that such numbers might be operated upon, and combined with other magnitudes, in all those ways which pure mathematics permits – ultimately suggesting derivative physical phenomena which might be sought in future experiments. There is no limit to the intricate ingenuity with which mathematical parameters can be interlocked, permuted and combined. This led to formal research which specialized just in that – the variety and diversity which can be shown by functions as they operate on any numerical magnitudes whatever. No one ever directly 'observed' a binomial expansion, whether in sweet peas or in rolling dice. But once appropriate observations are made, and recorded in numerically respectable fashion, then *it may be* that such numbers could be succinctly expressed, and summed over in, something like the binomial theorem. And there are infinitely more complex functional summaries which can encapsulate otherwise quite unintelligible clusters of data. The Theory of Functions provides an inexhaustible store of possible correlations for the data-clusters detected within experimental physics. Indeed, many formal operations within the Theory of Functions are so intricate that they surpass any conception of what might constitute their physical interpretation. One seems to be incapable of prophesying physical processes as unfolding in terms of such formal complexities. This led, in the later nineteenth century (e.g. with Hertz) to what we have come to call 'the Black Box attitude' towards a physical theory. Theoreticians, impressed by the unpicturability of so many operations within the Theory of Functions – operations nonetheless useful in the organization of, and 'reduction' of, chaotic heaps of physical data – such theoreticians have sometimes denied that it is any part of the physicist's business to provide pictorial correlates of what seem to him purely formal and instrumental transformations of observed data. This leads to the caricature of an experimental physicist: he feeds initial conditions in (like pennies) at the top of a 'Black Box Physical Theory'; then he turns the mathematical handle of that box as molded by the Theory of Functions such that new predictions and observation statements come tumbling out 'into the tray' at the bottom of the machine. Anyone enquiring after the physical operations *inside* that Black Box is scolded for asking improper questions. This attitude is very influential amongst theoretical physicists today, especially within Quantum Electrodynamics, and Meson Theory. And it is an attitude directly traceable to the reliance which nineteenth century physicists placed on proven operations within the Theory of Functions.

My 'postultimate' example requires little elaboration. That the work of Bolyai, Lobachevski and Riemann was considered *merely* formal (and hence slightly trivial) is a point that has often been made in histories of mathematics. That these individuals dared to play about with the properties of *hypothetical, unreal* spaces seemed somehow frivolous to the more established mathematicians of the nineteenth century; Gauss, Helmholtz, Weierstrass and Poincaré were exceptions. But, if that is how the work of these individuals seemed to professional mathematicians, it must have seemed completely irrelevant to the day-to-day problems of working physicists. Yet, consider the dawnings of relativity theory in the works of Clifford, Poincaré and Klein, as well as in the conjectures of working astronomers (e.g. Asaph Hall). Here is the ultimate embodiment of much that we recognize as theoretical physics today, and it lies well within the algorithmic creations of these nineteenth century mathematicians. This seems what one might have expected. Men like Clerk Maxwell, Einstein and Dirac – men who *created* their mathematical tools as they needed them in the working out of physical problems – these are the exceptional men who probe the historical rule. The large society of physicists looks gratefully to the established findings of pure mathematics for the provision of those complex inferential techniques, in the absence of which new observations might continue to lay in jumbled heaps; their 'pattern' and 'organization' and 'intelligibility' never being perceived in the sheer complexity of the phenomena. Pure mathematics has always provided simplifying lenses through which the data of nature can be comprehended by physicists.

IX

Let me italicize the semantical issue again. One can always argue *ex post facto*, that everything cited as contributing to the ultimate form of a scientific discipline – all that should be seen as part of that discipline to begin with. So much of alchemy was *really* chemistry. So much of astrology was *really* astronomy. And so much of nineteenth century chemistry, biology and mathematics was *really* physics. Why? Because it led to profound adjustments in physics itself.

Really, it hardly matters how one chooses to characterize the interrelationships. One could either note how physics spread its research-tentacles into subjects which had been thought walled within other disciplines. Or, one can choose to speak, as I have done, of the contributions these other disciplines made to nineteenth century physics. What matters is that we should be clear about the dynamic flow of ideas *into* the discipline which we have now come to recognize as physics – all of which can be recast into the following slogan:

No science is an island, not even when it is (like physics) the mainland.

References

Bernstein, Henry T. 1963. J. Clerk Maxwell on the history of the kinetic theory of gases, 1871. *Isis* 54 (2): 206–216.

Bochner, Salomon. 1963. The significance of some basic mathematical conceptions for physics. *Isis* 54 (2): 179–205.

Helmholtz, Hermann. 1847. *Über die Erhaltung der Kraft*. Berlin: G. Reimer.

Lilienthal, Otto. 1889. *Der Vogelflug als Grundlage der Fliegekunst: ein Beitrag zur Systematik der Flugtechnik*. Berlin: R. Gaertner.

Maxwell, James Clerk. 1873. Molecules. *Nature* 8: 437–441.

Maxwell, James Clerk, and John William Strutt Rayleigh. [1891] 1992. *Theory of heat*. London: Longmans, Green.

Mayer, J. R. 1842. Bemerkungen über die Kräfte der unbelebten Natur. *Annalen der Chemie und Pharmacie*. 42 (2): 233–240.

Chapter 8
The Trial of Galileo

The trial of Galileo has long been used by historians, by philosophers, and by scientists as a classic example of intellectual martyrdom, resembling somewhat the trial of Socrates. Since that is the way "the Galileo story" is usually represented, it turns out replete with villains and heroes. The scenario depends on who the scriptwriter happens to be.

If the narrator is a traditional eighteenth- or nineteenth-century Protestant historian, then Galileo's trial is taken as indicative of the Roman Catholic hierarchy's intolerance, dogma, suppression, popery, and worse. But if one encounters a Roman Catholic account, constructed by an equally eminent and skilful historian, it is Galileo himself who emerges as a rather impolite, brash, and arrogant prima donna. Some small support for this view may be found in the record books open to us all. What we must do here is recognize that the truth is far more complex and intricate than these one-sided Hollywood scenarios suggest. In fact, it was a tracery of confusion, of power politics, of hasty plots, of irrevocable outbursts and temper tantrums, crystallizing party lines and anxiety concerning the status quo.

Indeed, one way of appraising this awful moment in the history of thought at the beginning of the seventeenth century is to realize that the massive intellectual movements of religion and science were on a collision course. The status quo was represented by the ancient, traditional, orthodox, subtle, and detailed philosophy of Aristotle, as it had been transformed by patristic philosophers like St. Augustine and transmitted through the writings of Near Eastern thinkers like Averroës and Avicenna. All this encyclopaedic brickwork had been shaped into a massive cathedral of doctrine in the architectonic writings of such Schoolmen as Albertus Magnus and St. Thomas Aquinas. This was one of the greatest intellectual fusions of all time. In the fourteenth and fifteenth centuries it became solid, institutionalized, somewhat doctrinaire, and not easily displaced. Simultaneously flowering within the Church was something quite different: man's irrepressible curiosity about nature. The pursuit of this quickening interest was called "natural philosophy." It was concerned with the constitution of the world, with experimentation, with observation. These latter concerns were ultimately victorious over the absolute

© Springer Nature B.V. 2020

N. R. Hanson, *What I Do Not Believe, and Other Essays*, Synthese Library 38,
https://doi.org/10.1007/978-94-024-1739-5_8

authority of the Church, but some early battle lines within this conflict should be traced.

The Aristotelian world view is not something that can be summarized briefly. The thinking of Aristotle himself was based ultimately on an appeal to observation, and it seemed observationally quite obvious to him that men on earth were in a position quite different from hypothetical beings that might live on such other objects in space as the moon, the planets, or the sun itself. Because the earth is fixed, men don't feel themselves to be in motion as though they were on a boat; they are stabilized in the center of things, as the diurnal rotation of the sun and the Stars clearly seems to indicate. So in the Aristotelian world view the earth was thought of as immovable, and at the very hub of the whole cosmic system. The system was, in short, geocentric and geostatic.

The system, as Aristotle conceived it, was of no practical use. His rather vague statements on terrestrial centrality and immobility had to be fleshed out· with observational calculations – computations useful for devising calendars and navigational charts, guiding tax collectors concerning the date when they should next demand "tribute," instructing farmers about the times to plant, the times to harvest, and so forth. These practical, calendrical, computational interests required, in turn, a technique for calculating where the planets would be at any particular moment – present, past, or future. This technique was groped for by Eudoxus and Callippus. It was actually provided to a remarkable degree through the great mathematical achievements of that splendid trio of the Hellenistic period – Apollonius, Hipparchus, and most significantly, Claudius Ptolemy.

Ptolemy was a dedicated Aristotelian, at least in his cosmological pronouncements, and he conceived his work along geometric and geostatic lines. His *Syntaxis Mathematica* was, in effect, a logically articulated, geometric calculating device, which could be used both to predict with some accuracy where any one of the heavenly bodies might be at some future date, and to retrodict their positions in the recent past. This computing technique became conceptually tangled up with the Aristotelian philosophy of the constitution of the universe. Thus Ptolemy, just as Aristotle before him, distinguished between the kind of substance, matter, and nature that was celestial, divine, and heavenly, as opposed to the quite different types of matter, substances, and processes found here on earth. Ptolemy embraced Aristotle's contrast between superlunary existence – existence "above" the orbit of the moon – and sublunary existence, such as was found on earth. Everything "above" the moon partook of perfect motion, motion that in the Aristotelian framework was necessarily circular. Such movement did not terminate or come to loose ends. Rather, perfect motion turned back on itself with a complete rhythm and periodicity. Here on earth the natural paths of free-falling bodies, for example, are rectilinear. So freely falling bodies are necessarily sublunary, imperfect and incomplete – yet to achieve their "natural" termini.

Superlunary existence was perfect, moreover, in that objects above the moon's path did not decay, did not wax or wane, and had none of the flaws of terrestrial objects. It was in the perfect superlunary sphere that the products of divine creation were to be found. Such, according to Aristotle, and by derivation, to Ptolemy, was the constitution of the cosmos.

This grand cosmological scheme was developed with incredible subtlety and devotion through the Arabic and early medieval period. Ultimately, the entire theological framework of the Roman Catholic hierarchy came to involve the Aristotelian world picture in an essential way, for the constitution of God's universe, of God's creation, ultimately rested on some explication of the Aristotelian cosmology. The great achievement of St. Thomas Aquinas was examining alternatives to the Aristotelian picture, discussing them with great candor and logical acumen, and dismissing them as inadequate explanations of the problems. Heart and mind were joined in the work of St. Thomas Aquinas. His was not simple faith unstructured by reason; it was faith supported by reason, and that reason was a monumental tribute to the philosophy of Aristotle and his great Arabic followers. The entire thrust of higher scholarship and education within the fourteenth-century Roman Catholic hierarchy was toward working out in ever greater detail precisely these insights of St. Thomas Aquinas, Albertus Magnus, Jean Buridan, and other heroes of the medieval "schools." Indeed, what were known as universities in the thirteenth,.fourteenth, and fifteenth centuries were to a large extent simple Aristotle, institutionalized. The very business of pursuing the world of the mind, the world of reason, involved being in an institution, like the University of Paris, or the universities of Bologna, Pisa, or Padua, places where the complete and detailed study of Aristotle, in all his respects, in all his applications to the Catholic faith, could be undertaken virtually without restraint.

Nonetheless, even inside this monolithic philosophical edifice, criticism did develop. Individuals, like Aquinas himself, advanced many cautious criticisms of the Straight Aristotelian line. This was natural' enough; one cannot read and understand a philosophical work without becoming a philosopher to some extent. When an intellect like Aquinas sought to understand Aristotle, critical challenges were inevitably a result of comprehending the meaning of the original text. Also in the eleventh century, and certainly in the twelfth and thirteenth centuries, tension was developing between the Platonic interpretation of the Catholic faith and the Aristotelian interpretation, between Augustine and Albertus Magnus, between the Franciscans and the Dominicans. One important issue was that the Aristotelian tradition insisted that all natural human knowledge originates in sense experience, in observation of the world, whereas the Platonic tradition accounted for all knowledge as insight, gained through introspection and illumination. The difference can be seen reflected in the proofs of God's existence in the two traditions: the Augustinian Platonists take their start from a concept of God in our minds, the Aristotelians begin with the observed facts of causal processes in nature. The ascendance of the Thomistic philosophy constituted an Aristotelian victory over the Platonic elements which dominated the Augustinians. But even after Thomas' triumph, critical thinking and dissension appeared in the work of William of Ockham, Nicholas of Cusa, Jean Buridan, Robert Grosseteste, Nicole Oresme, and others. These scholars were concerned with the formal constitution of argument, the physical constitution of matter, and the divine constitution of the universe. In the late fourteenth and early fifteenth centuries they all began to raise doubts and ask pointed questions about the degree to which Scripture, and even Aristotle himself, could be relied upon to provide finely detailed truths about the complex phenomena of this world. Sometimes, for rational or logical exercise, these Schoolmen undertook searching examination of sacred and philosophical texts in order to explore more deeply to what degree consistent alternatives to Thomas, and to Aristotle, might be possible.

Many criticisms – e.g., of Aristotle's theory of the "solar system" (as we would now say) – began to appear not quite so ridiculous as the defenders of divine dogma argued. By the late fifteenth century doctrines and counterdoctrines ran parallel to each other: *primi inter pares*. At this time alongside Aristotle's cosmology existed another discipline called "calculational astronomy" or "mathematical astronomy," which, for all practical requirements of navigation, agriculture, calendar reform, tax collections, etc., had become of primary importance. While internal philosophical criticisms continued to mount, it was found that the practical Ptolemaic astronomy did not really work. Cosmology aside, the computations and predictions of arithmetic astronomy did not correspond exactly to the phenomena being observed. Navigators, tax collectors, and farmers were vexed.

It is in this framework that the great work of Nicolaus Copernicus should be considered; that work was called *De Revolutionibus Orbiam Celestiam,* "On the Revolutions of the Celestial Orbs." This treatise, published in 1543, was advanced in a totally medieval spirit, a fact that is often forgotten. The work of Copernicus is frequently hailed as the first ringing of the bell of modern times. This is not correct. The fundamental concern of Copernicus was merely to devise an astronomy that would do the work that Ptolemaic astronomy was trying to do, only do it much better. He found, therefore, that a number of revisions were absolutely essential. The virtues of some of these revisions had been recognized millennia before by Hipparchus, Apollonius, and Ptolemy himself. They had all realized that their calculations would proceed much more smoothly if the sun were considered to be the fixed center of the universe, but this hypothesis was generally rejected because it appeared to be refuted by simple observation of the sun's "movement." Copernicus' theory, in a sense, returned to this ancient conception, for it was a heliocentric and a heliocentric way of calculating the motions of the planetary system. System is the operative word here. The entire corpus of Copernicus' astronomy hangs together like a beautiful piece of machinery. Anything done at a "later" moment essentially depends on a previous event. And one cannot tinker with a small part of the system and expect the rest of it to keep running well. In this respect it was quite different from the Ptolemaic system, which resembled, in a sense, a plumber's toolbox; from this toolbox, depending on the problem, the particular calculation or device needed could be extracted and applied.

Copernicus was quite serious about the physical truth of his system. He did not purport to provide merely alternate means of computation to those provided by Ptolemy. Until recently, historians doubted this because of the clear distinction made in the Preface to the *De Revolutionibus* between mathematical hypotheses and physical descriptions. Mathematical hypotheses, according to the Preface, serve simply as a basis of calculation; they can be of any form whatever, and of any content – provided only that they are consistent. Their significance is as a basis for further inferences and computations, and they should never be confused with the sum of descriptive propositions, which really constitutes a physical theory. For many years it was thought that the Preface was written by Copernicus himself as an indication of what he really thought he was doing in *De Revolutionibus*. It is now known, however, that the Preface was written by another person, one Andreas Osiander, a Lutheran divine, and that it reflects his own theological views. Osiander wished to ensure that Coperinicus' work, which he had ushered through the press, received the most favourable reception possible. We can be confident that Copernicus did not subscribe to the views with which Osiander credited him. If Copernicus had been able to rise from his deathbed and really study the text that was placed in his hand on the last day of his life, instead of just glancing at a few pages, he would surely have protested against the spirit and the letter of the Osiander Preface.

Beyond the appearance of this signal work, Galileo's trial was prefaced by further trouble in scholarly ranks. Georg Joachim Rheticus, a clever and ardent disciple of Copernicus, was popularizing the heliocentric philosophy throughout Europe. Tycho Brahe, a Dane, brought out another anti-Ptolemaic theory, although in many respects it paid lip service to the Scriptures. Johannes Kepler, one-time assistant to Brahe and a great astronomer in his own right, shattered basic beliefs by showing that planets not only revolve around the sun but that they do so in "imperfect" elliptical orbits. Hence, although the Roman Catholic cosmology of the late sixteenth century appeared to be orthodox, it was increasingly under attack from within.

Convulsive unrest was thus transforming the schools from centers of indoctrination to the focus of intellectual search and anxious study. Externally the trembling tidal wave of Protestantism bore down upon the tradition. Within the Church, there were the Catholic Platonists and Franciscans, who refused to accept the doctrines of an Arab-modified Aristotle. There were also the mathematicians, who talked a new calculational language, largely misunderstood by the Aristotelians. The universities were no longer institutionalized retreats for sophisticated rehearsals of Aristotelian philosophy. Within these gray, cloistered halls seethed intellectual anxieties of all kinds – anxieties concerning the status of Scripture (myth or history), the status of the true conception of the universe (geostatic or heliostatic), the status of the organization of the Church (papal absolutism or intellectual freedom). Indeed, the anxieties which erupted under pressure during Galileo's lifetime did so in a way that should be a lesson for us all.

Galileo was born in 1564 (the same year Michelangelo died; the same year Shakespeare was born). He was precocious from the start. Slated early for a course in premedicine, he abandoned that field for mathematics. Then, after serving most effectively as a lecturer at the University of Pisa, he tired of that much honored position; he was, indeed, very poorly paid. He moved to the University of Padua, where in 1609 he happened upon a most remarkable discovery. He heard that the son of a Dutch lens grinder had put two lenses together and succeeded thereby in enlarging the visual images of distant objects. This rumor set Galileo's great mind into motion immediately. Within moments, it is reported, he blocked out the entire mathematico-physical framework of the telescope. He assembled one almost at once; it succeeded in magnifying an object's image three times. A later effort brought him up to a 32-times enlargement.

This was an incredible instrument in its time. It was as if Aristotle, for example, had been given an X-ray machine, so that he could perceive things beyond the senses of ordinary men. As a mathematician Galileo already realized that the heliocentric-heliostatic approach was going to prove more "rational" (i.e., well designed and deductively fertile), as an astronomical calculation device than the Aristotelian geostatic theory. Given these initial sympathies, he discovered many wonderful things by looking skyward through his telescope, things that tended to support his developing anti-Aristotelian position. He discovered the moon's surface to be rough, and not the smooth, polished marble it had been decreed to be. He discovered that the Milky Way was a constellation of stars, positioned at *all* possible distances from the earth, and not gemlike lights embedded within a single cosmic sphere centered upon the earth – the universe was not geocentric.

Galileo discovered four of Jupiter's many moons. He called them, in honor of his patron, "The Medicean Stars." Sunspots were perhaps his most important find. They appeared as gaping holes which looked like *imperfections* in what was universally thought of as *the* perfect celestial object. This made a deep crack in the Church's neo-Aristotelian position, although some professors were so committed to the invulnerability of the latter that they refused even to look through the telescope. Galileo was the first to see the phases of Venus, which proved that there is a dark side to that brilliant planet – the side turned away from the sun. This observation was totally incompatible with both unadjusted Aristotelian philosophy (wherein the planets were not construed as being "terrestrial" in any way), and unmodified Ptolemaic astronomy (whose basic geometry rules out *full* phases for Venus). Galileo was also the first to view the spectacular rings of Saturn. He did not construct the right analysis of them because his telescope was too small to reveal them clearly; he perceived them as "ears" or "handles" on the "mug" of Saturn. Nonetheless, his sketches were later duplicated in almost exactly the same form hundreds of years later by people who knew that Saturn had rings.

Galileo thus had succeeded in designing an instrument that brought closer to man's eye observational evidence supporting the antiorthodox position, the anti-Aristotelian position, the anti-Ptolemaic position. As a result his pro-Copernican leanings began to appear in his philosophical tracts and letters. In those days letters were circulated much as journals and periodicals are circulated today. They were the literary conduits through which new ideas were conveyed and circulated among intellectual peers. Although he did not articulate them fully, it appears from Galileo's writings that he pieced together Newton's First and Second laws, the *Law of Inertia* and the *Force Law,* in a somewhat restricted form.

Galileo's mathematical rationalism should be carefully noted. Within the Aristotelian Roman Catholic framework all discussion, philosophical analysis, cosmology, theology, etc., took place in logical or verbal form. The objective was to characterize the properties of certain *denotata* (the things that could be designated and attended to) in the most cogent and reasonable way. Thus man was a featherless biped; cubes were six-faced regular solids; water was a solid-liquid-gas. This is the framework of Aristotelian logic and analysis, within which there is no preassigned room for measurement, for number, for ratios, certainly not to the degree Galileo thought necessary for determining the properties and behavior of dynamic objects. Aristotle's philosophy was centered upon qualitative inferences. It called attention to the ways in which qualities can be categorized, or predicated of individual enti-ties. Thus it is of the nature of man to be featherless and bipedal. These must be qualities of anything properly called a "man." It matters not to what *degree* an entity is featherless, in one's reflections on manhood. That issue never arose in Aristotle's mind.

However, with Galileo, *quality* was not an important word; *quantity* was. It was he who made viable the distinction between the so-called "primary qualities" of matter and the "secondary qualities" – the primary qualities being things like weight, motion, shape, and the several dimensions. The secondary qualities were much less tractable – the softness, texture, smells, and colors of objects. Galileo distinguished these qualities, indicating his concern with only the first, although he was not wholly inarticulate about the secondary ones.

The signal consideration here is that Galileo, an accomplished mathematician, succeeded in using the concept of quantities in the understanding of ancient prob-lems about the nature of motion and about the relationship of the heavenly bodies to each other. He made coherent, intelligible analyses of eclipses and of lunar paths and the tides. He perceived the physical principles of ballistics. All these things he was able to approach in a totally new attitude, that of a numerical technician pre-pared not simply to argue about qualities, but to calculate. For Galileo, the claim to understand X had to be justified by providing precise predictions of X's future behavior. The latter (predictions) might not be a sufficient condition of the former (understanding), but it was certainly a necessary condition.

This was frightening for the orthodoxy within which his innovations were gener-ated. Each one of Galileo's sharp pronouncements seemed to negate the ideas that had become institutionalized in the framework of what his colleagues had learned in school and in church. Education of the young is an attempt to present and incul-cate the ordered knowledge which is at the fingertips of a wiser but older generation. The curriculum at any established university or school consists of the tried and the conservative learning that is *de rigeur* for children. When a threat to that kind of curriculum and to that kind of stability appears in such a person as Galileo, and when the "elders" learn to their dismay not only that he is intellectually agile – a magnificent tactician in argument – but also that when issues become complex he resorts to an unfamiliar symbolic technique, then a collision must predictably ensue.

The really ultimate blow to the conservative establishment was Galileo's failure to write in Latin. He wrote in Italian so that the man in the street could actually understand what he was saying. This was like taking the false front off tradition. Anyone who could read was suddenly in a position to see how twisted and unneces-sarily complex and archaic the philosophy and cosmology in the schools really were.

Galileo moved several times – intellectual gadflies are rarely stationary. He was given a lifetime appointment to the professorship of mathematics at Padua, but turned it down because he did not wish to harness himself to a fixed teaching schedule. He wanted to pursue original research, to explore the perplexities of physical science that meant so much to him. And so he accepted a much less secure appointment in Tuscany. The Grand Duke of Tuscany, Cosimo II, became his patron. In this station, supported by enlightened affluence, Galileo carried on the rest of his active life.

The old guard's distrust of Galileo was heightened by the existence within the Church of this period of a great deal of backbiting and ambition. Angry old men who were highly placed in the ecclesiastical structure were not prepared to have everything they held dear upset by the theories of someone like Galileo. What emerges from this particular chapter in the history of thought is that the real offenders in the "crime" against Galileo were not individuals such as Niccolò Lorini or Brother Tomasso Caccini, or Roberto Cardinal Bellarmine or Pope Paul V Borghese, or Maffeo Cardinal Barberini, the later Pope Urban VIII. These men were in a way only the tragic instruments of a monumental clash of ideas. The real offenders in this crime were ignorance, insecurity, failure of nerve, a breakdown of communications, and an attempt to preserve the familiar solidities against the incomprehensible new attacks from science and from experiment and from irrational theology.

Some of these weaknesses, although natural and even understandable to some extent, were nonetheless hard to forgive. In 1611 Lorini, an inconsequential preacher of Florence, feeling threatened and insecure (as his letters to his family indicate), lashed out against Galileo, as well as against his "highly overrated reed," i.e., his telescope. For Lorini, it was only a leaden pipe, with two bits of glass stuck in at either end. A "reed," indeed. As for the Dominican Brother Tomasso Caccini, who in the ecclesiastical scale of intelligence rates even lower than Lorini, he was a terribly ambitious man (again this is quite clear from *his* correspondence). In 1614 Caccini composed a complaint against Galileo, accusing him of heretical doctrine. The current pope, Paul V, had long since set his mind against the new scholarship, having declared that "the last thing we need in Christendom is a lot of wise scholars." Brother Tomasso sent his complaint of Galileo's objectionable doctrines to that pope. In this action he resembled somewhat an instructor at a great university writing secretly to the university president about leftism, radicalism, extremism, or atheism on the part of some annoyingly influential professor. What normally happens in such a case is exactly what happened at first in the "Galileo case": the complaint was identified as that of a near-crackpot and dismissed almost out-of-hand. Still, when viewed against the Church's general uneasiness about the future and the attacks upon orthodox doctrine, the dismissal of the complaint seemed to some highly placed churchmen as somewhat premature. At about this time occurred the scurrilous forging of a document by Brother Tomasso Caccini, a document most damaging to Galileo. In a letter Galileo had written, Caccini diabolically changed words and some phrases, and even added some expressions of his own. Triumphant scholarship has since discovered the forgery, but at the time the letter redounded to Galileo's dire disadvantage.[1]

At the same time, Galileo, always the disseminator of truth in no uncertain terms, presented his views somewhat untactfully. He attacked a pompous Jesuit, Father Christopher Scheiner, one of the senior fathers of the order – an individual who commanded great respect. This elder ecclesiastical statesman was haughty and rhadamanthine to an advanced degree. Galileo demolished him with many well-chosen words, knifelike epithets, sinewy arguments, and suggestive allusions. As this captain of Jesuits sank, beneath waves of Galilean debating devices, the entire order took offense. After all, the Jesuit order had become the virtual repository of all scholarship and theological doctrine within the Catholic Church. Now this insolent tormentor, aside from all his other publicized offenses, was actually savaging a senior hero of the order. This did not help Galileo's cause at all. Indeed, the Jesuits themselves, lacking then such a champion of real intellectual prowess as a Thomas Aquinas or a William of Ockham or a Nicole Oresme, began to feel the all-too-common, all-too-human weaknesses of insecurity. All Jesuits now felt the necessity of uniting against a common foe. Their interminable arguments with each other were suspended *pro tem*. All reckoned the real enemy to be the man beyond the chapter walls, Galileo himself.

Through a series of intricate intrigues, Galileo was called to Rome in December 1615 as a "consultant." His "expert" views concerning the possible heresy of Copernican doctrine were supposedly required. In fact this summons was a decoy, the real intention being to place Galileo's own views under the glass. He went to Rome, and with a light heart, inasmuch as no more confident individual lived. He knew his powers in dialectical debate; he was undefeatable, indeed virtually unchallengeable. Galileo's strength in argument was conceded by everyone, and his charm infused debate. He really *could* have been the Devil's advocate.

In Rome Galileo conquered everybody – even Cardinal Bellarmine, before whom Galileo was summoned, and from whom he received a letter freeing him from doing penance. The cardinal was old (74), and he was conservative. He was the living embodiment of the solidity, the organization, and the orthodoxy of the Church. Even the good cardinal, however, could find no fault with the manner in which Galileo addressed and attacked his problems, always in an objective, exciting, and stimulating way. But Bellarmine was not about to cut loose from his own theological moorings. In 1616, therefore, he found grounds for condemning the hypothesis of Copernicus as heretical – it went against Scripture in several well-known respects.

---

[1] Throughout this discussion, Hanson exchanged the names of Lorini and Caccini. Interestingly, though the 'triumphant' scholarship of Hanson's day had proclaimed that Lorini had added his own crude and incriminating interpolations to Galileo's *Letter to Castelli* (the document to which Hanson is referring in this discussion), contemporary scholarship has reversed that judgment. As it turns out, the version of the *Letter to Castelli* that Lorini sent to the papal authorities was a faithful copy of the original. Galileo saw fit to edit that original, and softened phrases that might have leaned toward the heretical, and scholars had mistook Galileo's edited version of the letter for the original. For details on this matter, see Fantoli (2012, 69–76). –MDL

The cardinal, however, was not totally intractable. Returning to the preface of Andreas Osiander, wherein occurs the distinction between a mere mathematical hypothesis and a physical truth, Bellarmine argued that Copernicus' treatise could be studied in the universities as a mathematical hypothesis, although it should not in any sense be construed to be propagated as a physical truth. He asked Galileo's opinion on this distinction, and indeed gained the scientist's consent, Galileo apparently recognizing at this point that the forces against him were too strong.

As a courtesy to Galileo the cardinal informed the scientist in advance of the Church's decision against Copernicus. A later bit of trumpery, perpetrated possibly by Brother Tomasso or more probably by some member of the Jesuit order, consisted of this: A record of this meeting between Galileo and Bellarmine was inserted into the papal files indicating that the scientist had been forced in 1616 to deny his position as a Copernican – to acquiesce, to abjure. This document was placed in the files sometime after 1616. It has been pointed out by many scholars that this is a spurious document, that it almost certainly had nothing to do with the events of the time.

Bellarmine's refusal to accept Copernicus' treatise as a physical truth was very disappointing to Galileo. He returned home chastened, a broken man. His own work in mathematics and astronomy had been developing toward a climax – it was to crest a wave of Renaissance activity in this new science of natural philosophy. All the vaults of knowledge were about to break wide open. This development was halted by the Church's refusal to place itself at the frontiers of learning.

So Galileo returned to Florence in defeat. Soon, however, he took up cudgels against another haughty Jesuit, Orazio Grassi, against whose views Galileo wrote a very exciting document called *I Saggiatore* (i.e., *The Assayer*). This particular piece pleased his friend Cardinal Barberini, who recognized that Galileo was one of the most magnificent philosophers of all time. Barberini and Galileo were members of the same discussion group, a philosophical academy called the Academy of Lynxes.

During Galileo's relatively quiescent retreat at home, Pope Paul V passed away. He was succeeded by Urban VIII, who was none other than Barberini himself. Galileo and his supporters rejoiced, for it seemed certain that the new pope would be more tolerant of new ideas and that he would accept what Galileo reckoned to be the true spirit of Roman Catholicism – to wit, that Church doctrine should never be incompatible with *the truth*. As a young man Barberini had concurred with Galileo in this persuasion. The enemies of the Church might be avarice, war, corruption, and vice, but never the truths of natural science. In any conflict between dogma and truth the former must give way. In a real sense, Galileo, during the early seventeenth century, was the devoted, the virtuous, the dedicated Catholic. He was fighting a kind of ossified orthodoxy within the vital organs of the Church, and his concept of the relation of dogma to truth was not to be accepted until the nineteenth century, when an encyclical was published under the title *Providentissimus Deus*, which in effect brought the Catholic Church full circle from Bellarmine to the pronouncements of Galileo.

Barberini as a Lynx and Barberini as the pope turned out to be two quite different people. As Pope Urban VIII he was not the same man with whom Galileo had discussed theories of magnetism and the tides and other philosophical matters. In those days argument had pursued its course, unhindered in its quest for truth, however shocking, unfamiliar, and hostile to tradition.

At first all appeared to be for the best. Scientific inquiry did not have to be concealed any longer, except from the Jesuits, who realized that things were not going their way with Barberini's appointment. They were further upset when two friends of Galileo – Niccoló Ricciardi and Msgr. Giovanni Ciampoli – were appointed as the papal censors. The latter were the individuals responsible for giving (or withholding) approval on any printed matter or work of scholarship generated within the extensive precincts of the Catholic Church.

Pope Urban, during an interview, was understood by Galileo to have *asked* for a completely neutral presentation of the two chief world systems: the planetary systems of heliocentrism and geocentrism. Urban requested this of Galileo – so that Galileo himself would not be proselytizing on behalf of a particular theory. Rather, the rival positions would be set up in a dialogue with each other. Galileo would not plead one against the other, but serve as an impartial referee above the fray.

Galileo worked on the assignment for 8 years; the result was *Dialogue of the Two Principal Systems of the World* (1632). It was an immediate success. Everybody seemed to applaud this clear, yet controversial exposition. It was pellucid; the rival positions were set out with great care – the geocentric view being articulated perhaps more cogently and persuasively by Galileo than by many of its most passionate advocates. Everyone who set eyes on this work felt elevated and ennobled by it, including the pope. Everyone, that is, but the Jesuits, who recognized themselves in too many places continually getting the worst of things in the dialogue. They screamed "heresy." Indeed, they went much further than this. In an audience with Pope Urban (by this time easily influenced and temperamental – almost mercurial) they persuaded him that he also appeared in the dialogue as one Simplicio, the Simple One, who was urging the orthodox, Aristotelian position. This enraged Urban, and his explosion of anger marked the end of all open-minded discussion of alternatives in the early seventeenth century. Scientific discussion simply disappeared because of the threat that Galileo seemed now to pose to persons and to powers and to politics.

The Jesuits succeeded in having Galileo summoned to Rome, although by this time he was aged and infirm, and his sight was extremely bad. The scientist asked for a dispensation, or at least a delay, so he would not have to come in the winter months. Urban, totally inflexible, steadfastly refused this request. In fact, he wanted to make an example of Galileo and his doctrines, particularly at this critical moment in the crosscurrent of ideas. It was a terrible moment for the human species, and for the development of Western thought in particular. (It suggests Senator Bilbo placing Einstein on the rack for having said a number of things that, if Bilbo understood them at all, seemed to him to be un-American.)

During Urban's fulminations the forgery of 1616 was "discovered," and was interpreted as constituting an earlier proscription against Galileo's ever teaching the Copernican heresy again. And yet, the Jesuits urged, this was precisely what he was doing in 1632. An Inquisitorial commission was appointed, and Galileo was forced to abjure; he was placed under house arrest for the final 8 years of his life. He retired, in effect. No more public firebranding! Galileo turned to work on the last great *opus* of his life, published in 1638 (not in Italy), *Dialogue of the Two New Sciences*. Here is found what is now known as "classical mechanics," worked out in its first exciting forms. By this advanced stage of his life Galileo was blind, and very sick. Nonetheless, even in this condition, he served as the intellectual inspiration of such great natural philosophers as Viviani and Torricelli.

The lesson to be learned from this tragic, sobering history is significant. Just as vivid today as it has ever been is the fact that doctrine – when it becomes institution-alized and vested in an organization; when it becomes stiffened against all possible attack because of administrators' lack of sufficient intellectual ability or vivacity, or the energy to fight off critical opposition; when its truth alone is not the only crite-rion for its evaluation – then such doctrine is no longer knowledge, it is dogma. When this happens, dialogue ceases. Free inquiry ceases. Communication collapses. Gossip and plotting, cloakroom intrigues – these take over. Small men make them-selves seem tall and strong with fighting stances and chest thumping. But history will record them ultimately as having assumed such postures in the cause of unreason.

The stiffening of the Jesuit and papal positions during the early seventeenth cen-tury made impossible the absorption, assimilation, and utilization of the new sci-ence – the glory of the Renaissance – whose effects are with us still. This rigidity made adaptation impossible, even for what had been the most flexible and absorp-tive organization of all time. The scientific revolution came ultimately to be set up *against* the Roman Catholic Church. It was anti-Christ to Rome, which in turn struck the newly enlightened as ignorance enshrined. Many of the writings by heroes of the scientific revolution seem to be antipapist at their inspirational sources. Copernicus was placed on the *Index Prohibitorum;* he remained there until the nineteenth cen-tury. Truly, the Vatican placed its imprimatur on very few major scientific works of the seventeenth, eighteenth, and nineteenth centuries. The breach cut by Bellarmine grew increasingly wider, so that in the twentieth century advanced scientific research and the Holy See have sometimes seemed to be at opposite ends of the universe.

It is interesting to speculate on the contributions Galileo might have made to sci-ence if his long, eventful life had been allowed to proceed without harassment from the Church. He is to us an Olympian even now; he is a towering giant because of his contributions to scientific thought, theory, experimentation, and instrumentation. If he had not had to spend his entire professional life with men of the stature of Bellarmine, Paul V, and Urban VIII, had he been able to devote himself exclusively to his work, he would probably have been a Newton, an Einstein, and a Darwin combined in one.

In a real sense the fault lies with all men, not just the Fathers of the Church. The human predicament was much then as it is now, a medley of elevating harmonies and cacophonous dissonance, and so pure a voice as Galileo's was certain to be muted, distorted, and silenced, just as it would most likely be today. Galileo reincar-nated in our time could not expect better treatment, because we the people probably have not changed much. All the more then should Galileo Galilei be admired. No man will ever sing his song again.

References

Berti, Domenico, ed. 1876. *Il processo originale di Galileo Galilei*. Rome: Cotta e Comp.

Brecht, Bertolt. 1963. *The life of Galileo*. Trans. I. Vesey. London: Methuen & Co., Ltd. [This is an excellent interpretation by a modern dramatist].

Brodrick, James. 1965. *Galileo; The man, his work, his misfortunes*. New York: Harper & Row.

De Santillana, Giorgio. 1955. *The crime of Galileo*. Chicago: University of Chicago Press.

Fantolie, Annibale. 2012. *The case of Galileo: A closed question?* Notre Dame: University of Notre Dame Press.

Galilei, Galileo. 1950. *Dialogues concerning two new sciences*. Trans. H. Crew, and A. de Salvio. Evanston: Northwestern University Press.

———. 1967. *Dialogue concerning the two chief world systems – Ptolemaic and Copernican*. Trans. S. Drake, with a foreword by A. Einstein. Berkeley: University of California Press.

Galilei, Galileo, Polissena Galilei (Sister Maria Celeste), and Mary Allan-Olney (compiler). 1870. *The private life of Galileo: Compiled principally from his correspondence and that of his eldest daughter, Sister Maria Celeste*. London: Macmillan.

Langford, Jerome J. 1966. *Galileo, science, and the church*. New York: Desclee Co.

Part III
General Philosophy

Chapter 9
On Being in Two Places at Once

In his early book *Reality*, Paul Weiss says interesting and important things. Some remarks however, besides being interesting and important, require further examination. The following quotation is a case in point.

> By identifying ourselves as concretely real with the 'X' of the law of contradiction, it is easy to substantiate the fact that the assertion 'someday in someplace I shall meet myself coming towards me' is absurd and material and that its denial is material and certain, significant and indubitable. But there is no need to bother with the demonstration. We know that we are unique individuals and that there is no sense in the supposition that we can be duplicated. This is a truth, fixed and deep enough to provide a satisfactory point about which our speculations may turn. My supposed repetitions are identified by myself and others as illusions, hallucinations, mirror reflections, images, and so forth, because we all know that I am an unduplicatable singular being, and that what is spatially distant from me is other than myself. No truth in logic or mathematics is more certain and no fact in daily experience or science is more significant than that we are unique individuals. If there were a choice between a dialectic which compels one to deny this, and the blunt affirmation to the contrary, it is the dialectic which must be put aside. (1938, 160)

That I shall never meet myself coming towards me is clearly true. It will never happen. But, Weiss notwithstanding, there is a real need to bother with a demonstration. For that he can say of the assertion above, that it is both absurd and material and that its denial is material and certain, significant and indubitable, this makes at least one of his readers feel unsure of the grounds for his claim. In one philosophical tradition Weiss would appear to have mixed the logical levels of his predicates. And while the error may turn out to reside in that philosophical tradition and not in Weiss's argument, nothing less than a demonstration one way or the other will settle the matter. If what he says is true, then a demonstration of his point should offer no difficulties. But, it might be argued that if Weiss's claim is a material one, then it cannot be logically certain. If, on the other hand, it is indubitable, then it cannot be material (i.e., contingent).

© Springer Nature B.V. 2020
N. R. Hanson, *What I Do Not Believe, and Other Essays*, Synthese Library 38,
https://doi.org/10.1007/978-94-024-1739-5_9

I

I propose to play the *advocatus diaboli.* It will be contended that Weiss's claim is indeed a material (i.e., contingent) claim – but just in the *de facto* sense that meeting oneself face-to-face never happens. That it should happen is inconceivable; inconceivable just as it is inconceivable that a man should flap his arms and fly to the moon, or that water should flow uphill, or that there could be a *perpetuum mobile*, or that I am not now writing these words. The claim is not inconceivable (so claims the devil's advocate) in that its denial is logically self-contradictory. If Weiss means to say that the possibility of my meeting myself face-to-face is inconceivable in some sense stronger than the first one delineated, and yet not inconceivable in the sense of being logically self-contradictory, then a demonstration is indispensable. In the absence of such a demonstration the following argument may be generated.

Consider the situation as the third person sees it. How does one ever know of a certain individual, X, that he has been or is in a particular place? We are usually said to know this, or not to know this, in terms of the reliability of our evidence. Suppose the evidence for X having been in New Haven yesterday is as conclusive as it ever can be with empirical claims. We have a sworn statement signed by the president and all faculty members of Yale University to the effect that X was there yesterday. We have photographs of X standing amongst familiar Yale landmarks, photographs which are dated and notarized. We speak to people who actually saw X in New Haven yesterday. From all this it would certainly appear as established that X was in New Haven yesterday. But now, is it *logically* inconceivable that exactly this same kind of evidence should have been produced to show that X was at Harvard yesterday? It is, of course, *de facto* inconceivable that such further evidence could be produced – it just never happens this way. But can it be argued that the production of verbal testimony, sworn statements, photographs, etc., *logically* could not be adduced in just as strong a fashion to show that X, besides being certifiably in New Haven yesterday, was also certifiably in Cambridge yesterday? Is it *logically* contradictory to suppose this, i.e., is the supposition of the form $P \cdot \sim P$? If it *is,* then the statement that X was in New Haven (only) yesterday is not itself an empirical claim – a bitter consequence.

If however, we simply dismiss one or other of these two clusters of evidence, on the grounds that one of them just *has* to be false, this would be a *petitio principii*; it would just be a covert way of restating (without further reasons) that a person cannot be in two places at the same time. But if *all* we have to prove that X was in *one* place at time *t* is duplicatable (logically) in just as strong measure to show that he was also in another place at *t,* then if we were right to rely on the evidence in the first place, we ought to rely on the second dose too. If, *ex hypothesi*, there is no difference in the evidence, we ought to accept it both times if we were prepared to accept it once. In point of fact, if such evidence could be produced why should we not boldly grasp the other horn and say that we now have evidence against the dictum that a person cannot be in two places at the same time? If it is true to say that X was in New Haven yesterday because of the evidence cited – and if it is not logically impossible that such evidence should be cited to show that he was also in Cambridge yesterday (after all, only natural laws are at stake) – then were such a thing to happen, the third person claim that X was both in Cambridge and in New Haven yesterday is precisely as strong as either single claim made separately. To deny this is to reassert what is at issue.

Certainly we would be astonished if such a thing happened. But the history of science, e.g., is full of such astonishment, discoveries of what had been thought impossible are suddenly shown to obtain. And once shown to obtain, it becomes clear that they never could have been logically impossible. Nothing in this third person example reveals such logical impossibility – though the psychological inconceivability is obvious.

It might be argued, as Weiss probably would argue, that the evidence in the third person is *different in kind* from what we have in the first person. It is just because third person evidence is what it is, that the situation imagined above can be entertained. But the first person is different. Could *I*, being what I am, know myself to be in New Haven and in Cambridge at the same time? Is this logically self-contradictory, or just psychologically inconceivable to an advanced degree? Do our logical concepts forbid it, or do our laws of nature forbid it? Is there *any* kind of situation which, though psychologically inconceivable, is none the less imaginable (i.e., not logically impossible), and which would raise with first person experience the kind of doubts we just raised with third person experience?

As a first approximation, consider Siamese twins, joined not at the hips, but at the temples. Let us suppose further, what sometimes does happen, that the brains of the twins are fused at this point. It is conceivable that two such human beings might share the same thoughts and even have the same perceptions. Their four eyes in combination would give an extremely complex composite vision; but nothing more complicated in principle than what obtains with, e.g., the chameleon, whose individually articulated eyes point in opposite directions from the sides of its head. In a sense then, these Siamese twins are different persons, with their own perceptual and conceptual equipment; yet in a sense they are the same person because *(ex hypothesi)* their total experience is the composite sum of the experience of each of them considered separately. Yet each one *has* the total experience. If one of them closes his eyes, the overall experience each has would be comparable to our overall experience when we have one eye closed. If twin X touches a stove, both X and Y feel the pain.

Several reactions to this example have been expressed thus: "Why X and Y are but one person, they are not different people at all." Precisely. This is exactly what is sought.

Now at this point we simply need a new empirical hypothesis, to get the two twins at spatially different points, yet leaving all other features of this example unmodified. We allow their point of contact, viz., the temples, to attenuate in an elastic fashion. The description of the twins' composite experience remains the same. They are still in a profound sense "the same person". But now their brains are connected by, and interfused with, a long neural conduit. The twins are now in different places. Yet they are the same person. Conceptually, it is only a minor empirical modification to this obviously empirical hypothesis, to let the conduit stretch

150 miles – the distance from New Haven to Cambridge. It is now a very fine neural fibre. Twin X is in Cambridge, and twin Y is in New Haven. Yet the composite perceptual and conceptual experience of both X and Y, considered together, is individually the total experience of X at Cambridge and the total experience of Y at New Haven. Thus if Y is sleeping in New Haven, X's "inner life" at that time would be as if the neural fibre connecting him with Y did not exist (assuming Y is not dreaming). But when Y is awake and perceiving and thinking, then X's total experience at Cambridge has been augmented by that much. At such times X and Y – who are, by the way, perfectly identical twins in every physical respect – would have precisely the same "inner lives". In a very important sense of the expression, they would be *the same person*. (Indeed some query giving them different names, 'X' and 'Y'.) The same person would be both in New Haven and in Cambridge, at the same time. All the physical characteristics of X and Y would be identical. All their perceptual, conceptual, and psychological characteristics would (because completely shared) also be identical. The *only* difference between X and Y would be the one on which Weiss ultimately rests, they are "spatially distant" from each other – which is precisely what this imaginary experiment has been meant to secure. X and Y are spatially distant, and yet they are the same person.

Now by a further hypothesis, again empirical, we simply suppose X and Y to keep in touch not by a neural fibre, but by an intangible *field* of some sort. The model of an electrical or a magnetic field will do. Everything that was claimed before for X and Y can still be supposed to obtain. The only difference now is that the mechanism of interaction between them has been changed to a kind of super wireless-telegraphy. Everything else however, is as before. X is in Cambridge. Y is in New Haven. Everything X perceives, Y perceives as well-in addition to his own perceptions – and vice versa.

In short, X and Y have the same inner-life, they are physically indistinguishable, and yet they are in different places. Although an extreme elaboration, this is not wholly unlike the stories we do sometimes hear about identical twins. When the two are separated, it is frequently claimed that crises affecting the one also affect the other, though they be miles distant. This is especially true when they have been, as it is said, "very close". Our present hypothesis is really the limit of a series of hypotheses concerning identical twins, each pair being in the sense above "closer to each other" than the succeeding pair. The only difference is that within the series one refers to pairs of people, while in the example above the "closeness" is so complete as to raise the question of whether it is adequate to refer to X and Y as different people at all. At this stage of abstraction to call X and Y the *same* person might do the descriptive job just as adequately.

All this is, I concede, a fantastic invention. It could never happen. But is it logically self-contradictory? After all, I could at least tell the story, something one cannot do when the "story" is logically self-contradictory. If it all did in fact happen, would we be confronted with a complete breakdown in our conceptual, linguistic, and logical machinery? Or would it just be the most unusual and inexplicable event of all time? Which claim one supports is clearly important for Weiss's thesis. He is opting for nothing less than the *synthetic a priori*. I say "opting" rather than "arguing", because of his opinion, "But there is no need to bother with the demonstration". If however, the strange example of X and Y can at least be entertained, then there is a need to bother with demonstration. For there seems to be an intelligible alternative account of the situation Weiss is considering.

Now, as they have been supposed to exist, X and Y *can* meet each other face-to-face. Nor is this the result of illusion, hallucination, mirror reflections, images, etc. – the only possibilities Weiss envisages. And, in their circumstances, it might be altogether natural for X or Y to say, "I shall soon meet myself coming towards me". Since the local experience of X at any time *t* and of Y at *t* is just the sum of what we outsiders would think the experience of each of them separately at *t*, then it certainly *will* be part of X's experience that he sees himself coming toward himself. Because Y's perceptions are part of X's own perceptions. And X's perceptions are part of Y's.

"But experience just isn't like that!" Exactly so. But that experience happens to be one way rather than another, is a *purely contingent matter.* That we are put together in the way in which in fact we are, this is not logically necessary. It is a mere matter of fact. There is no necessary reason why we could not have been constructed in just the way in which X and Y are in this imaginary example. After all, Aristophanes (in Plato's *Symposium)* does very little less than we have done here. And compare Melville (1930, 462–463):

> ... the monkey-rope was fast at both ends... an elongated Siamese ligature united us. Queequeg was my own inseparable twin brother... my own individuality was now merged in a joint stock company of two:... my free will had received a mortal wound...

So although X and Y are in two different places, it could still be natural to say that X and Y are the same person. Their inner-lives and their external appearances are indistinguishable. It is arbitrary which one we call X and which one Y; on what grounds could one be corrected if he said he saw X in Cambridge? If this state of affairs can be imagined, it shows that being in the same place is not a necessary condition for being the same person.

Nor is being in the same place a sufficient condition for saying of any X and Y that they must be the same person. Our X and Y began life within their mother. In the beginning they were as much a part of her as was her heart or her liver. As they developed, they remained "in the same place" as their mother, in some important sense of "in the same place". If one's heart or liver is "in the same place" as oneself, then by the same argument X and Y as embryos are in the same place as their mother. Of course, X and Y, although in the same place as their mother, are not in the same place as each other. Similarly my heart and liver are in the same place as *me*, but not in the same place as each other. Now even in this case, where X and Y and their mother are in the same place – in that sense of "being different people" which matters most in this connection – X and Y and their mother could be thoroughly different people. The mother is blond, blue-eyed, female. The twins are brunette, dark-eyed, male. The mother is calm, unimaginative, lethargic, and self-satisfied; the twins may be excitable and restless. How could people differ more than this? And if they are agreed to be in the same place, then being in the same place is not a sufficient condition for being the same person, since here, three different people are all in the same place. In other words, simply to lay it down by fiat, as does Weiss, that "What is spatially distant from me is other than myself", is to fail to see an essential point about the concepts "persons" and "selves" raised in examples like this. X and Y are spatially distant and yet they are the same. On the other hand, X in his mother's body is in the same place as she, and yet he could not be more different from her. So the importance of spatial distance in this problem is either that one forces it home as a definitional equivalent of *X and Y being different people* – and this is not untying the philosophical knot, it is cutting it through the middle – or else spatial distance is, as has been suggested here by the *advocatus diaboli*, simply one of the characteristics which, as a matter of fact, is always observed between one's self and other people.

II

One may reject the example of mother and foetuses as illustrating different people in the same place. Despite the fact that they are all biologically part of the same organism, it may still be felt that they are not in the same place. They have different coordinates,[1] even though they are fused at certain vital junctures. The example, nonetheless, is stronger than that wherein a large fish swallows two small fishes. Here the small fishes are contained, in a simple topological sense, within the larger fish. They are not part of the larger fish. When they become part – after digestion – they will no longer be small fish. X, Y, and their mother on the other hand, grow together and are interdependent for their health and for their existence. In short, they are one and the same organism, like the plant and its seeds. Yet it is not inconceivable that one might find it natural to distinguish them as different selves.

Still, if the example does not compel assent, let us treat it as but a crude first approximation. Let the case rest instead with the studies of Pierre Janet (1888) and Morton Prince (1890, 1908a, b, 1919, 1920). It is, of course, the classical example of the young lady whom Dr. Prince calls "Christine Beauchamp" to which the devil's advocate is referring now. Here, three distinct persons, or selves – this is the only natural and appropriate way of describing them-inhabit the same body; two of them at the same time! Prince succeeds in distinguishing

B I, Miss Beauchamp as he first met her;
B II, Miss Beauchamp under hypnosis;
B III, "Chris", later called "Sally";
B IV, "The Idiot", so named by Sally.

---

[1] Though even here we see how relative to context is deciding whether or not things have the same, or different co-ordinates. Radical atomism is the theory which reduces people to constellations of co-ordinates such that no two atoms can have the same co-ordinates. Other philosophical "isms" use their co-ordinates differently; their objectives are different.

These are not simply phases, or aspects, or modes, of a single person. Reading the cited works makes this immediately clear. These are *different people.* BI for example, besides having the marked traits of extreme vanity, religious scruples, uniform meekness and dependency, of never feeling anger, resentment, or jealousy, of being possessed of unending patience, of never being assertive, rude or uncharitable – besides all this she spoke and wrote fluently in French, was accomplished in mathematics, and had generally, for her years, attained a high level of culture. BIII, on the other hand, Sally, had no knowledge of French, or of any other foreign language. This enabled Prince to get information and instructions over to BI without interference from Sally who, being a thoroughly mischievous imp, missed no opportunity to embarrass and thwart the plans of Miss Beauchamp. Sally lacked any notion of responsibility or care. She was intensely jealous of BI, rebelliously independent, self-assertive, rude, merciless in her dealings with Miss Beauchamp – and yet a delightful, untutored child of nature in all actions which did not involve BI directly.

BIV was a strong, resolute woman; self-reliant, sudden and quick in a quarrel, easily angered and pugnacious, resenting interference in any form – yet a level-headed realist.

Further facts about this "family," as Prince calls them, is that BI has no knowledge of the existence of either BIII or BIV, although after years of being Sally's victim she becomes aware of the "devil within her". Sally however, knows every thought, emotion, and perception of BI. Still she remains a third person to it all. BI is always referred to by Sally as "she", or "her". BIII stoutly and resentfully denies any identity with Miss Beauchamp.

Consider the ways in which Dr. Morton Prince, an accomplished psychiatrist – closer than anyone else to "the family" – finds it natural and adequate to describe the strange phenomena he discovers in the course of over 6 years of examination. When referring to BI, BIII, and BIV, as three different people, Dr. Prince says (some italics are mine):

> I say three different, because, although making use of the same body, each, nevertheless has a distinctly different character;... different trains of thought,... different views, beliefs, ideals, and temperament,... different acquisitions, tastes, habits, experiences, and memories.... Miss Beauchamp, if I may use the name to designate *several distinct people*, at one moment says and does and plans and arranges something to which a short time before she most strongly objected, indulges tastes which a moment before would have been abhorrent to her ideals, and undoes and destroys what she had just laboriously planned and arranged... (1910, 1–2)

> I... asked... who "she" was. The hypnotic self was unable to give a satisfactory reply.
> "You are 'she'," I said.
> "No, I am not."
> "I say you are."
> Again a denial.
> Finally: "Why are you not 'she'?"
> "Because 'she' does not know the same things that I do."
> "But you both have the same arms and legs haven't you?"
> *"Yes, but arms and legs do not make us the same."* [This was the first appearance of

"Chris", later called "Sally" (27).] Further questioning as to why Sally was not the same as Miss Beauchamp brought the reply: "Because she is stupid; she goes round mooning, half-asleep, with her head buried in a book; she does not know half the time what she is about. She does not know how to take care of herself."

She [Sally] insisted she was [always] wide awake, *and resented in a way foreign to either BI or BII every attempt on my part to make her appear illogical in claiming to be a different person*... [Miss Beauchamp]... was in entire ignorance of the new self, Chris (BIII). *It was clear that there were three different selves*... (34)...

Miss Beauchamp under hypnosis:

Q. "What is the difference between you now and when you are not here?"

A. "I am asleep now."

Q. "Are you the same person?"

A. [Emphatically] "Of course I am the same person."...

Q. "Do you feel that you are exactly the same person?"

A. "Of course. Why should I feel differently?"...

Chris [Sally] appears:

Q. "Why have you suddenly changed?"

A. "I have not ch-ch-changed at all." [When she first appeared, Chris stuttered badly. Later this difficulty disappeared.]

Q. "You were not stuttering a minute ago."

A. "I was n-n-not t-t-t-talking a m-m-m-minute ago; 'She' was."

Q. "Who is 'she'?"

A. [Showing irritation and annoyance.] "I won't g-g-g-go through that n-n-nonsense again. I t-t-told you t-t-en d-d-days ago. If you d-d-don't know any better now, I shan't t-t-tell you."...

... BII [Miss Beauchamp's hypnotic self] never showed any evidence of persisting...

If you asked her what became of herself when Miss Beauchamp was awake as BI, she would answer she did not know. Did she exist at such times, as BII? No, she was waked up, that was all: She was BI: she was the same person. *The question itself, in her mind, implied an absurdity or wrong conception. She was BI; how then could she otherwise exist at the time, and as somebody else?* BI went to sleep, [was hypnotised] and we called her BII... With Chris, on the contrary, it was different. From almost the very first her language implied a concomitant existence for herself, a double mental life for Miss Beauchamp. She always spoke as if she had her own thoughts, perceptions, and will *during the time while Miss Beauchamp was in existence*... later, one of the personalities wrote short-hand in her diary so that Chris should not understand what she had written; and I was in the habit of using French to convey information which it was important should be concealed from Chris...

"We are not the same person," [Sally] would insist; "We do not think the same thoughts",... When asked... if she continued to exist as a separate and distinct self when BI was awake [Sally] asserted positively and unqualifiedly that she did... [Sally] observed things, when Miss Beauchamp was absorbed in thought, which the latter did not observe, and remembered much that had been forgotten or never known by her...

[Letter from Sally to Dr. Prince.] "You are most absurd and idiotic to waste your time and sympathy on such a perfect chump as our friend is. [That is, Miss Beauchamp.] ...Our friend is going to weep salt tears when she knows I have written you..."

I felt certain that it was Sally trying to pass herself off as Miss Beauchamp... Charged with the fact, and put to the test of reading French, which this personality could not do, she at first evaded, but soon, seeing that she was caught, burst out laughing...

For two wills to contend against each other *they must coexist*. Sally, then, did not simply alternate with Miss Beauchamp, she coexisted with her... [All this] can only be interpreted... as a struggle between two co-existent minds in one body.

[Sally] was jealous of Miss Beauchamp's superior attainment, of her culture, and above all of her popularity with her friends, and of the care and solicitude shown for her... "Nobody seems to care what becomes of me," she would complain, when a plea was made that Miss Beauchamp's life should not be interfered with...

[Sally] is a distinct... character, [her] trains of thought, memories, perceptions, acquisitions, and mental acquirements, [are] different from those of BI... her personality, her perceptions, her thoughts, and her will, co-exist with those of BI... Sally maintains... that she knows everything Miss Beauchamp [and BII] does at the time she does it – knows what she thinks, hears what she says, reads what she writes, and sees what she does; that she knows all this as a *separate co-self* and that her knowledge does not come to her afterwards when an alternating self, in the form of a memory.

In the *Journal of Abnormal Psychology* (1919–1920) Prince discusses this case again:

Sally, besides alternating with the others, had a co-conscious existence, in that she persisted as a self, i.e., as a *separate mental system* possessing a differentiated self-consciousness.... Thus there were two *I*'s in existence.

I have quoted at length from Prince's work. His mode of expression is vital here. If the facts had been adequately expressible in a different way, Prince would have found that way. His work is a serious, technical, exacting piece of description; he certainly is not hunting for striking metaphors or for a picturesque or emotive style. He seeks only to set out the facts. And the facts concerning Miss Beauchamp are best described by talking of *different people in the same body, distinct persons in the same place at the same time.* Now just to lay it down that Prince's language *must* be in some way indirect and metaphorical, is again to reassert (still without reasons) that if X (BI) and Y (BIII) are in the same place at time *t*, they must be the same person. *Petitio principii:* this is what is at issue. And the challenge is to describe a state of affairs which would naturally be described in a way which clashes with this part of Weiss's dictum. Dr. Prince has met this challenge for me. Having the same body is not sufficient for being the same person. "Arms and legs do not make us the same," said Sally. "How dost thou know that some entire, living, thinking being may not be invisibly and uninterpenetratingly standing precisely where thou now standest...?" (Melville 1930, 677).

The twins, X and Y, provided us with an example which disclosed that "being in the same place" was not a necessary condition for "being the same person". BI and BIII, give us an example which shows that "being in the same place" is not a sufficient condition for "being the same person". *So "not being in the same place" is neither a necessary nor a sufficient condition for "not being the same person".* X and Y are not in the same place and are the same person; BI and BIII are in the same place and are not the same person.

That we shall never meet ourselves face-to-face therefore, is materially true (and perhaps even *necessarily true*, by an argument to be given in the next section), but it is not logically true. Its denial cannot be self-contradictory, not even for Weiss when he identifies himself "with the 'X' of the law of contradiction".

III

What point was Weiss trying to make? It is a duty of anyone who has proceeded as in the foregoing to point out not only where another philosopher may have failed to make his case, but also to try to restate what he was aiming at.

At this moment I am conscious, slightly thirsty, and am sitting in my chair typing. My confidence in this is as great as any confidence as I could ever have in any claim. Indeed if I were asked to entertain what it would be like for this to be false, I should not know what to entertain at all. Because if confidence at this empirical level goes, confidence at all levels goes. Whatever could make me doubt that I am sitting here, conscious, thirsty, and typing, could more easily make me doubt any alternative state of affairs. Hence the doubts I am asked to entertain may be dismissed as presenting no tenable alternative to what is now quite certain to me. It is for me inconceivable that I am not at this moment sitting here, conscious, thirsty, and typing. And if one allows that a situation which is inconceivable is the negation of some situation which is necessary, then it is necessary that I am sitting here, conscious, thirsty, and typing. It is necessary, because it is simply inconceivable for me that anything should count as evidence against it. And if nothing can count as evidence against it, then it cannot be false. And if it cannot be false, then it must be true. And if it must be true, then it is necessarily true. So it is necessarily true that I am sitting here, conscious, thirsty, and typing. Nothing could dissuade me of this. [In this droplet of argument is contained the whole cloud of Cartesian philosophy.]

But this way of putting it involves just the problem encountered in Weiss's own exposition. Because although it is necessarily true that I am sitting here, conscious, thirsty, and typing, it is not necessarily true in the same sense that it is necessarily true that every Euclidean equilateral triangle is equiangular. To deny that every such equilateral triangle is equiangular is to assert what is palpably self-contradictory, in the form $P \cdot \sim P$. For me to assert that I am *not* here at this moment, conscious, thirsty, and typing, is not to assert anything of the form $P \cdot \sim P$. It is just to assert something for which I can form no conception of supporting evidence. This is a profound difference; and it is just this difference which Weiss's reflections fail to reflect.

If Weiss had argued that it is necessarily true that he will never meet himself coming towards him, and meant by that that evidence to the contrary was simply inconceivable for him – just as evidence against my now sitting here conscious, thirsty, and typing is inconceivable for me – then his claim would have been unambiguously clear and unexceptionable. The examples of X and Y, and BI and BIII, in that case would only have been calculated to suggest what the forming of such a conception, even for Weiss, might be like. Such examples may not work, of course. Weiss may remain unconvinced that any evidence whatever should count in favor of his meeting himself face-to-face. But then at least the nature of the discussion thenceforth would be clear. When Weiss goes on to say however, "No truth in logic or mathematics is more certain and no fact in daily experience or science is more significant than that we are unique individuals," there is suggested, to me at least, a possible confusion between the *a priori* of logic or mathematics, and what might be called the "*a priori* of epistemology". I am not sure that I understand what it would be to claim that the fact that every Euclidean equilateral triangle is equiangular is *more certain* than the fact that I am now sitting here conscious and typing. (I just had a drink.) Of course, nothing can disconfirm for me my claim that I am now sitting, conscious, and typing. And nothing can disconfirm for me the claim that every equilateral triangle is equiangular. But the reasons for this are totally different in each case. A Euclidean triangle which is equilateral but not equiangular is not possible, i.e., the denial of this constitutes a simple inconsistency in the use of one's language and symbolism. Evidence which would prove to me that I am not now sitting here, conscious, and typing is not possible either, i.e., I could not possibly *accept* any evidence which went against what I now am quite certain I am doing. But the logical gulf between these two types of necessity needs no further italics; and no example yet brought forward in all the long and eloquent history of philosophy from Descartes, through Leibniz, Kant, Hegel, Bradley, Whitehead, and last but not least, Weiss, need force one to think the gulf narrowed by 1 millimeter. Because all such examples exploit – by failing to distinguish – two senses of "inconceivable," two senses of "could not be false", two senses of "necessarily true".

Weiss's contention therefore is not established, because he has not made it clear in what sense of "necessarily true" it must be necessarily true that the assertion "'some day in some place I shall meet myself coming towards me' is absurd and material and that its denial is material and certain, significant and indubitable". He is, after all, championing a candidate for a *synthetic a priori* statement. His predicates make that undeniable. But surely he cannot hope to do *that* without a demonstration! Kant himself did not attempt such a thing. Particularly inasmuch as circumstances are conceivable [i.e., not logically self-contradictory] in which a person, our X for example, may wish to speak of meeting himself coming towards him; a possibility which is not met by observing that human experience is not like what the example supposes. This was the point of the example, namely, to show that, as he put it, *Weiss's contention is contingent on the way human experience in fact is.* This was the point of Sect. II as well. If however, Weiss's contention about the impossibility of a person's being in two places at the same time, and its corollary concerning two persons' being in one place at the same time, if this was meant to concern impossibility in the sense that it is impossible that I should accept evidence against my now sitting here, conscious, and typing, then one can only wish that Weiss had said more about this. Because this kind of necessity, the kind which it is logically possible to deny but which is empirically inconceivable to disconfirm, well deserves the kind of thorough analysis which Weiss might be prepared to give it. Because presumably Weiss would not be content simply to agree with the *advocatus diaboli* and call this "psychological necessity". If however, this is too weak a characterization – and I readily concede that it is – we must be told clearly and forcefully why this is so.

References

Janet, Pierre. 1888. Les actes inconscients et la mémoire: pendant le somnambulisme. *Revue Philosophique de la France et de l'Étranger*. T. 25: 238–279.

Melville, Herman. 1930. *Moby Dick; or, the whale*. New York: Random House.

Prince, Morton. 1890. Some of the revelations of hypnotism: Post-hypnotic suggestion, automatic writing, and double personality. *Boston Medical and Surgical Journal* 122 (20): 463–467.

———. 1908a. My life as a dissociated personality. *Journal of Abnormal Psychology* 3 (4): 240–260. [Anonymous letters compiled by Morton Prince].

———. 1908b. An introspective analysis of co-conscious life, by a personality (B) claiming to be co-conscious. (My life as a dissociated personality). *Journal of Abnormal Psychology* 3 (5): 311–334.

———. 1910. *The dissociation of a personality*. New York: Longmans, Green.

———. 1919. The psychogenesis of multiple personality. *Journal of Abnormal Psychology* 14 (4): 225–280.

———. 1920. Miss Beauchamp: The theory of psychogenesis of multiple personality. *Journal of Abnormal Psychology* 15 (2–3): 67–135.

Weiss, Paul. 1938. *Reality*. Princeton: Princeton University Press.

Chapter 10
Copernicus' Rôle in Kant's Revolution

In opposition to common sense I dare to imagine some movement of the Earth;... since mathematicians have not (yet) agreed with each other, I was moved to think out a different scheme...; by supposing the Earth to move, demonstrations more secure than those of my predecessors (could) be found for the revolutions of the... spheres... all (celestial) phenomena follow from this (supposition). (Copernicus 1566, iib)[1]

Like Copernicus, Kant sought to explain the properties of observed phenomena by postulating a kind of activity in the observer. This is the "Copernican Revolution". Nonetheless, in expositions of Kant's metaphysics the expressions "Copernican Revolution" and "Copernican Hypothesis" have come to assume a perhaps unwarranted role. Commentators and historians of philosophy suggest that Kant himself actually *used* these phrases and that there is one and only one meaning in Kant's mind for such language.[2] Though these distinguished Kantian scholars intimate both that Kant used the expression "the Copernican Revolution" and also that he meant to compare his revolution with that of Copernicus in one and only one way,[3] the following analysis aims to show that it is still worth inquiring whether this is an adequate account of the connections between Copernicus and Immanuel Kant.

Nowhere in either edition of the *Kritik der reinen Vernunft* does the phrase "the Copernican Revolution" occur. Nor does the expression "Copernican Hypothesis" occur either. In his translation of the *Kritik,* Dr. N. Kemp-Smith renders *mit den ersten Gedanken des Kopernicus* as "Copernicus' primary hypothesis" (1950, 22). But he is at least cautious enough to temper this mistranslation by adding the

---

[1] Praefatio Authoris; from "... ac propemodum contra communem sensum..." to "... illorum phaenomena indesequantur..." [The translation Hanson offers here closely follows the C.G. Wallis translation of the Preface and Dedication to Pope Paul III (Copernicus [1939] 1995, 5–6). – *MDL*]

[2] Thus H. J. Paton (1936, 75) writes: "Kant *compares* his own philosophical revolution with that initiated by Copernicus." (my italics). A. C. Ewing (1950, 16) says: "But Kant *means that he resembles* Copernicus in attributing to ourselves, and so classing as appearance, what his predecessors had attributed to reality." (my italics).

[3] "...it is this doctrine and this doctrine alone..." says N. Kemp-Smith (1918).

© Springer Nature B.V. 2020

N. R. Hanson, *What I Do Not Believe, and Other Essays*, Synthese Library 38, https://doi.org/10.1007/978-94-024-1739-5_10

German as well. In Kemp-Smith's *Commentary*, however, the reader is led to suppose that Kant himself used the expression:

(Kant's) 'Copernican hypothesis'... *he claims* (is) merely a philosophical extension of the method (of positive science).... (Upon) the 'Copernican hypothesis'...Kant *dwells* at some length. Kant's *comparison* of his new hypothesis to that of Copernicus... The apparently objective movements of the fixed stars... are mere appearances, due to the projection of our own motion into the heavens... it is this doctrine *and this doctrine alone* to which Kant *is referring*..., in thus *comparing* his critical procedure to that of Copernicus... etc.. (1918, 19–22, my italics)

Compare S. Alexander (1909–1910, 49): "… Kant *himself signalised* the revolution which he believed himself to be effecting as a Copernican revolution". Lindsay writes (1936, 50–51): "This new way of conceiving the possibility of *a priori* knowledge Kant *compares* to the revolution brought about in astronomy by Copernicus." Lindsay then goes on to quote Kant as saying "… Copernicus' primary hypothesis…".

How these otherwise scholarly writers can so wantonly render *Gedanken* as "hypothesis" is baffling to me, unless, of course, they are simply forcing this English word on Kant to strengthen their own general interpretations of his philosophy. But Carl J. Friedrich is more careful and more respectful of his native language, and of Kant's ability to write in it, than are the aforementioned Britons. Friedrich (1949, xxvii) translates *den ersten Gedanken des Kopernicus* correctly as "the first thought of Copernicus". Hence he shows no tendency to make Kant characterize his own philosophy as a "Copernican revolution" or a "Copernican hypothesis". Friedrich lets Kant's references to Copernicus serve only to indicate Kant's dissatisfaction with a chaos of existing theories, and his decision to abandon them and make trial of another.

Now what exactly does Kant say? In the 1787 Preface to the second edition of Kant's *Critique of Pure Reason*, we read:

The example of mathematics and natural science, which by a single and sudden revolution have become what they now are, seem to me sufficiently remarkable to suggest our considering what may have been the essential features in the changed point of view by which they have so greatly benefitted. Their success should incline us, at least by way of experiment, to imitate their procedure.... We should then be proceeding precisely in accordance with *the first thought of Copernicus*.[4] Failing of satisfactory progress in explaining the movements of the heavenly bodies on the supposition that they all revolved around the spectator, he tried whether he might not have better success if he made the spectator to revolve and the stars to remain at rest. A similar experiment can be tried in metaphysics....(Kemp-Smith 1950, 22; Bxv-xvii)

Similarly, the fundamental laws of the motions of the heavenly bodies gave established certainty to what Copernicus had at first assumed only as an hypothesis, and at the same time yielded proof of the invisible force (the Newtonian attraction) which holds the universe together. The latter would have remained forever undiscovered if Copernicus had not

---

[4] Hanson here follows Kemp-Smith's translation with the exception of this sentence. As Hanson mentioned earlier, Kemp-Smith had "We should then be proceeding precisely on the lines of Copernicus' primary hypothesis." – *MDL*

dared... to seek the observed movements, not in the heavenly bodies, but in the spectator. (Kemp-Smith 1950, 25, Bxxii, note)

The two expressions "Copernican revolution" and "Copernican hypothesis" do not occur in indices to the other two *Critiques*, and particularly not in the fully indexed edition of Kant by F. Meiner.[5] In fact, the entire Kantian corpus makes no reference to Copernicus other than the two occurrences (quoted above) in the *Vorrede* to the second edition of the *Kritik der reinen Vernunft*, added 6 years after the completion of the first edition. The *Kant-Lexikon* of R. Eisler (1930), usually reliable on the first *Kritik*, reveals no further references to Copernicus anywhere.

This is not only a matter of Kantian philology. Reference to the "Copernican revolution" has carried the burden of the most important expositions of Kant's philosophy. That so much weight should for so long have been placed on so tiny a textual foundation may encourage further questions about interpretation. In any case, philosophers have a duty initially to read Kant's words as Kant wrote them: "What did he say?" is prior to "What did he mean?" Concerning Kant's references to Copernicus, these two questions have been thoroughly confounded.

The origin of the expression, of course, lies in the *Vorrede*. Perhaps the mischief consists in using "revolution" in the first sentence of the paragraph beginning at the bottom of Bxv (quoted above). But the preceding discussion leaves no doubt that the "revolution" referred to here has *nothing whatever* to do with Copernicus. The *Vorrede* is addressed to a discussion of the affinities and differences between mathematics, physics, and metaphysics. The first two disciplines, after a period of groping (*Herumtappen*) certainly became sciences. They entered upon the *sicheren Gang einer Wissenschaft* as the result of the "revolution". When referring to mathematics (in Bxi) the word "revolution"[6] is emphasized ("gesperrt"); Kant does not say upon whom the light of mathematics suddenly broke *(ging ein Licht auf)*. With physics, the world had to wait longer for its revolution. Only 150 years before Kant, Bacon had "inspired fresh vigor in those who were already on the way to [the discovery]" (Bxii). It was with Galileo, Torricelli, and Stahl that a light broke upon all students of nature: *"so ging allen Naturforschern ein Licht auf"* (Bxiii). They all had the vision to cast old theories aside in order to test some bold new hypothesis. When (at Bxvi) Kant refers back to "the examples of mathematics and natural science which by a single and sudden revolution have become what they now are", he is not making any reference to Copernicus and his heliocentric doctrine *per se*, but rather to the successful foundation of experimental physics by the great scientists of the seventeenth century.

In Bxv, Kant asks whether a change in the method of metaphysics, corresponding to these revolutions in mathematics and natural science, might not end *its* random groping (cf. Bxv, *Herumtappen);* whether or not this can be done can only be discovered by a trial (Bxvi). Similarly Copernicus, when he found he could not achieve satisfactory results by assuming one hypothesis, made trial of (*versuchte*)

[5] Hanson here refers to the works of Kant published by the Felix Meiner Verlag. – *MDL*
[6] Where Hanson offers translations of Kant, they are from Kemp-Smith 1950. –*MDL*

another.[7] In metaphysics, it is possible to make an analogous trial (*auf ähnliche weise Versuchen*). This much is the main point of Kant's argument. The name "Copernicus" is brought in here only to illustrate the propriety of making trial of an untested hypothesis, particularly when extant theories seem fruitless. Any of a number of other scientists could, and in fact *do*, illustrate this point for Kant. Further parallel between Copernicus and himself are not central to Kant's exposition at this point; which is *not* to say that Kant *never* conceived of further, and perhaps more important parallels. But whether or not this is so surpasses the *letter* of Kant's own writing, something one would never gather from the commentators quoted earlier.

Consider further that in 1759 and in 1760, Kant lectured on mechanics. For this purpose he used Wolff's *Elementa Mechanicae*.[8] Appended to this work is a dissertation on scientific method, the 'Commentatio de Studio Matheseos Recte Instituendo'. Kant would have been very familiar with this tract.[9] In sections 309–311 of Wolff's dissertation, there is a discussion of the very point at issue – the uses of novel hypotheses as a means of scientific progress. The example given is the hypothesis of Copernicus together with its subsequent verification by Kepler and Newton. Kant's reference to Copernicus in the *Vorrede* may thus have been introduced with this passage in mind, and not necessarily as a more comprehensive reference to the effects of "activating" the observer in astronomy and in epistemology. At least this possibility ought not to be dismissed out-of-hand in favor of the more orthodox exegesis.

Note also an allusion which Kant makes (Bxxii note) to a further parallel between *De Revolutionibus* and his own *Kritik*. There Kant argues that what he is setting out purely hypothetically in the *Vorrede* will be established "apodeictically, not hypothetically" in the body of the *Kritik*. There is a very similar relation between the preface of Copernicus' *De Revolutionibus* and the body of that treatise itself. Because where in the *Praefatio Authoris* the heliocentric principle is asserted only as an hypothesis, in the body of the work its use is taken for granted.[10]

I have so far been attempting to show (1) that in the main texts of both editions of the First *Kritik*, Kant never spoke of a "Copernican revolution" or even of a "Copernican hypothesis"; (2) that Kant was never concerned, even in the Preface to

---

[7] See Copernicus' own words at the head of this paper, and compare Bxvi quoted above.

[8] Part of the *Elementa Matheseos Universæ* (1746).

[9] See Adickes (1924–1925, I.11n.)

[10] "... what I am saying may seem obscure here, nevertheless it will become clearer in the proper place." Copernicus, *De Revolutionibus*, Dedication to Pope Paul III (my translation). In Copernicus' *magnum opus* we must, of course, distinguish the Dedication to Pope Paul III from the very first foreword to the reader. The latter was almost certainly the mischievous work of Andreas Osiander, as is made clear beyond doubt in Gassendi's *Life of Copernicus* appended to his *Tychonis Brahei* (1654). The Dedication, however, is indisputably by Copernicus himself; these were facts *definitely* established only in 1873, but hinted at in the mid-seventeenth century. Professor Kemp-Smith mistakenly refers to the Osiander portion in the name of Copernicus, in order to show how the latter regarded his "hypothesis". The hypothesis-talk was Osiander's invention, calculated to save *De Revolutionibus* from an early Papal death. Copernicus' claims were really much stronger.

the second edition, to stress the doctrinal similarity between his own epistemological teachings and the astronomical theses of *De Revolutionibus*; (3) that Kant's main reason for referring to Copernicus in the 1787 Preface concerns his intended contrast between the hypothetical and the established (or demonstrated) stages of a scientific discipline and to point up the periodic need of new departures in science when old theories have lost their vitality; and finally, (4) that in the *Vorrede* (with its elaborate historical parallels, so conspicuously absent in the present case) Kant's reference to Copernicus in Bxvii may not stand in any primary relation to the main thrust of his argument.

Now that this much has been said, we must take stock. Even if it is clear that Kant nowhere uses the expression "Copernican Revolution", and that such reference as is made to Copernicus need not be viewed solely by the one interpretation which commentators have supposed, it still remains for us to inquire just what illumination the expression "The Copernican Revolution" *does* shed on the main corpus of Kant's metaphysics. For even though they are wrong in suggesting that Kant *explicitly* made this comparison of his own philosophy with the astronomy of Copernicus, Kantian scholars are correct in assuming that there is a fruitful analogy between these two great works.

Kant openly asserts a similarity between himself and Copernicus in but one respect; each of them made trial of an alternative hypothesis when existent theories proved unsatisfactory. The revolutions in thought with which Kant explicitly compares his own revolution have nothing specifically to do with Copernicus. But how are we to understand the last reference to Copernicus quoted above in Bxxii note? A further analogy between Kant and Copernicus is implied here. It is this which to some extent justifies the tradition according to which commentators speak of "Kant's Copernican Revolution". What is implicit in this last reference suggests that the revolutions in mathematics and natural science of which Kant speaks, in expounding his own metaphysics, are not *merely* revolutions; they are revolutions of a quite special variety. These were not revolutions simply because a fresh hypothesis was substituted for prior theories. They were also a revolution in ways of thinking (*Revolution der Denkart*). The demonstration that every equilateral triangle is also equiangular must have been carried out initially by some geometer who discovered that it was useless merely to follow with his eyes what he saw in the triangle, or even to trace out the elements which are thought in the concept of 'equilateral triangle' by itself. That is, neither empirical observation of equilateral triangles, nor an analysis of the concepts involved in speaking of such geometrical entities, will serve to demonstrate any mathematical truth. What must be employed is rather what Kant calls "the construction" of concepts; we must exhibit *a priori* the intuition corresponding to our concept (B741). What this hypothetical ancient geometer discovered was that it was necessary to produce the figure of an equilateral triangle *by means of what he himself thought into it*. He thus exhibited *a priori* its equiangularity, as is in accordance with the geometrical concepts we now possess. To have had certain *a priori* knowledge the geometer must have attributed nothing to the equilateral triangle except what followed necessarily from what he injected into it in accordance with his geometrical concept, i.e., its equiangularity (Bxii).

What Kant takes to be essential to this revolution is that the geometer's mind is not concerned just with the empirical object, some particular equilateral triangle; or even with the concept *equilateral triangle* derived by abstraction from such objects. It is concerned rather with its own act of construction, with what is put into the figure in accordance with the concept. *A priori* knowledge in mathematics arises from the mind's awareness of its own special operations.

Special difficulties arise when Kant tries to give a similar account of the genesis of natural science. Because here the revolution (once again a *Revolution der Denkart*) is the introduction of the experimental method. What *is* it to discover the experimental method as Galileo, Torricelli, and Stahl are said by Kant to have done? They discovered that reason has insight only into what it produces in itself in accordance with its own plan (Bxiii). Here again, a superficial inspection of objects will never give us a binding law of nature. But reason will never be satisfied with anything less than such a law. Reason confronts nature with its own ultimate principles, e.g., those set out in the *Analogies*, and with the experiment thought out in accordance with these principles. Reason is the judge who compels witnesses to answer questions which he himself formulates. The revolution whereby natural science ceased to be groping, was due to the realization (by Bacon, Galileo, Torricelli, and Stahl presumably) that our researches into nature ought to conform to what questions and principles the scientist's reason itself puts and applies to nature.

So much Kant actually claims. The value of all this as a piece of history of science is, of course, extremely dubious. Galileo is struggling every moment for greater and greater objectivity: thus he dispenses with the subjective reactions to heat by inventing a publicly observable thermometer. And he attempts a similar shift of emphasis in the case of time where he sought an effective pendulum-clock. To characterize the essence of such discoveries as Galileo's realization that his researches into nature had to conform to what his own reason put into nature is, to say the least, mildly shocking. Galileo never makes such a claim for himself; in fact, the case is quite the opposite. Similarly with Copernicus. This need not matter, of course; Kant may be telling us something about these great scientists which even *they* did not know. This is rather unusual as a technique in history of science. Nonetheless, it must be granted that for Kant the revolutions in mathematics and physics had something in common, over and above their being disciplines in which bold new hypotheses took the place of older, unfruitful theories. In each case, the mind was somehow attending to what it itself had put into its objects.

This doctrine is only *implicit* in the Preface to the second edition of Kant's *Kritik* (1787); it is not openly stated in the words which Kant actually uses. But in view of this doctrine, the special situation of metaphysics itself may now be considered. Mathematics and natural science had become what they were in Kant's day by a tremendously rapid advance, remarkable enough to make Kant reflect upon the essential character of this new way of forming conceptions. Can metaphysics imitate mathematics and physics in this manner? From the structure of the first *Kritik*, it seems clear that Kant is looking not merely for some sort of metaphysical revolution in the weaker sense, some new hypothesis which will extricate the philosopher from the chaos of previous epistemological theories. He is looking for a revolution which has the same fundamental character as that which he had implicitly outlined for mathematics and natural science.

Before Kant, metaphysics had proceeded on the assumption that all knowledge must conform to objects *(sich nach den Gegenstanden richten)*. But on this assumption all attempts to acquire *a priori* knowledge of objects (so necessary if physics and mathematics are to stand on what Kant felt to be a firm foundation), all such attempts must end in failure. Kant therefore suggests that we at least try *(versuche)* the hypothesis that objects must somehow conform to the structure of our knowledge. The proposed revolution in metaphysics therefore is to follow the line suggested by the revolutions in the methods of mathematics and physics. Not only will a new hypothesis be put to trial in place of the older enervated theories, but now we may consider that perhaps the mind, in all these cases, "puts something into" its objects, imposes certain properties upon them necessarily.

Here (Bxvi) appears the first reference to Copernicus. He too, swept aside older theories and tried a relatively new hypothesis. This Kant makes quite explicit. But submerged and implicit in this example may also be the obvious point that Copernicus sought to account for the properties of observed celestial phenomena by investing the observer with a certain activity.[11] Kant thinks the metaphysician can make an analogous experiment: *"In der Metaphysik kann man nun... es auf ähnliche Weise versuchen"* (Bxvi).

Kant's thought is something like this: in explaining the movements of celestial bodies Copernicus rejected the natural assumption that the movement was in the stars themselves; he tried instead the view that this movement was in the spectator. The movement is "put into" the stars by the spectator. That is the way Kant construes the Copernican hypothesis, and his own philosophical parallel to it is definite, and important. But, and this is the real issue here, not all of the parallel is explicit in Kant's work. That Copernicus tried a new hypothesis in place of older theories is explicit in Bxv-Bxxii. But that Copernicus (like Kant) had hit on an hypothesis whose main point was to take what had been regarded as characteristics of the observed object and explained these in terms of the characteristics of the observer himself – this interpretation of Copernicus is not at all explicit in Kant's own exposition, Professor Paton and Kemp-Smith notwithstanding.

---

[11] This also seems suspicious as a piece of history of science. *De Revolutionibus...* seeks primarily to show that, as a matter of physical geometry, all the data which gave rise to the astronomical computing system set out in Ptolemy's *Almagest* can equally well be accounted for (i.e., explained and predicted) by shifting the primary reference point of the ancient system from the Earth to the Sun. The geometry which resulted would be much tighter and more elegant, the introduction of *ad hoc* (i.e., unsystematic) hypotheses would be minimized, and one's physical imagination would be less offended. But Copernicus was essentially a medieval astronomer. He thought himself to be working within the old framework of ideas more effectively, by making certain formal and systematic alterations. Almost certainly he was not aware of the full implications of his geometrico-physical modification. And Copernicus never expresses himself as I suspect Kant would have liked him to do, by stressing that his hypothesis consisted in "investing the observer with a certain activity."

So Kant was urging that, like Copernicus, metaphysicians must make trial of a new hypothesis. Moreover, the new hypothesis is to be of a quite definite kind.

In light of all this, it appears that while we are justified in following the tradition of Kantian scholarship in saying that the Königsberger effected a Copernican revolution in metaphysics, we must, in the interests of scholarship, distinguish the explicit from the implicit features of Kant's own claim.

We must certainly refuse to allow commentators to obliterate the distinction between what Kant said and what he "must have meant" in their zeal to establish the latter. And, in fact, Kant's understanding of what Copernicus actually did can only be ascertained by comparing the texts of the *De Revolutionibus Orbium Coelestium* and the *Kritik der reinen Vernunft*.

References

Adickes, Erich. 1924. *Kant als Naturforscher*. Berlin: W. de Gruyter.

Alexander, S. 1909–1910. Ptolemaic and Copernican views of the place of mind in the universe. *Hibbert Journal* 8: 47–66.

Copernicus, Nicolaus. 1566. *De Revolutionibus Orbium Coelestium, Libri VI*. Basel: Heinrich Petri. English edition: Copernicus, Nicolaus. [1939] 1995. *On the revolutions of the heavenly spheres*. Trans. C. G. Wallis. Amherst: Prometheus Books.

Eisler, Rudolf. 1930. *Kant-Lexicon: Nachschlagewerk zu Kants sämtlichen Schriften, Briefen und handschriftlichem Nachlass*. Berlin: E.S. Mittler.

Ewing, A.C. 1950. *A short commentary on Kant's Critique of pure reason*. London: Methuen.

Friedrich, Carl J. 1949. *The philosophy of Kant: Immanuel Kant's moral and political writings*, by Immanuel Kant. New York: Modern Library.

Gassendi, Pierre. 1654. *Tychonis Brahei*. Paris: M. Dupuis.

Kemp-Smith, Norman. 1918. *A commentary to Kant's Critique of pure reason*. London: Macmillan.

———. (trans.). 1950. *Immanuel Kant's Critique of pure reason*, by Immanuel Kant. New York: Humanities Press.

Lindsay, A.D. 1936. *Kant*. London: Oxford University Press.

Paton, H.J. 1936. *Kant's metaphysic of experience: a commentary on the first half of the Kritik der reinen Vernunft*. London: G. Allen & Unwin.

Wolff, Christian. 1746. *Elementa Matheseos Universæ*. Genevæ: Apud Henricum-Albertum Gosse & Socios.

Chapter 11
It's Actual, So It's Possible

Socrates: Can men of your century fly faster than the earth turns?

Contemprates: We've already *done* it!

S.: And because you have done it, you say you can do it?

C.: Of course.

S.: Can you fly to the moon?

C.: We *can*, in that nothing in our scientific theories prevents it. We lack only the technology for that feat. We can do it, but not just now.

S.: And because present concepts permit it, you reason that it can, in principle, be done?

C.: Yes.

S.: Can you men fly faster than light?

C.: No – well, yes and no.

S.: What do you mean?

C.: Flying faster than light is not logically impossible like squaring a circle, or encountering a quadrilateral triangle. A different universe might contain velocities in excess of *c* and a science which might meaningfully describe such speeds; traveling faster than light is *logically* possible. But our entire system of physical concepts would crumble were anything discovered to exceed *c*. "Flying faster than light" expresses what is inconceivable, but not inconsistent.

S.: But if by some as-yet-unimagined experimental techniques, I could actually show you an object traveling faster than light, then you would say that this was possible?

C.: I would have no choice but to do that.

S.: Tell me, Contemprates, can twentieth-century men fly along the legs of a quadrilateral triangle?

C.: No, no more than could the men of any age. For "… is a quadrilateral triangle" is a self-contradictory epithet. Whatever is quadrilateral cannot be triangular: this is necessarily true. Its denial is inconsistent. And so it is logically impossible for any man to fly, or bicycle, or walk, along the legs of a quadrilateral triangle.

© Springer Nature B.V. 2020

N. R. Hanson, *What I Do Not Believe, and Other Essays*, Synthese Library 38,
https://doi.org/10.1007/978-94-024-1739-5_11

S.: But if, as a result of circumstances I cannot now describe, I could actually show you a married spinster riding a three-wheeled bicycle along the legs of a quadrilateral triangle, would you then admit that such a thing is logically possible?

C.: I doubt that *any* conceivable circumstances could serve you in the ways you require. But, allowing that this *could* happen, this case would then be no different from the previous ones, where we reasoned from the actual occurrence of an event to the technical, physical, and logical possibility of that event occurring.

S.: In other words, Contemprates, you are saying that if *P* actually happens, then *P* is *logically* possible (as well as technically and physically possible)?

C.: Yes, *P* always entails that *P* is *logically possible*. This, Socrates, has the force of a philosophical principle.

S.: But you have often stated another principle which seems, to my ancient head, to contradict the present one: *no contingent state of affairs ever entails a necessary state of affairs*. Nothing is necessary in nature; nor is anything necessary entailed by what actually happens. From what has happened, one cannot infer what *must* happen. In short, you have taught that a proposition whose negation is consistent cannot entail another whose negation is inconsistent – save in that 'degenerate' case wherein a necessary proposition is said to be strictly implied by any proposition whatever. (Here, ordinary entailment is not in question.) You have propounded all this, Contemprates. Yet now you argue from the actual occurrence of *P* to *P is logically possible*. This contradicts your earlier dictum. For any contingent value of '*P*' this entailment will run from what is contingent to what is logically necessary. Because the proposition *P is logically possible* is, for contingent values of '*P*', itself logically necessary.

C.: Is it, Socrates? Suppose I choose a self-contradictory contingent proposition as '*P*''s value? Then, *P is logically possible* would be logically false.

S.: A self-contradictory proposition cannot be contingent, since its negation is necessary. Nor can it be necessarily true, since it is the denial of a necessary truth. But I concede your point to this extent. Let me say that *P is possible*, if true at all, is necessarily true (even when *P* is contingent). If it is false, it is absurd.

C.: What prevents me from saying that *P is logically possible*, if true, is contingently true? After all, as Descartes conjectured (1964, vol. 1, 145, 149, 151) another Creator might have designed the class of logically possible propositions so as to exclude *this* value of *P*. *That* this *P* is logically possible is contingent on the way our world is, and what counts as logic in our world.

S.: Descartes comes after my time, but if he says that, he speaks absurdly. He is presenting us with no real alternative to the logic we have, and which determines what is and is not conceivable. He is talking words. Besides, I can prove that *P is logically possible* is a proposition which, if true, is logically necessary.

C.: How will you do that?

S.: The easiest way is to assume, hypothetically, the negation of *P is logically possible*, where *P* is a contingent proposition. If this leads to a contradiction will you concede that the proposition in question is logically necessary?

C.: Yes. If *P̄* is, or leads to, a contradiction then ~*P* is logically necessary.

S.: Well, then, let us begin:

(1)   *P* is logically impossible   Supposition

And let it be given that

(2)   *P* is contingent   Premise

C.: You mean by this, I take it, no more than that '~*P*' expresses a self-consistent
   proposition.
S.: That is correct: *P* is contingent if and only if ~*P* is self-consistent.
Then

(3)   ~*P* is contingent   from (2); if *P* is contingent,
                                          then ~*P* is contingent
(4)   ~*P* is no tautology   from (3); no contingent proposition is a
                                          tautology since its negation is consistent
(5)   *P* is not self-consistent   from (1); what is logically impossible is
                                          not self-consistent
(6)   ~*P* is a tautology   from (5); the negation of what is not
                                          self-consistent is tautologous
(7)   *Reductio ad absurdum*   (4) and (6)
(8)   That *P is logically impossible* (1),   whatever leads to a contradiction is
      when *P* is contingent (2),   itself self-contradictory
      is necessarily false

Therefore, I conclude:

(9)   The proposition *P is logically*   The negation of what is logically
      possible (when *P* is contingent)   impossible is logically necessary
      is itself logically necessary

Hence *P is logically possible* is always a logical proposition. It is logically necessary when *P* is contingent or necessary, and logically false when *P* is inconsistent. And since no proposition can be both necessarily and contingently true, *P is
logically possible* cannot be (as you claim) contingently true.
C.: I felt this proof to be spurious, Socrates. What you have proved is that the supposition of *P is logically impossible* leads to a contradiction, *reductio ad absurdum.* But I never asserted that *P is logically impossible*; my claim was only that
   P is logically possible, if true at all, is contingently true. Have you spoken to this
   assertion?

S.: I have indeed. And the proof just set out destroys this your claim, Contemprates. But if the issue remains unclear, let me begin, as you suggest, with

(1)   *P is logically possible*, if true at all, is contingently true   Supposition

This is your contention; if *this* devolves into a contradiction, Contemprates, will you concede its denial, namely, *P is logically possible, if true at all, is logically necessary?*

C.: Yes.
S.: All right; let it be given that

(2)   *P* is a contingent proposition   Premise
(3)   So *P is logically possible,* and   from (1) and (2); every contingent
      thus *contingently true*   proposition is self-consistent and
                                                  hence logically possible
(4)   So '*P* is logically impossible' also   from (3); if *Q* is contingent, ~ *Q* is
      expresses a contingent proposition   contingent
(5)   So '*P* is self-contradictory'   from (4) 'is logically impossible'
      expresses a contingent proposition   *means* 'is self-contradictory'
(6)   *Reductio ad absurdum*   It is a necessary fact about *P* that it
                                                  is self-contradictory, and in no sense
                                                  contingent, as in (5)

C.: *Petitio Principii!* This is what is at issue. I claimed that *P is logically possible, if true, is contingently true.* Yet you 'refute' me by inferring the 'absurdity' '*P is self contradictory*' *expresses a contingent proposition.* But this 'absurdity' is my supposition – as it had better be if your demonstration is logically sound. Just calling this 'absurd' is no more compelling than had you simply rejected my supposition as 'absurd' at the start.

I still want to say that whether or not *P is logically possible* is contingent on the ways in which homo sapiens uses languages. Other beings, with different types of communication, might not have regarded *P is logically possible* as we do, Socrates. This suggests that your way of construing *P is logically possible* is but one of several conceivable alternatives – and hence not absolutely invulnerable? 'Logically necessary' is a significant designation only within our linguistic frame of reference, hence contingent upon that frame.

S.: Aha! so the Cartesian word-nonsense *has* possessed you, Contemprates. That the *sentence*, or *marks*, or *sounds* '*P* is logically possible' can be used to express a necessary proposition – this is contingent. But that the *proposition* expressed by this sentence, these marks, these sounds – and by these, '*P* c'est une possibilité logique', '*P* secundum artis logicae regulas est,' and '*Π* κατα τη γογικήν ἐστιν' – that *this* is a proposition which (if true) is logically true, is logically necessary.

Its denial (1) leads to what is absurd ((4) and (6)). The argument above is in no way contingent upon the habits of sentence users. It is purely an analysis of the logical structure of propositions.

Yet you claim both that *P is logically possible* is entailed by *P is actual* and also that nothing contingent can entail anything necessary.

C.: Yes, I concede that I confused necessary facts about propositions with contingent facts about the sentences expressing them. More, I agree that there may be a fundamental incompatibility between at least two of my philosophical principles.

Could the perplexity be resolved in the following way? Suppose we deny that *P* is ever merely contingent. Consider these candidates for the status of 'necessarily true, contingent propositions': "I am now conscious, discussing possibility with you, Socrates." This statement is, for me, inconceivably false. It could not but be true in the present circumstances.

S.: Surely not, Contemprates. I could cite possible evidence which, if it obtained, could conceivably shake your confidence in the claim you have just made.

C.: No, if my confidence dissolved at *that* level, I would have no reason for accepting any account of such counter-evidence. That I am now conscious and engaged in discussion is a fact so fundamental to the organization of my experience that it determines my capacity to accept anything at all *as* evidence. Nothing could dissuade me that I am now conscious and engaged in discussion, (just as it often happens that within a dream nothing could occur to persuade the dreamer that this was a dream). Nothing now conceivable could count against this claim. It is (in its context) completely invulnerable, and hence unfalsifiable. Thus, uttered in this context, it can only be true. And if it can only be true, it must be necessarily true.

S.: Yes, but…

C.: The same may be said of "No *perpetuum mobile* exists", and "Nothing travels faster than light". While factually informative, these assertions are yet in practice treated as unfalsifiable. A Ph.D. thesis describing a *perpetuum mobile* would not even be read. An article speculating on experimental velocities in excess of *c* would have difficulty getting published. Because to entertain such things as true leads directly to conceptual breakdown.

Now perhaps every contingently true proposition to some degree shares this characteristic: that entertaining its falsity crumbles related ideas. Ask me to entertain the thought of an ancient Greece without a Plato and I simply do not know what to think next about classical civilization. And could I really be wrong about Alexander, Attila, Charlemagne, Napoleon, and Hitler? If I could be, what surety have I that I am correct about anything at all, other than when playing purely formal symbol-games – if such things there are?

S.: Just so. You cannot be *sure* of anything contingent.

C.: That is an excessive statement, Socrates. There is no point in describing a contingent proposition as 'uncertain' unless it makes sense to speak of one which *is* certain. Otherwise, nothing would contrast with the epithet "uncertain contingent proposition"; it would be vacuous.

S.: I could contrast the certainty of formal symbol-games with the uncertainty of informal, contingent statements.

C.: But we already *know* that physics is not mathematics; we cannot discover the world by reflection alone. We know that establishing certainty in mathematics is typically different from establishing certainty in natural science. To insist that the latter be subjected to the former's criteria, barred forever from having its own standards of *proof* and *certainty*, this seems a questionable procedure. If it is ever informative to say of some contingent proposition that its truth is doubtful, then it must be both possibly informative and possibly true to say that some contingent propositions are indubitable, in a genuine, nonderivative sense of 'indubitable'.

S.: I have you now – by your own argument. Because if it can be informative and true to say of a contingent proposition that it is indubitable (and I am not conceding this, but merely entertaining it on your behalf), then *some* contingent propositions must be thoroughly dubitable. To deny this would (by your own turn) be to rob "indubitable contingent proposition" of any descriptive force; it would no longer contrast with anything. So there must be some corrigible values for '*P*' (for example, "some philosopher is twenty feet tall") which entails the incorrigible assertion "It is logically possible that some philosopher is twenty feet tall." There *must* be such values for '*P*', Contemprates, otherwise the force of your Kantian-type examples is lost.

C.: Yes, I lose that point.

S.: Besides, Contemprates, even supposing your consciousness, perpetual motion, and the speed of light, to figure in invulnerable statements which are somehow typical of all contingent statements, these will not solve the perplexity in the inference from *P* to *P is possible*. Although the negations of your statements express inconceivable states of affairs, these are not 'inconceivable' in the sense required, that is, demonstrably self-contradictory. The negation of *P is possible* when *P* is contingent, is not merely unentertainable (as is a *perpetuum mobile*, a velocity > c or my present unconsciousness). *P is impossible* leads to something of the form $Q \cdot \sim Q$.

C.: You mean that the class of necessarily true statements contains two subclasses: (1) propositions whose negations are conceptually untenable (for example, *no perpetuum mobile exists*) and (2) those whose negations are self-contradictory (for example, *no quadrilateral triangle exists*).

S.: Precisely. Even allowing your candidates, where *P* is factual yet necessarily true in sense (1), this does not help; it only reposes our difficulty. *This P* still entails *P is logically possible* which, if true at all, is necessarily true in sense (2). But your second philosophical principle might as easily read "A proposition whose negation is not self-contradictory cannot entail a proposition whose negation is self-contradictory". The problem thus remains, even supposing you are correct in your interpretation of the nature of contingent propositions.

C.: So the tension between the principles *If P is actual then P is logically possible* and *From what is contingent nothing necessary follows* is not relieved either by denying the logical character of the statement *P is logically possible* or by denying that *P* can be merely contingent.

S.: So it seems.

C.: Well, there must be something unusual about the entailment between *P is actual* and *P is possible*.

S.: What do you suggest, Contemprates?

C.: Suppose an individual, call him 'J.S.M.', when challenged for his reasons for asserting $2 + 2 = 4$, offers this:

"When very young I played with blocks, marbles, apples, and the like. I learned this much: that when I put any two blocks into a box with any other two, the result was always four blocks in the box. Similarly marbles. And for any duet of apples placed with another duet, one quartet of apples resulted. Indeed, *any* couple of any entities (of any logical type) when bracketed conceptually with any other duet of 'nameables', always issued in a quartet of designata – for example, a block, $\sqrt{2}$, Oedipus, and triangularity. Any duet, plus any other duet, is always observed to form a quartet. Therefore, $2 + 2 = 4$."

S.: Oh? Surely the statement expressed by *this* use of '$2 + 2 = 4$' is entirely different from that expressed when '$2 + 2 = 4$' figures in the logic of arithmetic.

C.: That is just the point.

S.: J.S.M.'s statement that $2 + 2 = 4$ just summarizes his (not unusual) experiences. It may be difficult for him, or anyone, to entertain supposed exceptions to this summary. But because it *is* a summary, of J.S.M.'s finite experience at that, exceptions must be logically possible. What is expressed by this use of '$2 + 2 = 4$' must be, in principle, vulnerable.

C.: Precisely. We know, for example, that two cubic feet of mercury mixed with another two cubic feet will not fill a four-cubic-foot vessel. (If it did, fluid physics would need revision.) Were this known to J.S.M., it would have affected his mode of generalizing that $2 + 2 = 4$, since it would have constituted a counterinstance to a merely empirical law.

S.: But *this* would have no effect upon the arithmetical statement that $2 + 2 = 4$.

C.: None whatever. The sentence '$2 + 2 = 4$' may be used in these different contexts – as a contingent summary of J.S.M.'s experiences and as a statement of what necessarily follows from the axioms of elementary number theory. But *that* the contexts are so dissimilar shows that the statements expressed in these audibly indistinguishable utterances are disparate.

S.: What '$2 + 2 = 4$' can be taken to express then, is a function of our knowing in detail from what it follows, and what follows from it.

C.: Yes. Thus 'It is possible that *P*' can express many different types of statement. Your first questions forced me to distinguish technical, physical, and logical possibility, Socrates, and there are other varieties of possibility as well, for example, 'his credentials mark him as a possible candidate', 'harpsichord and trombone are an impossible combination', etc.

Philosophical employments of '*P* is possible' do not always make it clear what proposition is being located by this sentence. Sometimes, as for example in 'metaphysical' discussions of actuality and possibility, being and essence, several propositions are run together under the same sentence all at once, making comprehension difficult.

S.: It occurs to me here, Contemprates, that we may not be clear enough about the details of the simple statement '*P*' – from which we take '*P* is possible' to follow. That *P* is consistent is an essential part of the meaning of '*P* is contingent'. Concerning the inference from *P is contingent* to *P is logically possible*, the truth of *P* is not necessary. Thus, from the false contingent proposition that 'threnody' appears thrice in this sentence, it certainly follows that it is possible that 'threnody' should appear thrice in this sentence. Because, since *it is logically possible that P* tells us no more than that *P* is self-consistent (that is, entails nothing of the form $Q \cdot \sim Q$), then, whether or not the situation described by *P* actually obtains is not relevant to the question of the logical possibility of *P*. That *P* is a contingent proposition, even a false one, is all we need know to infer that *P* is logically possible.

C.: *P is logically possible*, therefore, is just shorthand for saying that, for any conceivable situation, if '*P*' is used to describe it, then the statement that *P* is not self-contradictory. This is a minimal commentary on the structure of statements. Since it is axiomatic that any contingent proposition, as well as its negation, *must* be consistent, then the statement *that* it is consistent (that is, logically possible) follows necessarily.

S.: Yes. The further remark 'the contingent statement *P* is true', or (what is the same) '*P* actually obtains', adds nothing new from which *P is logically possible* follows. The latter follows irrespective of *P*'s truth.

C.: Of course, Socrates, *something* important does follow from P's being contingently true, namely that *P* is physically and technically possible, or describes a physically and technically possible situation. But this is of no direct logical interest.

S.: My main point is simply this – the assertion that *P* actually obtains constitutes a triple claim, from only one part of which does it follow that *P* is logically possible. It is (1) the claim that some undescribed, but merely designated event has come about, making the world to that extent different from what it had been before, and (2) that although this need not have been so, the mark(s)/sound(s) '*P*' does actually serve to express the proposition *P*, which in its turn describes the state of affairs (noted in (1) to obtain), and (3) the claim that *P* is a contingent proposition. '(3)' tells me that *P* is logically possible. '(1)' and '(2)' tell me that *P* is true, and hence technically and physically possible as well. For the inference *P therefore* P *is logically possible* to hold however, *P* need only be understood as claim (3). Only so far as (3) *is* understood in the assertion *P is actual/obtains/is true* can *P is logically possible* be said to follow.

C.: If, however, *P is possible* meant also that *P* is physically and technically possible, as well as logically possible (as it often does), *this* could not be inferred just from *P is contingent*. One needs to know that *P* is true.

But I agree that the claim *P* is really tripartite. It is not clear to me, however, how this will resolve our perplexity.

S.: Well, Contemprates, if it is correct to say that the 'contingent assertion' *P* really constitutes a tripartite claim, and if it is also correct to say that *P is logically possible* follows directly only from (3), the claim *P is contingent*, then our perplexity about inferring what is necessary from what is contingent may be transformed. Because, Contemprates, although *P is logically possible* is, if true at all, necessarily true, we can now see that (3) (from which this follows), is also necessarily true, if true at all.

C.: But '(3)' was '*P* is a contingent proposition'; are you saying that *this*, if true, is necessarily true?

S.: Yes I am.

C.: But *P* need not have been contingent. In a different world, with different language users... oh, wait a moment! I have traveled this road before – and you showed me where it leads. See if I've learned my lesson.

That the *sentence* '*P*' can be used to express a contingent proposition, this *is* contingent. But that the proposition expressed by the sentence '*P*' – and '*Π*' (in Greek), etc. – that this *proposition* is itself contingent is a claim which is either necessarily true, or self-contradictory. The claim that *P* is contingent is really the claim that ~*P* is consistent. But then, to deny this claim is just to allow that *P could be* self-contradictory. The credentials of this contention are assessed by logical means alone. We require no experiments to determine whether or not a proposition is consistent. If *P is contingent* just is the claim that ~*P is consistent*, then the whole issue is a purely logical one. If it is true that *P is contingent*, then it is necessarily true.

C.: Yes, this parallels your earlier argument against me, that if *P is necessary* is true, it is necessarily true. Both contentions become clearer as one distinguishes the contingent facts concerning a sentence's uses from the necessary facts concerning a proposition's structure.

S.: Yes. That *P is true* is contingent; it depends on (1) how the world is, and (2) our actual uses of language. But that *P is contingent* is necessary or absurd; it turns only on (3) the question of logical structure of the proposition ~*P*.

C.: I see; and *since P is contingent* is (if true) necessarily true, and since it is only from this part of the claim *P actually obtains* that *P is logically possible* follows, then the apparent strangeness of our initial inference dissolves. For this is now but an inference from what is necessarily true (or absurd) to what is necessarily true (or absurd).

The other elements in *P actually obtains*, or *P is true* are contingent; the proposition expressed by '*P*' (or '*Π*') need not have correctly described the world, but as a matter of mere contingent fact it does. But *P is logically possible* does not follow directly from this, but only from the fact that the proposition expressed by '*P*' (or '*Π*') has a self-consistent negation.

This, Socrates, parallels what I sought to say about J.S.M.'s use of '2 + 2 = 4'. '2 + 2 = 4' cannot be known to express either an empirical or a necessary proposition without knowing whether the reasons for the utterance were of the J.S.M.-type, or the elementary number theory-type. So also, deciding in practice whether '*P* is possible' expresses something contingent or necessary is feasible only when one knows from what this claim follows, and what follows from it. You have shown, Socrates, that when it means *P is logically possible* it follows from *P is contingent*. Only insofar as *P is true*, or *P is actual*, contains the claim that *P is contingent* can *P is logically possible* be said to follow from it. And you have also shown that nothing follows from the fact that we do infer from *P is actual* to *P is logically possible* which forces one to qualify one's acceptance of *the* fundamental principle of contemporary philosophy, namely, that a contingent proposition (that is, one with a consistent negation) cannot entail a necessary proposition (that is, one with an inconsistent negation).

S.: Yes, that about sums it up, Contemprates.

C.: Still, Socrates, despite the fact that it is a tripartite claim, only one part of which is necessary, *P actually obtains* does entail *P is logically possible*. If *a* entails *α*, then *a and b* entails *α*, even though *b* may be contingent and *α* necessary. If *Xanthippe is a wife* entails *Xanthippe is female*, then *Xanthippe is a wife, and we are discussing possibility* entails *Xanthippe is female*. So similarly, if *P is contingent* entails *P is logically possible*, then also *P is contingent, 'P' expresses the proposition that P, and P is true* entails *P is logically possible*.

S.: But then P *is contingent* conjoined with any utterances whatever – necessary or contingent, significant or insignificant, well formed or ill formed – would have to be said to entail that *P is logically possible*. This conclusion strikes me as unacceptable, Contemprates. There is not now time to discuss it, but I would be inclined to say that although *P is contingent* does entail *P is logically possible*, I would like some other connective-word to signify the relationship between, for example, *P is contingent and all men are blond* and *P is logically possible*. It is not the terminology but the distinction I am concerned with.

C.: Well, we'll leave it at that. Xanthippe seems to feel that we've been talking long enough.

Reference

Descartes, René, Charles Adam, and Paul Tannery, eds. 1964. *Œuvres de Descartes*. Paris: J. Vrin.

Chapter 12
On Having the Same Visual Experiences

A. "When you see a stoplight and say 'Red', it can never be known whether you are having the same visual experience which I have when, simultaneously, I see the same stoplight and also say 'Red'."

B. "Can't it? Why not? Surely it is possible for us both to submit to various optical and ophthalmological tests; this would decide the matter. Our verbal responses can be ignored; but by pairing and discriminating colours and shapes it can certainly be determined whether or not we both see red or see the shapes in the same way. What would a question about seeing which could not be settled in this way be a question about?"

A. "No, this misses the point. We both may perform identically on these pairing and discriminating tests; there may not be one case in which our verbal or other behaviour differs when confronted with a stimulus like a stoplight. We may indeed both be supposed to see a colour, or a shape. Yet we can never be sure that just because the behaviour is identical the 'internal' colour perception is the same."

B. "You mean then, that nothing whatever could decide the matter? But if this is so, then your worry is not a genuine *empirical* worry at all. It can only be a disguised piece of linguistic legislation, at best – or a conceptual muddle at worst. A worry which cannot possibly be resolved is not the kind of worry I propose to let myself have. If I have no idea what would solve the puzzle, I cannot persuade myself that I really know what the puzzle *is*"

There, too often, the matter is left, B feeling that A is profoundly unclear about what he is asking for, and A reckoning that B's refusal to have his difficulty indicates only a lack of depth and a narrowness of technique. A's position is quite shocking, despite the fact that many discussions (both inside and outside philosophy)

© Springer Nature B.V. 2020

N. R. Hanson, *What I Do Not Believe, and Other Essays*, Synthese Library 38,
https://doi.org/10.1007/978-94-024-1739-5_12

often come perilously close to it.[1] Because if you and I can never know whether we are having the same experience in those situations where we both confront, e.g. a stoplight, and respond by saying "Red," then we can certainly never know whether we are having the same experience when we are looking at triangles, or chairs, or tables, or other people, or any of the other philosophical furniture of the external world. To have the doubt which A expresses just is to entertain the most radical kind of solipsism, although many seem to have had the former without explicitly entertaining the latter. It is hardly satisfactory to point out here, as some recent philosophers, like B above, seem to have thought it sufficient to do, that the question is not a soluble one. *This* is precisely what A is claiming. Discussions often conclude therefore in agreeing that the difficulty is insoluble and then proceed to construct theories as to how it is that A and B ever manage to agree upon so much in the world, despite their disagreements about "basic" visual experience. Here words like "convention", "agreement", "stipulation", and "ostensive", are exercised rather heavily. We can, apparently, never really know that we have the same "inner" visual experiences. But we can, by various sorts of arbitrary agreements and conventional decisions, proceed to act as if we do have the same experiences. The value of the agreements and decisions is assessed pragmatically. If we can all succeed in managing our affairs in the world by agreeing that X and deciding that Y, then in so far do we rely on X and Y.

Quite often, of course, A is *not* making an empirical claim. His way of expressing his worry is frequently just a way of announcing how he proposes to use words like "experience", or "colour", or "behaviour", etc. But certainly there are some cases in which A would feel himself to be making an empirical claim. He might think that he was in some way delineating more rigorously than is usually done what we are actually entitled to claim is "going on" when you and I say "Red" on confronting a stoplight. One feature of that experiential situation, says A, is that there is a limit to the amount of comparison possible between what you and I undergo.

In this case however, we can shake A from his claim by some hypothetical description of a test which would enable us to say of α and β, or of you and myself, that we are having exactly the same experience. Or, in a different jargon, it ought to be possible to imagine a test which would settle whether or not we are having the same visual sense data.

When discussing sense data it is essential that two different employments be distinguished. Theorists like Broad and Price have regularly spoken of "sense datum experiences". This must at least refer to *something which happens* in percipients. The claim to have had a sense datum experience in this sense is thus an empirical claim. How to verify or falsify the claim may raise other difficulties, many of which are conceptual in nature. Nonetheless the locus of sense data here is inside experience. Distinct from this is the more purely logical use of references to sense data, as "the limit of a series of everdiminishing empirical claims". In this role the sense

---

[1] E.g. "...How do you know you see the same color I do?... Since 1660, when Isaac Newton discovered the properties of the visible spectrum, we have slowly been learning the answer..." (Land 1959, 84).

datum is nothing that ever *happens*. It is rather the logical residuum of a certain analysis of perceptual experience. Epistemological discussions suffer when the sense datum theorist slides from the first employment of this technical term to the second without warning. Thus the datum seems to be indisputably something which all percipients experience (if they experience at all) until "the opposition" begins suggesting empirical tests which might check the theorist's thesis. Then sense data are marvelously transformed into "logical destructions out of ordinary experience". It is one of the purposes of this paper to trace the tracks of the sense datum theorist's claim as he slides off the empirical highway onto the logical verge.

If such a test can be imagined, then it will be obvious in any particular case whether or not the worry being broadcast is just a covert means of setting out definitions, or does really constitute a perplexity concerning the facts.

Imagine this experiment: α and β are sitting side by side facing an illuminated wall. Mounted on the wall is a square of graph paper, whose grid marks are in fine, but definite, black lines. Drawn on this graph paper is a bright red triangle, to which we shall refer as Δ.

Do α and β have the same visual experience; are they having the same visual sense data? A said we can never know the answer. But surely we can know. For let us give both α and β a piece of graph paper identical to the one hanging on the wall. Provide them both with a bright red pencil of exactly the hue of Δ. By hypothesis, we suppose that both α and β are master draughtsmen. They are so skillful in hand and eye that they can and do reproduce on paper exactly what they see before them. α and β are now asked to draw on their allotted graph papers precisely what is drawn on the illuminated piece. We can eliminate by a relay of bi-prism lenses any perspectival differences due to the fact that the eyes of α and β cannot have identical spatial coordinates at the same time. After they have finished sketching, we superimpose α's drawing on β's drawing and hold them up to the light. If there is any difference whatever between the two drawings, then this cannot, *ex hypothesi*, be the result of inaccurate drawing. It must indicate that α and β had not had the same visual experience, had not seen Δ in the same way. We can easily make a decision in this case: if the drawings are congruent, then α and β saw the same thing, had the same visual experience, entertained the same visual sense data. If there is any detectable difference whatever between the two drawings, then α and β did not have the same experience.

There can be no difference between what α draws and α's visual sense data – at least, not on any orthodox account of visual sense data (e.g. those of Broad and Price). Similarly with β's drawing and what β experiences. For what could such a difference consist in? The entire content of the visual sense data is set out here in the drawings which α and β produce. *The Mind and Its Place in Nature* (Broad 1925), and *Perception* (Price 1932) suggest nothing geometrically present in the former which is absent in the latter. Nor does the *Critique of Pure Reason* for that matter. Because whatever α and β may be supposed to add to the raw data before their "private images" become drawable, they either add the same ingredients as we in the third person must add, or they add different ingredients. This will be decidable when we compare Δ with α's and β's drawing of it, or when we compare the latter with each other. Colours, shapes, configurations of lines, in the drawing these can be precisely what they are in the sense experience. Indeed, there seems to be a sense in which we can now say that the drawings of α and β are their sense experiences. Because since there is nothing discernible in the one which is lacking in the other, the two must be identical in every respect which can matter for the understanding of what visual sense data are. To have confronted the one just is to have seen everything which composes the other. If there were a difference, this would either go against the hypothesis that α and β are master draughtsmen, or go against the thesis that there is nothing in a visual sense datum other than lines, shapes, colours and their several configurations.

There are no odours, or pains, or pangs in visual sense data – at least not as the latter are classically represented. Odours, pains, and pangs are certainly not drawable or paintable. It is assumed here (not implausibly, I hope) that every constituent of a visual sense datum is drawable or paintable, even its organization. Hence, everything contained in the visual sense data of α and β is drawable or paintable. By the master draughtsman hypothesis then, what a and P draw just *are* their visual sense experiences.

One effect of this *Gedankenexperiment* is to make everything that matters about visual sense experience perfectly public (as I suspect it always has been anyhow). This is analogous in a very rough way to Galileo's transformation of the previously private experiences of heat and cold and the passage of time, into operations amenable to public assessment; the invention of the thermometer and the first attempts at constructing a pendulum clock. So too here, we can now decide in principle the previously unanswerable question of whether or not two observers are having the same visual sense experience or different ones.

Should A object, saying that there could be no such experiment in that the master-draughtsman hypothesis is unfulfillable, the logical point is nonetheless established. Because though as a matter of empirical fact there are no expert draughtsmen, it is possible that there could be; the supposition is not self-contradictory.

Of course A may claim not that the hypothesis of the master draughtsman is empirically unfulfillable, but that it totally begs the question. He may argue that the whole complexity which he first expressed is buried right inside this assumption. He could develop his argument in the following way: α and β are both observing Δ on grid paper. So of course it is logically possible that their drawings should be congruent drawings of Δ on grid paper. But this proves nothing. The heart of the matter is untouched, and the above argument is a *Petitio Principii*. Because suppose that when α is confronted with Δ, he verbally identifies this as a red triangle and indeed draws on his grid paper a red triangle; suppose however, that the actual "inner" experience which he has is of a figure which, if we could but inspect it, we should identify as a green circle. For α, the external red triangle is completely "coordinated" with his internal green circle. So naturally, when he seeks to draw what he is internally aware of, he will have to draw on paper what we see as Δ. How could what he draws be a drawing of that figure on the wall unless the two were more or less congruent? But by A's new supposition an inner perception of what others would call a "green circle" can, with α, only be drawn by him as what we see as a red triangle, a reports, and draws, a red triangle; but he *has* a green circle "internally coordinated" with it. A says that *we* can never know of anyone else whether or not his inner visual life is coordinated with external objects as ours is, or as α's is. Indeed α cannot even know which is the case – of himself – nor can we know which is the case with ourselves. All this still leaves the middle term (i.e. α's internal visual experience of what we should call a green circle) unexamined and undiscussed. And this is simply to return to the difficulty which A says he felt in the first place. The same criticism has been made of Mr. Patrick Trevor-Roper's thesis concerning the effect upon the work of an artist of the state of his eyesight. In a letter to the *Manchester Guardian*, Mr. J. Boumphrey remarks "… presumably Mr. Trevor-Roper did not overlook the fact that in painting the shapes which he saw with astigmatic eyes El Greco would reverse the optical process and the images on the canvas would theoretically end up the same as the originals!"

It now begins to appear that the perplexity is really irresolvable. Still at this stage A might claim with some plausibility that this in itself does not make the perplexity a non-empirical one. The very fact that we tried to suggest an empirical test shows that we were understanding A's perplexity in the way in which he claims to have been proposing it. What A now points out is that our recommended test fails completely. The reason it fails is because of a further empirical supposition which A makes; between an observed object Δ and an observer's behaviour with respect to Δ (even his perfect drawings of Δ) there could be, logically, an entire alphabet of symbols (obscurely related, or even unrelated) with which the observer unerringly co-ordinates the same external objects and the same behaviour (drawings) on all occasions. Of this supposition, A might reasonably urge that it is not in itself self-contradictory, nor is its negation self-contradictory. So it is an empirical hypothesis. Yet it seems, with A, to lead to a perplexity which is empirically irresolvable.

There is still one further suggestion which we can make however, which may meet A's fully developed argument. And the suggestion is indisputably empirical in nature, that is, not self-contradictory, nor possessed of a self-contradictory negation.

A's dissatisfaction with the hypothetical empirical test described above consisted in this: α and β succeeded in getting themselves from an observation of Δ to the production of the drawing of Δ on the paper presented to them, only through the intercession of an inner visual experience (an "internal apprehension" of their own visual sense data). On any orthodox account of sense data this "interceding variable" is not publicly inspectable; which need not by itself mean that it is not inspectable at all or does not really consist in an empirical operation at all. So, apparently, A will be assuaged by nothing less than an experiment which will allow the public inspection of one's private sense data. In other words, it must be possible for a third person to observe directly, not just what is on the wall and what the subject of the experiment draws, but what is actually "in the subject's head".

But this requires no more than a refinement of *Gedankenexperiment*-technique. If by hypothesis one can allow that α and β are both master draughtsmen, then also by hypothesis one can now imagine at hand all the laboratory equipment necessary for making the internal sense experiences of α and β publicly observable. The proposed experiment will now do something like this: α and β will sit facing Δ on the wall. But instead of the drawing performance required in the earlier example, we now suppose a complicated cluster of electrodes and optical and electronic equipment, whose purpose is to make it possible to project α's and β's "private" visual experiences of Δ out from their centres of visual awareness, back through the optic nerve, tunica retina, and through the anterior part of the subject's eyes – back on to the wall itself. This hypothetical equipment then, will simply turn α and β into animated projectors. Whatever it is they "have" as private sense images, these are flashed back on the wall. By a further complication, we make it possible for the subjects to see only the original red triangle. The image projected through them back on to the wall can only be viewed by a third person wearing special spectacles.

In this case then, it ought to be easy to decide whether α and β are having the same visual experience, the same sense data. If the figures projected out from their eyes are absolutely congruent with each other (it does not matter whether or not they reproduce the Δ on the wall) then α and β had exactly the same visual experience, the same private visual sense data. If there is any difference between the two projected figures, then we simply decide the opposite way. Now no one can say what the exact details of the experiment imagined above would be like. But this does not matter; the suggestion is an indisputably empirical one. And if its description makes sense, in general, then the fact that we cannot perform this experiment tomorrow indicates only an engineering deficiency, not a logical one. The limitation in principle supposed originally by A would now seem to have disappeared.

This is not absolutely precise. The question is whether, at any point along their optico-neurological relays, α and β are getting exactly the same signal. This is not finally answered by the above hypothesis. However the original stimulus is transformed while coursing along the intra-cranial reticula of α and β, our hypothesis above suggests only that these changes could (as a matter of physical principle) be run through "in reverse". This would but reconvert the signal as it obtains at any point in our subject's reticula back into what it was as the original stimulus. In fact, our imaginary apparatus, by working in this way, could not even disclose the properties of an intra-cranial signal, since these very identifying traits would have been retransformed into those of the original extra-optical stimulus. Our arrangement above would only have the subjects shooting the original stimulus back out at the wall whence it came, – like a couple of slot-machines rejecting bad pennies. No. A more refined hypothesis would consist in interrupting at a point p in the neural relay, and then conveying the signal-as-discovered-at-p directly to public observation. Rather than shooting the signal back out through the eyes of α and β we might convey it out e.g. through the subject's temples into instruments capable of rendering the signal thus detected available to public observation. This would make it possible (in principle) always to decide whether such a signal as picked up at p in α and at the same location in β was the same for the two subjects, or different. The point is that, although this description is rather more complex than that of the *Gedankenexperiment* described above, it remains indisputably possible from a contingent point of view. The question "Do α and β have the same signal at p?" remains in principle decidable.

Suppose that the perplexity has not disappeared, however. The *Gedankenexperiment* projected their sense data either back through the visual reticula of α and β, or else directly out of the subject's crania. This assumes, as A seemed at first willing to assume, that α's and β's private experiences were somehow to be construed as the final stage in a series of physico-chemico-neurological events, a view often encountered in orthodox sense datum theory. Indeed the alternative to this is either a purely logical theory in which sense data cannot ever be said to *happen*, or a very obscure empirical theory which must encompass a crude psycho-physical parallelism; the latter could not explain, or serve as an analysis of, any epistemological perplexity simply because it is not really intelligible at all. If then it is meant to be understood as empirically true that the physical events occur sequentially, ultimately generating the visual sense datum, then it follows that it is empirically possible to consider the sequence in reverse. Thus light reflected from Δ can etch a photographic plate, which can then in its turn itself have light projected through it from behind to throw Δ back on to the wall.

It is, moreover, conceivable that a tiny mirror could be surgically introduced into the back of the eye which would reflect out what had entered through the front. To go further along the assembly-line, we could imagine the introduction of an intricate electronic receptor at the terminus of the optic nerve which would re-route all the afferent impulses back whence they came, and with a suitable electrophysical conversion at the retina shoot out through the iris just exactly what had previously come in. An even more complicated mechanism would be required at the visual cortices, and so on as far as the sense datum theorist wishes to go.

Similarly, some fantastically intricate but not logically inconceivable device could project the final events in these series as they actually occur in α and β back out along the perceptual assembly-line whence they arose, and throw the resultant image back on to the wall. If the claim to have a visual sense datum is meant to be an empirical claim, and if it is a claim in any way or at any time connected with having open eyes, and visual stimuli – if in short there is any sense in speaking of a "visual sense datum *experience*" – then the possibility of this suggested sequence-reversal follows necessarily.

A could argue here however, that one cannot simply identify the ultimate "inner" visual experience with the last event in any physico-chemico-neurological series of events. On other grounds, of course, this is in general a perfectly correct position to adopt. The "last thing to happen" in such a series is the detection, or observation, of "the first thing to happen", e.g. the introduction of an object into the observer's field of vision. But this final detection is not itself detectable, anymore than observing is observable or seeing is seeable. Neither does 'checkmate' simply name the last move in a particular chess game. Had this position been adopted initially as A should have done, it would have made his worry even more difficult to express, and even more difficult to have. But beginning as A did, this new announcement that he now proposes to disqualify each newly-refined experimental suggestion as always falling short of some ineffable experience which cannot in principle be got at, is somewhat startling. For this shows either that A's initial worry was not after all an empirical one, despite appearances, or that A has slid from that level of discourse in which sense datum experiences can serve as the subject matter of genuine empirical talk, to a quite different level where the sense datum is merely an analytic invention introduced for the purpose of discovering what we are right about even when our perceptual claims are totally wrong. It also shows that A's original way of expressing his worry may have been misleading, or false, or both. Such *Gedankenexperiments* as those suggested (and their subtlety can be increased just as readily as A's talk of sense datum experience can be) would certainly decide whether or not α and β were having the same visual experience (sense data) in any context wherein the question arose as a genuine empirical doubt, and not as part of a rigorously inflexible epistemological theory, or as a systematically elusive metaphysical uneasiness. If every attempt to present an empirical solution to A's ostensibly empirical difficulty is met by his making the subject of discourse less and less vulnerable to testing, we should soon gather that his perplexity was not really the sort of perplexity we had thought it was, and perhaps not even the perplexity A had thought it was. A's doubts are now metaphysical in the worst sense. Though he poses as needing help with the problem, he will not *allow* us to help with any suggestions of experiments. À *la bonne heure.* But now that respect in which α and β cannot be known to have had the same experience has now been reduced by A himself to the logical vanishing point. It is no longer possible to tell *what* A's trouble is.

One thing which follows from all this is that the reason we can never in fact say of two percipients confronting the same visual stimulus that their "internal" visual experiences are the same, is not that though the question is empirical we are logically forbidden from answering it. The reason is either (1) that in order to get into a position where we could reasonably decide we would require equipment much in advance of anything now envisaged (assuming the question to be empirical), or (2) that there is something fundamentally wrong with the whole drift of the question, i.e. that assuming the question to be an empirical one, "having a visual experience" is not related to the chain of physico-chemico-neurological events which precede it as the developed photographic negative is related to the chain of events which preceded it. It may be different in kind. In short, the making of such a decision with respect to A's initial question is not impossible in principle, it is only very difficult as a matter of practice. A's attempt to elude the latter conclusion in favour of the former seems to result either in some arbitrary stipulations concerning the definitions of "colour", "experience", "test", etc., or else in the obscurest of metaphysical doubts.

The conclusion then is this: that it is in principle possible to decide of two subjects whether or not they are having the same visual experience, or the same private sense data, *if* the claim that they are or are not having the same experience or sense data at a particular time is an empirical claim in the sense in which A proposed it. But though these two expressions, viz. "same visual experience", and "private sense data", have been used (as in the tradition of the sense datum theory) as virtually synonymous, it has really been with the purpose of distinguishing them that all the foregoing has been undertaken. Because suppose that instead of Δ, our experiment involved one of those multiple-aspect figures familiar in Gestalt psychology (e.g. the duck-rabbit, the wife-mother-in-law, the cup-faces, the reversing-staircases). In this case it ought to be possible to say of α and β that, despite their having exactly the same sense datum experiences, in the sense that the figures projected through their eyes on to the wall are perfectly congruent, they may nonetheless have different visual experiences. If α sees the duck and β sees the rabbit, nothing in orthodox sense datum theory will explain this difference so long as it can be argued that their sense data are identical, geometrically indistinguishable. If I see the wife and you see the mother-in-law, our sense data need not differ when our visual experiences do. So long as "α and β have the same/different visual experiences" and "α and β have the same/different visual sense data" both remain empirical claims and are not converted by the theorist into an equivalence when the argument begins to go against him, then *it is meaningful to speak of situations in which two observers having indistinguishable visual sense data nonetheless have disparate visual experiences.* The sense datum theory then fails completely as an analysis of visual experience.

References

Broad, C.D. 1925. *The mind and its place in nature*. London: Routledge.
Land, Edwin H. 1959. Experiments in color vision. *Scientific American* 200 (5): 84–94. passim.
Price, H.H. 1932. *Perception*. London: Methuen & Co. Ltd.

Chapter 13
Mental Events Yet Again: Retrospect on Some Old Arguments

Abstract Contemporary Psychological Behaviorists argue against the existence of mental events in ways that are but little more refined than those of Watson, Lashley, and others 40 years ago. The orthodox case stated by today's psychologists against the possibility of a science of mental events is inconclusive. This is because their attack is usually based on *factual* considerations, instead of being, what it ought to be, a conceptual analysis of the *idea* of a mental event. This article seeks to disclose some weaknesses in the standard pattern of attacks on Introspection. It does this by defending the latter (i.e. Introspection) against such factually-orientated criticisms as those of Watson, Lashley, Hull and Skinner. It then challenges Introspection as it should be attacked; not externally with counterfacts, but internally with demonstrations of what is untenable in the very concept of a science of private events called "mental".

I

Although a modified Behaviorist myself, this paper makes me feel like a lion about to be thrown to Christians. My purpose is to remark that talking about mental events and processes *is* objectionable; but not for the reasons advanced by Behavioral psychologists. The usual arguments against the "data of consciousness" are inconclusive. Let's establish this and then consider the *only* effective attack on introspectionist psychology.

Most of us, when not psychologizing or philosophising, use and understand words like *know, feel, mood, intelligent, habit, conscious, sensation, imagine, pretend, remember, think, motive*… etc. Our conversations do not limp along incomprehensibly when such terms arise. We use this language naturally, – to convey just what we wish. So, in an important sense of "know the meaning", we all know the meaning of these "mental" words.

But at the question "just *what* do you mean?", ordinary conversation *does* come to a halt. We feel the embarrassment Augustine felt about Time: "What then is

© Springer Nature B.V. 2020

N. R. Hanson, *What I Do Not Believe, and Other Essays*, Synthese Library 38,
https://doi.org/10.1007/978-94-024-1739-5_13

time?" he asks. "If no one ask of me, I know; if I wish to explain to him who asks, I know not". (1876, XI.xiv.17).

People who have not thought about "mental" words might also be embarrassed if similarly challenged.

Academicians however, have overcome this embarrassment. The great names, beginning arbitrarily with Descartes, are Leibniz, Locke, Berkeley, Hume, David Hartley, the Mills, Kant, Herbart, Lotze; and, in a different current of the same conceptual stream, Fechner, von Helmholtz, Wundt, Hering, Brentano, Stumpf, G. E. Müller, Ebbinghaus, Mach, Avenarius, Külpe, Titchener, and James. Deep differences divide these thinkers. Yet here is a confluence of doctrine concerning the mind and consciousness. Short of a history I can only outline the details of that doctrine.

Every man has both body and mind. Bodies are in space; subject to mechanical laws. Bodily processes can be observed externally. But minds are different, – unweighable, unbeatable, ungoverned by mechanical laws. They are not witnessable by external observers. But their contents are nonetheless witnessable – by the subject. One can do this with increasing objectivity, until, as with Wundt, experimental techniques become essential to the description of one's inner life.

There are, thus, two kinds of existence. *Physical* existence, and *mental* existence. All physical happenings connect mechanically. But mental happenings occur in insulated fields, – "minds". Normally, there is no causal connection between what happens in any two minds. Moreover, the streamlike nature of consciousness entails that the mind (whose life *is* that stream) must be aware of what passes down it. Nor can the mind err about its contents; it is immune from illusion and confusion. Perceptual judgments may be confused. But consciousness and Introspection cannot be. How could I misjudge my own sense-data? The things thus known, – sensations, images, memories, *Bewusstseinslagen*, etc., – have attributes indefinable in spatial, temporal, or even in quantitative terms. Sensations and images have, besides attributes like intensity, extensity, also *quality* and *clearness*, – properties nowhere implicit in physics or chemistry. Nor are they mathematically describable, – despite Herbart's attempts.

Further, consciousness unites past, present and future; it includes not only the explicitly known, but also a penumbra of implicit meanings, – all contributing a distinctive organic quality. These contents of consciousness are organized. Thus the mind is selective; – this is without parallel in the inorganic world. Elements are fused into a unique whole, – more than a coexistence of parts. Objects are seen not as constellations of color-patches; but as men, or as cars, or as houses. A house is not just a pile of bricks; so too a perception is more than a momentary awareness of colors, or noises. The perception is *of* these things *organized* in a certain way. And this organization it is the mind's function to effect. The knowledge one has of his own consciousness, is not on the same logical level as the contents of that consciousness, any more than the shape of a house could be just another object produced in a brick kiln.

The mental life, moreover, coheres in an awareness of the self as the object of all internal perception. Through all this analysis of mental events and consciousness runs the hint that most unsophisticated people would assent to just this account. It seems continuous with what ordinary folk mean by *mind,* and by *consciousness,* – a claim the boldest Behaviorist has not yet made in behalf of his own position(s). Nor are these old-fashioned answers to an old-fashioned problem. Ordinary people cannot quickly be brought to see the truth of doctrines which cast doubt on this view of mental events. When put as critically as McDougall puts it[1], the issue demands a carefully-drafted rebuttal:

> Any day, I suppose, anyone of you may be called to serve on a jury to try a criminal charge. Suppose it to be a charge of murder. One man has shot another and killed him. The testimony of witnesses establishes beyond doubt that the prisoner held the pistol and fired the fatal shot. The witnesses give you a full Behavioristic account of the incident. From this you may be able to conclude with confidence that the prisoner *intended* to shoot his victim. There you have reached from the description of behavior a conclusion which goes beyond the purview… of Behaviorism…. But the court will be interested in a still more difficult psychological problem, namely – granted that the prisoner intended to shoot his victim, did he intend to kill him? The fullest possible Behavioristic account would fail to throw light on this question. But there is a still deeper psychological problem which the court must solve, before it can do justice to the prisoner. Granted that his intention was to kill his victim, what was the motive of that intention? There you have the most essential problem of the case, and one before which the… the behaviorist… is perfectly helpless. (287).

Just on this question of mental events, Behaviorism (from Watson even to the present day) has opposed the psychology of Wundt, Külpe, and Titchener. The Behaviorists adopt a range of positions expressing their disapproval of "mental events". They believe that Titchenerian speculations would *not* become effective just by the redoubling of Introspective effort. The most definite of Behaviorist positions is to say: *Mental events do not exist.*

This is the position Watson (1920, 94) takes: "The Behaviorist 'ignores' mental events… in the same sense that Chemistry ignores alchemy, astronomy, horoscopy, – and psychology telepathy and psychic manifestations". Weiss (1925, vii) says the same thing: "The factors which traditional psychology vaguely classifies as conscious or mental elements merely *vanish* without a remainder into the biological or social components of the Behavioristic analysis".

Lashley's position (1923a, 246) barely differs at times: "Grant me" he says, "the postulates of the physical sciences and I can show you how the phenomena of mind may arise within a system which has no other attributes than those which the physicist ascribes to his phenomenological world". And again "The description of a rat opening a problem box is as complete an account of the *process* of thinking as can be given from introspective data".

Another Behaviorist position is: *Mental events may exist, but the behaviorist is not interested in them.*

This position is adopted by Bekhterev (1913). It is also a way out of controversy sometimes adopted by Watson[2], by Lashley, and by more recent writers grown

---

[1] In a paper called: 'Purposive or Mechanical Psychology?', which the editor of the *Psychological Review* 30 (1923), artfully inserted between the two halves of Lashley's famous article: 'The Behavioristic Interpretation of Consciousness'.

[2] See Watson (1913).

weary of battling metaphysicians. But not only does this concede the issue, it grants also a psycho-physical parallelism inconsistent with Behavioral analysis.

A third position from which Behaviorists have reacted against mental events is this: *Mental events may exist, but they are scientifically intractable; hence they are not a suitable subject matter for experimental psychology.*

This "Methodological Behaviorism" is (again) clearly stated by Watson. He wrote (1913, 175): "Will there be left over in psychology a world of pure psychics...? I confess I do not know. The plans which I most favor... lead practically to the ignoring of consciousness... I have virtually denied that this realm of psychics is open to experimental investigation..." Weiss also occasionally took this line. "The...(Introspectionist) point of view *can*, of course, be consistently maintained. There is justification for inferring the existence of conscious correlates for at least some of our actions, but the heuristic value of this assumption seems doubtful... the behaviorist disregards the entity which the functionalist calls *consciousness*". (1917, 314)

An alternative position: *Mental events may exist but are too boring to command attention.*

The spirit of Lashley's article (1923a, b) is behind this formulation. Though he raises other considerations for and against Introspection and uncovers weaknesses in the program of Wundt and Titchener, Lashley realizes that none of his observations are conclusive. His rejection of the traditional approach rests on this: that his own work is exciting, fruitful, and promises clear answers to well-formulated questions. Titchener's stuff however, and that of Driesch, Dunlap, Fernberger and McDougall seemed dreary and tortuous in contrast. Doubtless this explains the attraction to the Behaviorist persuasion of younger psychologists. Quickly bored with metaphysical terminology they prefer to get to work with rats and tachistoscopes, reaction times and data recorders.

Another stand Behaviorists can take regarding mental events is this: *Mental events exist, – but only when correlated with observables; only when a first-person description can be correlated with third-person observations of the subject's external or physiological behavior.*

This is implicit in Lashley and more well-formulated in Tolman, Hunter, Skinner and Hull. Some psychologists have even claimed that when the above conditions are fulfilled, then a report that a particular mental event has occurred in a subject *means the same thing* as reporting overt or physiological behavior in the subject. We will return to this later.

II

Behaviorists have adopted any one, or several, of these positions regarding mental events and Introspective Psychology. How would a Titchenerian reply? Let's first put the Behavioral program into its strongest armor. Let's suppose:

1. That the utterance "I am hungry" is *completely correlated* with observations of saliva flow, stomach contractions and the like. (If stomach contractions are not in fact correlated with first-person reports of hunger, as Davis and Garafolo seem recently to have shown, then assume something else is.) No instance wherein a normal subject sincerely says "I am hungry" is one where laboratory correlates are lacking; – and no case where correlates are observed is one where the subject would deny he was hungry.

A word about "a normal subject sincerely says". The Cartesian gymnastics some Behaviorists undergo to avoid possible dishonesty, caprice, illusion, or hallucination in a subject make it sound as if we could *never* ascertain whether a person is lying, or hallucinating. But most of us, and especially magistrates, counselors and doctors do this all the time. What reasonable grounds would there be for the psychologist to be dissatisfied with reports of consulting physicians that a subject is normal, – and reports of other qualified individuals that he is sincere? These reports can have all the certainty any empirical statement could ever have. Why this show of skepticism? It *ought* to lead to complete Solipsism, and the end of Behavioristic Psychology as a science. This stricture seems to be used only against first-person testimony. Third-person testimony is not supposed open to such doubts, because of "public verifiability".

But when a psychologist makes a distinction on principle he should follow through, – even if similar suspicions arise about third-person testimony. Why not accept that there *are* standard ways of determining whether a subject is sincere and otherwise normal? If competent authorities satisfy me of this at 11:55, then at noon I am free to proceed with experiments *sans* Cartesian encumbrances. The alternative to this is a doubt which *no* empirical findings can alleviate, – i.e. a metaphysical doubt. The risk of error never vanishes when proceeding as I suggest. But neither does this happen in any other scientific discipline.

To continue: I have supposed an absolute correlation between a subject's utterance "I am hungry", and laboratory correlates, such as his stomach contractions. Next I suppose

2. that no case wherein a normal sincere subject (henceforth, simply "subject") says "I am thinking of x" is a case lacking in vocimotor behavior.[3] And no observations of the characteristic vocimotor behavior occurs when the subject would deny he is thinking of x. Suppose further

3. that whenever a subject says "I am feeling X (or that X)", we always detect characteristic glandular activity; and that in no case where this activity is observed would the subject deny that he feels X, or feels that X. Next, suppose

4. that whenever a subject associates two ideas, words, or images, the psychologist finds some correlated reflex of the Pavlovian type, although more complex. Moreover, whenever such a reflex is discovered, the psychologist also notes in the subject behavior in accordance with some particular association of ideas,

---

[3] E.g. tiny laryngeal movements, as discussed by Watson (1920).

words, or images. So all sensations, images, and associations of ideas are supposed wholly correlated with the subject's reactions. Finally suppose

5. that every sensation of X is absolutely correlated with some discriminatory case where the subject would deny having sensations of the specifiable sort.

In short, allow the program of the early "naive" Behaviorists (and of the not-so-early-and not-so-naive Behaviorists) as *completely substantiated*. Let every "mental event" be assumed correlated with laboratory observables. The correlation is non-random, positive and exact. The observer can, from the report of a mental event, infer with highest probability to the concurrent activity of the subject's bodily organs. And from observing such activity the investigator could infer with highest probability to the concurrent readiness of the subject to describe some "mental event". The Behaviorist's program is thus set in the strongest terms – exceeding the most extravagant claims of Holt, Tolman, Lashley, Weiss, Hunter, Skinner or Hull, in the hey-day of optimism about Behavioristic psychology. Against this, how ought mental events to be approached?

III

Would such success in the program sanction the dictum of Watson, Weiss, and Lashley that mental events *do not exist* at all, that they *vanish* without remainder into the biological and social components of behavioristic analysis? Not at all! Introspectionists need not regard the success of Behaviorism in finding observable correlates of mental events as proving the non-existence of those events. Is one justified in concluding from the fact that man can be considered a physical-mechanical entity, or a physiological or neurological entity, that this rules out his existence as an economic entity, a socio-anthropological entity, or a music-composing entity? No, all these *complement* each other. Each makes some contribution to our total understanding, the whole forming an interrelated theory. The believer in mental events might be surprised at our hypothesized success of the Behaviorist program, but he need not conclude that therefore no mental events exist. He could embrace the Behaviorists' discoveries as complementary to his own. These he may continue to regard as irreducible to anything like observable behavior, the latter being different in type from anything with which he concerns himself.[4]

---

[4] Compare:

Münsterberg: "Psychology thus presupposes… a most complicated transformation (of our mental life), and any attitude… which does not need or choose this special transformation may be something else, but it is not psychology" (1899, 112).

McDougall: "Psychology can have no bearing upon, and no application to, the problems of human life, the problems of voluntary conduct, the acts of men proceeding from desire and will" (1923, 279).

Moore: "… the term 'psychology' means and can only mean the science of mind (consciousness, mental life, or other more or less equivalent expression),… some other term must be invented for the science of behavior, and for that comprehensive science which covers the study of consciousness and of behavior in their mutual relations" (1923, 235).

Now for the second possible way in which a Behaviorist, flushed with success, might express his distaste for mental events. The Behaviorist might take a softer line, allowing that "mental events" may exist, but arguing that they cannot constitute a scientific subject matter. This is said *ad nauseam* about History, Ethics, and even the Social Sciences. Apparently these very subject matters rule out laws of the Newtonian type. Observations familiar in Chemistry and Physiology, − the metrical and statistical techniques which now almost define scientific enquiry − in disciplines like History, Ethics, and even (say) Anthropology, these are impossible. The contents of consciousness form a similar subject matter. In principle these must remain intractable, resisting formulation in Newtonian-type laws, and orthodox observation or experiment.

Again, the Introspectionist need not faint in defeat. He can argue that only on one narrow definition of "science" will his subject matter, − and those of History, Ethics and Anthropology − be ruled out as unscientific. If Newtonian type laws, controlled experiment, and observation of the metrical-statistical type, are the *sole* desiderata of an enquiry being conducted scientifically, then none of these constitute sciences. But is one obliged thus to define a science? Why should techniques already employed limit what procedures are to be allowed? Had this argument always obtained, new scientific techniques would never have arisen; exit all the metrical and statistical techniques of the last 200 years. These were constantly challenged. If *they* qualified as sciences apparently all objectivity was lost. Hence outbursts against Darwin, Weismann, DeVries, and Morgan, Einstein, De Broglie, and Schrödinger, Heisenberg, Born, and Dirac. All introduced techniques making formerly intractable areas amenable to objective treatment. Often this was against the protests of contemporaries who urged that certain types of phenomena *must* elude objective treatment. Every psychologist has heard learned outsiders say that where man is concerned, really scientific treatment is impossible.

Two issues intertwine here. (A) Some claim of a subject matter that it cannot become an object of scientific enquiry by arguing that no now-recognized technique could ever illumine it. This constitutes an empirical prediction based on slender evidence, if any at all. How can one *know* that in the next century History and Ethics will not be pursued with the techniques Physics now employs? It may seem unlikely; but any more so than it seemed to our ancestors that Energy, Life, and Intelligence could be dealt with in terms now familiar? Consider the techniques of Biophysics and Biochemistry; and many biological and medical types of enquiry have been transplanted *en bloc* into orthodox domains of psychology and social psychology. What then supports the claim that established scientific techniques will never be employed in History, Anthropology, Ethics, or the science of Mental Events?

(Mind you, there *are* considerations which make optimism about "scientific" futures for these disciplines out of place. But these are not empirical considerations. For the moment, however, the argument that certain disciplines will *never* employ now-orthodox techniques, appears to have little to support it.)

(B) The second issue concerns whether "suspicious" disciplines may develop their own techniques yet to become tools of scientific enquiry. Probability theory, the theory of statistical approximation, the operational calculus, and unnumbered other techniques have transformed intractable data into respectable sciences. To predict this will never happen with the studies above, is empirically risky.

Suppose the Behaviorist's claim is contracted. He says now only that enquiry into consciousness and mental events cannot *at present* be called "scientific". But this is not worrying. The Introspectionist can now point to problems which have yet to meet their scientific solutions. Why a counsel of despair? Now the confirmed Introspectionist will mark his brotherhood with other researchers who have unsolved problems on their hands, – the New Cosmologists, the Meson theorists, the Business-cycle Analysts. Nor would the Introspectionist wilt at Lashley's claim that mental events and consciousness are boring.[5]

Clearly Wundt, Külpe and Titchener, none of them stupid, were not bored by, or superstitious about, their "new" psychology. People get bored, Titchener argued, because they lack the patience every science demands when problems arise.

Perplexities in the analysis of mental events get intricate, and the Behaviorists quit! This even Lashley (1923a, 238) calls "the egotistic fallacy". Behaviorists changed the subject. Because *they* could not answer questions about mental phenomena, the questions must be unanswerable; they turned to easier questions, answers to which were quick.[6] But (the argument continues) these easier questions are *different* from those which perplexed the Introspectionist. Does it follow that the latter are not genuine questions, that they refer to a non-existent subject matter, that they can never be answered objectively?

So even where the Behaviorist program is supposed fulfilled to the limit, nothing need force the Introspectionist to retreat. He can still say he is discussing something different from the Behaviorist, that this is in no way less amenable (in principle) to possible scientific enquiry than other empirically discoverable phenomena; nor need he be dismayed when others do not find his interests exciting.

IV

The fifth Behaviorist reaction to mental events is more moderate, yet in some ways more radical. It proposes that mental events exist, but only insofar as they are correctable with laboratory observables. This constitutes a *redefinition* of *mind* and *consciousness*, like that proposed by Bawden in 1918: "... mentality, or mind, is a name for the fact of the control of the environment in the interest of the organism

---

[5] "The conception of mind has undergone a long course of evolution and many of its supposed attributes are only vestiges of... superstition, religious dogma, and false psychologizing..." (1923a, 247).

[6] "[They omit] a whole universe of phenomena, which have been supposed to constitute the chief realm of psychology" (239).

through the interaction of inherited capacities and acquired abilities". Contemporary psychologists often add a significant rider: when the observer says of someone else "X has just uttered 'I am hungry'", and also says "X's stomach is contracting in the specified way", then the two statements *have the same meaning*. In other words, "Hanson says 'I am hungry'" (henceforth proposition 'H') and "Hanson's stomach is contracting in the specified way", (henceforth 'S') mean the same, i.e. have the same use, have all the same consequences, are verified in exactly the same way.[7]

If *this* contention is correct then Introspectionism is really dead. Because then no Introspectionist's statement will convey anything beyond what the Behaviorist conveys in describing what is correlated with introspective reports. They are discussing the same thing, – in different words which (in third-person speech) have the same meaning. Two such statements make the same assertion. Hence neither can convey information the other does not.

"John is the male sibling of Jane" means the same as does "Jane is John's sister". The words are different. But the same assertion is made. I could substitute one for the other and convey the same news. So no special kind of facts is expressed by one of them not expressed by the other. The Introspectionist cannot now say he is dealing with a subject matter different from that of the Behaviorist. There can no longer be an independent subject called "Introspective Psychology". The subject matters of the two disciplines are equivalent. So if this account can be sustained, the Introspectionists have propounded but a prolix way of stating facts which Behaviorists can state succinctly, and without metaphysical overtones. Introspectionists asked a host of misleading questions because they failed to appreciate the content of their own assertions. (This may be true even if the present Behaviorist version is *not* correct).

What could the Introspectionist do when confronted with this account? The obvious counter-attack is this: that a person's utterances are correlated with his behavior (bodily or organic) is something we discover empirically. Were there a perfect correlation between a subject's utterance "I am thinking about X", and his vocimotor behavior, this would constitute an important factual discovery; important because, while it *could* have been false, it is in fact true. This discovery excludes a real possibility, a state of affairs imaginable, but not actual. Is saying that H and S *mean the same thing* an acceptable way of describing this factual discovery? Consider: "John is the male sibling of Jane" means the same thing as does "Jane is John's sister". Assert one and you cannot meaningfully deny the other. Both statements make the *same* claim, have the same meaning. This fact we discover not by experiment and observation, but by understanding what is being claimed in either case. Equivalence of meaning is not a factual issue. It is logical or conceptual in character. But that utterances are discovered correlated with observed activity, this is factual. So assume the sentences "Hanson says 'I am thinking about X'" and "Hanson's vocimotor apparatus is acting in the specified way", are both discovered to be true. This is a factual discovery whose denial is false, – but not meaningless. Occasions may

---

[7] This is an almost verbatim rendering of remarks made in lectures and in print by Professor D. Ellson of Indiana University.

not exist wherein the Behaviorist would say "Hanson says 'I am thinking about X'"
where the facts do not allow him also to say "Hanson's vocimotor apparatus is acting in the specified way". But the very idea of such an occasion cannot be self-
contradictory. To state S (i.e. "Hanson's stomach is contracting in the characteristic
way") when one must deny that Hanson is saying, or could say, or would say that he
is hungry, this may never in fact happen. But it is not impossible to imagine what it
would be like for it to happen, as it is impossible to imagine what it would be like
for John to be the male sibling of Jane whilst Jane is *not* John's sister. And if one can
imagine such a distinction between S and H, then it is a distinction in the meaning
of the expressions being employed.

The contention seems to be this: whenever phenomena are invariably correlated,
their descriptions are really different ways of describing one, single phenomenon.
Thus the descriptive sentences, H and S, since they describe two events which (we
are supposing) never occur separately, are therefore but different word-forms by
way of which the same assertion about the same event is made. Hence the two sentences have the same meaning.[8]

Let us carry through the argument in the medium of *third* person talk. Because it
seems clear that H and S do not have the same meaning in the first person, i.e. when
I am doing the talking. Because should I say of myself "Hanson says 'I am hungry'", I would not mean to say anything about my stomach contractions. I may not
even know that my stomach *is* contracting, – as must be the case with most people
who say they are hungry. And if I do not *know* that my stomach is contracting, how
then could it be claimed that when I say I am hungry I *really mean* to say something
about my stomach? I mean no such thing. I mean I am experiencing pangs of a
characteristic sort for the remedy of which quantities of food inserted in the appropriate place is required. That is what I mean when I say I am hungry.

But some contemporary Behaviorists disavow first person-experience in a way
that can be likened only to portions of Descartes' *Discours de la méthode*. They
imagine that there are no ways of distinguishing whether one's perceptions are
veridical or the results of hallucination. Hence, so it is argued, one must either go
Cartesian (and build up experience out of indubitable verities) or go Behavioristic
(and abandon first person talk as insignificant, or unscientific).

This is a philosophical howler of the textbook variety. It represents the most
confused kind of metaphysics. *What* is being doubted when one doubts that his
perceptions are veridical? This cannot be a genuine scientific doubt. If it were it
would be possible to assure the psychologist and his subject at 5 minutes to noon
that they are not in any non-normal state; their perceptions at noon time ought then
to be counted as veridical. But that this question is not of this type seems clear.
Nothing would really satisfy such a doubter that his perceptions were non-halluci-
natory. If the Behaviorist *could* be satisfied about this (as most non-philosophers
could be), then the argument from possible illusion would not oblige one to agree
that only third person talk is reliable in the laboratory. But if we can convince a

---

[8] "… the statement 'I am conscious' does not mean anything more than the statement that 'such and
such physiological processes are going on within me'" (Lashley 1923a, 272).

person that he is not hallucinating, then his first-person talk can be just as reliable as his third person talk, even when he is describing mental events like the feeling of hunger pangs. If nothing can convince the subject and the psychologist that hallucination is *not* dominating their experiment, then their doubt is not a genuine empirical doubt, but only metaphysics. In principle this is like my schooldays-discovery that I cannot know that when I see the color I call "red", I am not seeing the color you see when you say "green"; all reference to oculist's tests were dismissed as irrelevant. But then what could be relevant? Answer: *nothing*. Hence the Schoolboy's doubt is not the kind we occasionally have about someone's vision, but a metaphysical (i.e. a factually unanswerable) doubt. Besides, a systematic perceptual demon would mislead the experimenter as much about the behavior of others as about his own internal states.[9] Let us glide over these difficulties and allow the Behaviorist to speak in the third person, ruling out first person talk as publicly unverifiable. (The Introspectionist will nonetheless note this precisely as the decision *not* to discuss what interests him). What now? The contention is that when the psychologist says H he means exactly what he means when he says S.

If this is only a decision to use old words in a new and technical way, there can be no quarrel. It's a free-ish country, so the Behaviorist can use language as he pleases. But from this it could not be concluded that H and S meant the same thing in any ordinary use of "mean the same thing". Because I decide to use "eclipse" to mean the same as "a can of beans", it does not follow that "eclipse" means "can of beans". But the Behaviorist occasionally argues that when two events (e.g., my utterance "I am hungry" and the experimenter's observation of characteristic stomach contractions) are invariably correlated, then H and S mean the same. What must the classical Introspectionist say to this?

He could note a range of invariably correlated events whose descriptions remained semantically distinct.

Begin with some first approximations:

1. Suppose that the correlation between the occurrence of fire and the appearance of smoke was taken to the absolute limit. Would the two reports "There is fire twenty feet ahead" and "There is smoke twenty feet ahead", mean the same? Not on any ordinary account. Even were the statements mutually inferrable, with minimal risk of error, that fact alone would not force one to think that whenever it is appropriate to make the one assertion it is equally appropriate to make the other. At night smoke may obscure visibility, whereas fire might increase it. Only an arbitrary recommendation will force these statements to have the same meaning.

2. Again, lightning and thunder are highly correlated. (If lightning occurs at T, thunder will be heard at $T + S/1100$ s (where "S" is the foot-distance of the observer from the lightning).) $\Theta = T + S/1100$ is to be ruled out because of its

---

[9]Lashley (1923a, 251) demurs: "The behaviorist may study a behaviorist in the act of studying a behaviorist, and is justified in concluding that his own processes of study resemble those of the other".

complex nature, then something more than the word *correlation* must be employed. I am going to take lightning and thunder as non-randomly, positively and exactly correlated for any observer by the function Θ. Yet it is ludicrous to suggest that "There is lightning twenty feet ahead" means the same as does "There is thunder twenty feet ahead". Their consequences are entirely disparate. Yet the two events are absolutely correlated by Θ, – as indeed are any two finite sets of apparently random data if one sets up no restrictions on the mathematical complexity of the correlating function involved.

3. A book has pages. Moreover, it is tangible, has weight and is combustible. On no occasion will I discover any of these properties lacking; all are invariably correlated. Yet the sentences, "That book has pages", "That book is tangible", "That book is heavy", "That book is combustible", all have different meanings. We can imagine one obtaining when the others do not: just as I can *imagine* H being true and S being false. The consequences of each are different. The verifying procedure for each is different. Thus one cannot conclude from an invariable correlation between A and B that a statement referring to A will mean the same as a statement referring to B.

4. There are hundreds of such examples: an object's color is always co-present with its dimensions. Yet no one would take a statement about an object's color to mean the same as a grammatically similar statement about its dimensions. Electromagnetic radiation of $\lambda = 7000-7500$ Å is seen by normal observers in specifiable conditions as red. Conversely, red (in the same circumstances) is correlated with optical exposure to radiation of this wavelength. But it would be naive to conclude that therefore statements about radiation of 7000–7500 Å, and grammatically similar statements about the color red *mean the same.* This would be tantamount to saying that the color red *is* electromagnetic radiation of this wave length, – an elementary howler. Complete correlation does not obliterate conceptual differences.

The Behaviorist may return to the charge. He may state that H ("Hanson says 'I am hungry'") means the same as S ("Hanson's stomach is contracting in a characteristic way"), because *not only* are the two events invariably correlated; *they cannot be distinguished.* There is no difference because no difference is detectable. But the Introspectionist (and I) will stoutly reassert that the two events, and hence H and S, certainly can be distinguished. Movements of my mouth, and noises emanating therefrom are perceptually distinguishable from stomach movements. The differences *are* detectable. They are described at length in Titchener.

It may be countered that this difference is just a carry-over from first person talk, which is already ruled out. But now we see just *what* has been ruled out: everything that goes against the Behaviorist's contention. So he wins his argument by his own fiat. He proves that two sentences mean the same because they cannot be distinguished; but when someone tries to distinguish them the means for doing so are struck from the permissible moves in the Behaviorist's game.

The contemporary Behaviorist first rules out Introspection, and first-person talk, for the same reasons his predecessors had. Hence he is able to prove that the two third person statements H and S mean the same (i.e. cannot be distinguished), when he himself makes it impossible for Introspection to distinguish them.

Is it even true that in the *third* person no distinction can be made by the Behaviorist himself between H and S? It may be a fact that nothing which followed from the occurrence of the one could fail to follow from the occurrence of the other. But how is this fact learned in the first place? *You* can establish that I have just said "I am hungry" by listening. And that my stomach is contracting is established by watching instruments. It is, however, a dogma of philosophical analysis that if one establishes the truth of statements in different ways, then those statements have not the same meaning. So even the disinfected third person statements favored by the Behaviorist (i.e. H and S) cannot have the same meaning.

V

In general, someone like Fernberger[10] need not concede defeat, − not when the attacks are these Behavioral reports of experimentation, or recommendations as to the uses of a technical language. The reason is clear. *Introspectionism is really a program of research*, − not a set of conclusions. Behaviorism is also a program, a highly successful one. It has announced conclusions about behavior, and even about intelligent behavior. But a program's conclusions are not identical with its prospectus. The conclusions of a science are true or false, or probable or improbable. Programs are not true, not false, not probable nor improbable. They are promising or unpromising, fruitful, unfruitful, developed or abandoned. All that can be done by conclusions running counter to a program's prospectus is to contract the scope of the latter. So the imputation that the *empirical* findings of Behavioral psychology obliterate Introspectionism can forever be resisted. The Titchenerian will insist that the Behaviorist discusses something different from his own interests, − consciousness and its contents. Nor will he concede that these interests are unempirical or objectively untestable. Disguised methodological recommendations which can always be shown to rest on implicit rules of the game – making it impossible for the Introspectionist to win, − need never force him into this concession. (It must be possible for one's opponents to win an argument if one's own victory is to be construed as resting on facts, as opposed to logic.)[11] The Introspectionist can withstand all this because the Behaviorist's scrutiny of a "science" of mental events reduces to: "What we are doing is fruitful; so any other approach, – like Introspectionism, – must be concerned either with non-existent entities, or entities which can't be "scientifically" studied, or which are boring, or which are exactly equivalent to the entities the Behaviorist studies. The Introspectionist can deny each of these in turn. Each rests on empirical considerations whose very logic ensures that they cannot

---

[10] See Fernberger (1922).

[11] Compare Lashley (1923a, 245):"... by changing the rules of the game innumerable self-consistent systems can be developed".

deliver the *coup de grâce* to a *program*. Facts can force us to modify programs, but not to say that they are false. Programs, like plans, invitations, and prescriptions, are not of the right logical type to be true or false. When fought for stoutly, more than a recitation of facts may be required to cause a full retreat from a program like Introspectionism.[12]

So far as psychologists, philosophers, novelists and lawyers are concerned, the utter rout of dualism suggested in some Behaviorist-oriented textbooks has just never occurred. It is too doctrinaire to say that only the uninformed, antediluvian, unintelligent and obscurantist believe that there is in every rational person an experiential state, − "consciousness", "inner life", "mind" − inaccessible to the observations of others. Changing the subject, concentrating on the paraphernalia of the laboratory, reading and writing ultra-statistical papers, − none of this is going to persuade ordinary people that what they think they encounter every day of their lives does not exist, or is unworthy of study, or is equivalent only to what some third person might say about the subject's behavior. No *new facts* will make the believer in mental events abandon his old facts as mythical, boring, incapable of objective scrutiny, or equivalent to third-person behavior statements.

VI

Yet introspectionism ought to be abandoned: all the good people just mentioned ought to reappraise the content of what it is they believe to constitute the facts of their inner life.

Because, though the program of Introspection and mental events cannot be shown to be *factually* false (just as dualism cannot), − it is nonetheless conceptually unsound (just as dualism is). The people who defend mental events, consciousness, and their own inner life may misunderstand the logic of what they are wanting to assert; this despite centuries of intelligent employment of such terms. Many people fail to appreciate what they themselves must do to get information about themselves. The Introspectionist may argue he is discussing a different kind of fact from the Behaviorist. But he can scarcely say he is employing a different kind of logic. Facts will not shake him. Contradictions in his own program will. It is this latter attack which the Behaviorist must undertake if the opposition is to be disabused of ideas concerning mental events and consciousness which have haunted philosophers and psychologists for centuries.[13]

Behaviorist arguments have rested on facts; they ought to have rested on logic. Polemics have argued that mental-event talk was false; but much of it is *not even*

---

[12] On this matter Lashley errs badly: "The controversy between behaviorism and dualism is not a question for philosophy, but one to be answered strictly in the light of empirical evidence provided by psychological study" (1923a, 246). To me this indicates a failure to have appreciated the nature of the controversy between behaviorism and dualism.

[13] Thus Lashley observes: "... the behaviorist seems to have failed to strike at the root of the dualistic systems".

false. I can only sketch some of these arguments, *sans* the details found in the original, careful treatments in Wittgenstein's *Philosophical Investigations* (1953) and in the recently published *Blue Book* and *Brown Book* (1958).

The Behaviorist must not put himself into a position like that of the old fashioned Atheist. Many Atheists wish to assert *both* (A) that there is no God, and also (B) that to claim that there *is* a God is itself nonsensical. But an assertion, *P,* must make sense even to be false. If *P* is nonsensical, then the counterassertion, *not-P*, is equally nonsensical. "The number two weighs four pounds" is nonsense; it is equally nonsensical (and for the same reasons) to say "The number two does not weigh four pounds". If the Atheist says "It is not the case that God exists" he ought not also to claim that "God exists" is nonsensical. When the Behaviorist says "... mental events don't exist" or "... are equivalent to behavioral events" he ought not also to say[14] that Introspective talk about mental events is nonsensical. One cannot have it both ways.

The "nonsensical" claim is the stronger one. It says that talk about mental events is confused and confusing, *not even* false. Citing *facts* against the existence of mental events can at most show that claims concerning mental events are false. The Introspectionist can then contract his claim and preserve the significance of his enquiry. If the Behaviorist wishes to clear away Titchener's camp-followers, he must do what seems rarely to have been done, – attack the logic of Introspection. This means not just characterizing the evidence for mental-event talk as question-able, nor showing that the methods by which Introspectionism proceeds are scien-tifically suspect. Wundt, Külpe and Titchener could easily sidestep such charges. Nor does it mean doing what Ellson aspires to do, – namely, invent a new technical vocabulary in which talk of mental events cannot arise. The Introspectionist can refuse to play that wordgame. Wundt, Titchener and McDougall would liken this approach to the cutting of the Gordian knot. It does not solve the problems of mental events, but invents a terminology in which such problems cannot be stated.[15] No, the man who thinks he has problems will not be impressed by ways of suppressing, rather than solving, those problems. What *will* put him off, is the demonstration that he has not really got a problem at all; that his very questions entail unacceptable conclusions. Introspection-talk will be quieted only if it can be shown that on his own premises no answer which would satisfy the Introspectionist can be derived.

Consider how a novelist would describe the mind of his hero, or villain. "I have told you how Joe checks and rechecks his conference notes, tries out his friends' advice by countless and ingenious tests, ties his shoes in seven distinct motions of the hand, has worn out two copies of *Who's Who* during official interviews, invests only in gilt-edged bonds, checks the Mrs.' budget every Friday and Monday, and

---

[14]As have most Behaviorists (including Tolman, Lashley, Weiss, Hunter, Skinner, Hull, and Ellson).

[15]This is analogous to the attempts of modern philosophers, like Carnap, Reichenbach, Hempel and Bergmann, to crack old conceptual chestnuts by creating an artificial language in which tradi-tional philosophical problems cannot arise. This is like fixing a flat by resolving to walk, or like patching a lover's quarrel by joining a monastic order.

only buys his children toys of the *Monopoly* and *Cash-Register* type, – now that I have told you all this I am going to tell you about Joe's mind".

Never! Such an ending would not occur. Because in so describing Joe's propensities to behave, the novelist *is* telling us about Joe's mind. Writers, magistrates and counselors are so busy *describing* people's minds that it never occurs to them to set out a special section on, e.g. Joe's inner life. To do this would be analogous to saying "Well, you have shown me all of the rooms in your house, the attics, the basement, the garden and the grounds, your sons, your daughters, your wife and the dog, – now show me *your home*". It is obvious that seeing all these things *is* seeing one's home, – not just the physical correlates of one's home. I was once stopped in Cambridge by an American soldier who had obviously visited and photographed thoroughly every old college within 4 square miles; he asked me where he could find *the University*. My answer disappointed him. It seemed to puncture his expectations of something better yet to come. As a little boy I had the same reaction on learning that the parade of platoons moving past me was not going to be followed by something even better, – the march-past of the Regiment itself.

These are confusions of levels of discourse. They are easily noted in cases when e.g., someone asks where the average American lives, or supposes that the probability of the law of gravitation might be raised if he just begins dropping things. But, the confusions are not easily noted in the cases of *consciousness*, and *minds*. These latter are umbrella-words. Their logical function is to make terse, economical references to an intricate network of information, description and prediction about the ways people carry on their daily activities. This is precisely how people without philosophical-psychological theories defend and support their assertions about other people's minds.

I might say of a colleague that he was ruthlessly ambitious, not seriously reflective, too quick in discussion, prone to slant a quotation, rather flamboyant in the presence of a visitor, aggressive when his own views are subjected to criticism, tireless in his industry over new causes he feels he ought to champion; and with all this, he is yet disarmingly pleasant on chance meetings, makes the undergraduate feel confidence with a few well-put questions, and exceedingly generous. What would you be asking me to say if, after all this, you asked me for information about the man's mind? Have I not just given this? This is what all of us do when we say significant things about other people's minds, mental capacities, intelligence, wit, sensitivity, humor, emotional steadiness, industry, acuity and honesty. These "mental" words just denoted, are generalized notions which we employ when discussing things people actually do, and the things we expect they would do in certain situations. "Don't leave him alone in the room with the cash register open", "Don't expect her to take the responsibility if the headnurse leaves", "Good old Jane, still at it, – what devotion", "Watch out, here comes John's big fish story again"; – these are *the* ways in which we describe, and discover other people's minds. They are just the ways ordinary people will recognize themselves as employing. Any theory which makes a different supposition must make it clear why ordinary concepts fail for specialized, extraordinary purposes.

Most of us, when not straining to say something philosophically or psychologically novel, gather knowledge about *ourselves* in ways not in principle different from the ways I learned about my colleague's inner life. It is not even true that we learn about our own minds, and capacities and dislikes, preferences and inclinations, – in ways different from the ways in which we learn such things about others. Certainly the data are easier to come by. But that doesn't make them different in type. How e.g. does one know of himself that at times he is counter-suggestible, quarrelsome in discussions that might have been amicable save for his incursion? Not by turning an inward eye onto some internal region wherein it is set out in long-hand what sort of a chap he *really* is. One learns these things of himself just as he would learn the same things of another; by noticing how often discussions in which he participates soar into high metabolism, how naturally he plays *Advocatus diaboli*, by observing how his colleagues seem to him to err more and more. That is how in an autobiography he would set out such revelations about himself. One rarely verbalizes such information for his own benefit, nor must he devise special ways of extracting the truth about himself from himself. This only means that the data are for him more readily acquired – not that they differ in principle from what one requires to make similar pronouncements about him in a biography. Being me, I am extraordinarily alert to patterns of such data within myself. My point of vantage *vis-à-vis this* data is better than yours is likely to be. This is no scientific revelation, but rather a loose account of how people naturally proceed to talk about the capacities, dispositions and intellectual excellences of others, and indeed, of themselves.

So, the Introspectionist account is *not* logically continuous with the ways in which ordinary people ordinarily think and talk about mental capacities and mental events. Oh; it is compatible with what ordinary people will say if you stop them on the sidewalk and ask them what *mind* is. I have done this, and the messages of over-reflective schoolteachers and overconscientious persons comes through to me as a report of what would otherwise have been described in quite different terms.[16]

This is clearly a thesis to be defended in detail and I have already written too much. But as Wittgenstein (1953) has argued so well, there are myriad activities which *directly* display qualities of mind. They are not themselves covert intellectual operations. Intelligent practice is not necessarily the *result* of prior internal theorizing. Rather, internal theorizing is but one practice amongst many others which we regularly characterize as having been intelligently or stupidly conducted. That there *must* be some internal process for every justifiable use of a mental-conduct word is a prejudice of late nineteenth century logicians, philosophers and psychologists, a prejudice which has wrecked the psycho-physical theory from within.

The result is a menagerie of shadowy mental processes like *inference;* how strange that Introspectionists spent so much energy trying to describe such a process without asking themselves whether expressions like "John is half through inferring that every equi-angular triangle is equi-lateral" really *sound* all right. But such logical questions eat at the heart of talk about internal mental processes. It seems a pity

---

[16] This stresses Bridgman's dictum: "If you wish to know what a man means by a term, don't ask him, watch how he uses it" (1927).

that such talk was met by Behaviorists not by logical examination, but by the assertion of *counterfacts*.

Make no mistake, the story of the demise of Introspectionism and dualism must be a logical story. It consists of taking the psycho-physical theory and contrasting it with the concepts we learned to manipulate at mother's knee, — and then tracing in detail the impossible consequences of the theory's presuppositions. For it is on the grounds of logic and logic alone that a logically spurious theory must be attacked. The circle-squarers were squelched by the logical *proof* that their position was impossible, not by observations of the past failures of circle-squarers. So must the Introspectionist finally be squelched, — not by demonstrating how little his predecessors have done, but by showing the logic of the entire dualistic theory to be broken-backed.

References

Augustine. 1876. *The Confessions of St. Augustine*. Trans. J.G. Pilkington. Edinburgh: T. & T. Clark.

Bawden, Heath. 1918. The presuppositions of a behaviorist psychology. *Psychological Review* 25 (3): 171–190.

Bekhterev, Vladimir Mikhaïlovich. 1913. *La psychologie objective*. Paris: Librairie F. Alcan.

Bridgman, P.W. 1927. *The logic of modern physics*. New York: Macmillan Co.

Fernberger, S.W. 1922. Behavior versus introspective psychology. *Psychological Review* 29 (6): 409–413.

Lashley, K.S. 1923a. The behavioristic interpretation of consciousness I. *Psychological Review*. 30 (4): 237–272.

———. 1923b. The behavioristic interpretation of consciousness II. *Psychological Review* 30 (5): 329–353.

McDougall, W. 1923. Purposive or mechanical psychology? *Psychological Review* 30 (4): 273–288.

Moore, Jared S. 1923. Discussion: Behavior vs. introspective psychology. *Psychological Review* 30 (3): 235–235.

Münsterberg, Hugo. 1899. *Psychology and life*. Boston: Houghton Mifflin.

Watson, John B. 1913. Psychology as the behaviorist views it. *Psychological Review* 20 (2): 158–177.

———. 1920. Is thinking merely action of language mechanisms? *British Journal of Psychology. General Section* 11 (1): 87–104.

Weiss, Albert Paul. 1917. Relation between structural and behavior psychology. *Psychological Review* 24 (4): 301–317.

———. 1925. *A theoretical basis of human behavior*. Columbus: R.G. Adams & Co.

Wittgenstein, Ludwig. 1953. *Philosophical investigations*. New York: Macmillan.

———. 1958. *Preliminary studies for the "Philosophical investigations", generally known as: The blue and brown books*. Oxford: Basil Blackwell.

Part IV
Logic

Chapter 14
Imagining the Impossible

I

If I can imagine X, then X is not logically impossible. If X can be thought, then *"X"* is consistent. If one could draw a picture of X, then *"X"* is not self-contradictory.

What justifies these claims? A statement as to what is or is not possible, is a logical statement. It tells us whether or not some purported description is of the form $P \cdot \sim P$. The claim that X is logically possible just is the claim that X's description is not of the form $P \cdot \sim P$. And such a claim, if true, could not but be true. The claim that a quadrilateral triangle is impossible, is the claim that "X is a quadrilateral triangle" is a description which can be reduced to the form $P \cdot \sim P$. If this claim is true, it necessarily could not be false.

Prima facie, however, it does not seem that a statement about what one can and cannot imagine, or think, or picture, concerns logic. It is in some sense contingent to say of me that I can imagine that X, while you cannot imagine that X. And so it is contingent also to say of everyone either that they can, or cannot, imagine X. The negation of this seems not to be reducible to the form $P \cdot \sim P$.

But then the claim that if we can imagine, or think, or picture X, then X is possible, – this looks like a claim consisting in the inference of a logical statement (concerning X's possibility) from a contingent statement (concerning what we can or cannot think, imagine, or picture). Clearly, this had better be wrong. It could be wrong in any of three ways: (1) It may be that the statement "'X is a quadrilateral triangle' is of the form $P \cdot \sim P$; i.e. X is impossible" is *not* a logically true statement at all, but a mere statement of fact, i.e., the fact that "X is a quadrilateral triangle" *is* of the form $P \cdot \sim P$. That is, we do in fact use "X is a quadrilateral triangle" as being of the form $P \cdot \sim P$, – but we need not have done so. (2) It may be that "No one can imagine a quadrilateral triangle" is not a contingent statement at all, but a necessary one. That is, it may be that "Someone is imagining a quadrilateral triangle" is self-contradictory, and not merely factually false. Or, (3) it may be that "If one can imagine X then X is not impossible" is not the entailment we are inclined to suppose it is.

© Springer Nature B.V. 2020
N. R. Hanson, *What I Do Not Believe, and Other Essays*, Synthese Library 38,
https://doi.org/10.1007/978-94-024-1739-5_14

II

Let us consider these alternatives in turn:

1. Is the statement "'X is a quadrilateral triangle' is impossible, (i.e., of the form $P \cdot {\sim}P$)", – is this statement itself a logically true statement or only a statement of fact? We really do use "X is a quadrilateral triangle" as being of the form $P \cdot {\sim}P$. But need we have done so? Is not the fact that we *do* do so contingent on our linguistic habits and conventions? Does not the statement itself just describe what is factually the case?

   Well, let *this* be *given* as true: *that X is a quadrilateral triangle is impossible. How* is this true? Are further observations still relevant as they are with "To swim around the world is impossible"? Certainly not. If it is true that S is impossible, – then it is logically true. There is no chance of future observations changing the verdict. If it is true that 'S' is of the form $P \cdot {\sim}P$, then to deny this would be not merely false, but self-contradictory. The matter is settled by reflecting on the logical structure of negation of "'S' is of the form $P \cdot {\sim}P$", – not by further observations of in action. So "S is impossible (or possible)", if true, is logically true.

2. Concerning whether or not "No one can imagine a quadrilateral triangle" is contingently or necessarily true it might be argued: "But the contention that if X is imaginable it is therefore possible, has nothing to do with you or any other individual, or all individuals. It does not concern what particular persons can or cannot imagine. It concerns what is, or is not, imagin*able*. It is concerned with the nature of what can or cannot be thought or pictured. That some people in fact have restricted imaginations, while others have boundless imaginations – *this* is contingent, and irrelevant. What is at issue is the very structure of experience. It is of the essence of imagining, thinking, and picturing, that we cannot imagine, think, or picture what is logically impossible. What is one denying who denies this? Here is an axiom about imagination and thought if ever there was one. This does not require further analysis; it is the basis of further analysis. This insight is itself what justifies our uttering other propositions concerning what is or is not possible. And the insight is expressed in a statement which, although it recounts a basic fact of experience, is nonetheless necessarily true. Its negation may not be of the form $P \cdot {\sim}P$, so it need not be a tautology; yet the statement could not be false. What would it be *like* for it to be false?"

What we have *really* just been told is that we justify that (x) $(THINKABLE\ x \supset {\sim}IMPOSSIBLE\ x)$ by inferring it from the equivalent principle: $(x)(IMPOSSIBLE\ x \supset {\sim}THINKABLE\ x)$. But whence came we by this prior knowledge, which, although it be conveyed in a statement formally equivalent to the first, yet differs from it in that it appears to consist now in an inference from a logical statement to a contingent one? Why then should it be conceded that *this* inference is necessarily true although not tautologous? *Must* it be granted that no deeper understanding of our acceptance of such an axiom is possible; that it is something simply seen or not seen, but not something for which one can give, or expect, arguments? *Must* we agree that this is an unquestionable condition of offering justifications and analyses for other philosophical problems? Even were this true, must we concede that nothing more can be said?

III

Is it part of what we *mean* by saying that *X* is logically impossible, that we cannot think it? This does not seem to be so. To say that a quadrilateral triangle is impossible is to say that its description is of the form $P \cdot \sim P$. One can say *this* however, without any reference to thought; anyone's or everyone's. In fact, if *X is* logically impossible, we *cannot* form a mental picture of *X*, e.g., a quadrilateral triangle. But then the connection between *X* being impossible and *X* being unthinkable may only be an empirical one. There simply never has been a case of anyone thinking, imagining, picturing the logically impossible. *Prima facie* however, this is not in principle different from saying: there never has been a case of a *perpetuum mobile*. No one has ever succeeded in building one. And, given *our* physical world, no one ever will. Given *homo sapiens*, no member of that species has ever built a *perpetuum mobile*, nor thought/imagined/pictured the logically impossible. But it need not be self-contradictory to suppose either of these circumstances to obtain; it would just be false.

Perhaps we do not even *have* the concept of building a *perpetuum mobile*. But this again is a statement about what kinds of concepts we *do*, in fact, have. "I just built a *perpetuum mobile*" may be not conceivably true, but it is not logically false. And there never has been a case of anyone thinking, imagining, or picturing the logically impossible; "I just imagined something which is logically impossible" may be not conceivably true, but is it therefore logically false? "I just imagined something whose description is of the form $P \cdot \sim P$" is not *itself* a statement of the form $P \cdot \sim P$, although it may indeed be a false statement.

So "is unthinkable" need not be considered part of the meaning of "is logically impossible". Nonetheless it sounds far too weak to say that therefore (x) $(IMPOSSIBLE \ x \supset \sim THINKABLE \ x)$ expresses only an as-yet-un-falsified contingent regularity, or perhaps just a "psychological inconceivability". For we cannot form any notion of what would count against this principle. We cannot even find a hypothetical value for "x" such that $(\exists x)(Ix \cdot Tx)$. We cannot form any conception of what would count against $(x)(Ix \supset \sim Tx)$. If we cannot form such a conception though, then $(x)(Ix \supset \sim Tx)$ cannot but be true. And if it cannot but be true that (x) $(Ix \supset \sim Tx)$ then it is necessarily true that $(x)(Ix \supset \sim Tx)$, even though $(\exists x)(Ix \cdot Tx)$ does not reduce to the form $P \cdot \sim P$. "I have a mental picture of a quadrilateral triangle" is, if false, contingently false, even though we can form no idea of what it would be like for this to be true.

It is already long overdue that we should carefully distinguish two senses of the expression "is necessarily true". For these are regularly confused in discussions of inconceivability, possibility, and necessity – and they have been slipping in uncontrollably in the foregoing.

It is necessarily true that I am sitting here writing these words. And yet the statement that I am doing this is an empirical statement. How then can it be necessarily true? It is necessarily true, for me, because in fact no evidence I could now entertain could possibly shake my present belief in the claim that I am sitting here writing these words. If at this moment I had any reason whatever to doubt this, then I would have thereby been robbed of any reason to think any other empirical statement at all reliable. My ability here and now to entertain evidence for other propositions (one way or the other), depends on my present inability to entertain evidence against my being now conscious, sitting here at this table writing these words. If the latter goes, everything goes. So no evidence can count against this proposition for me. It cannot, for me, be false that I am now sitting here consciously writing these words. So it cannot but be true. If it cannot but be true however, then it is for me necessarily true that I am sitting here consciously writing these words. This was the sequence entertained just a moment ago.

But although it is now necessarily true for me that I am sitting here writing these words, it is not logically impossible that I should not be doing so.[1] That is, the negation of the statement "I am sitting here consciously writing these words" does not itself reduce to the form $P \cdot {\sim}P$. So although it cannot but be false that I am not sitting here and consciously writing these words, this statement is nonetheless not self-contradictory. An exactly similar analysis can be given for such a proposition as "A *perpetuum mobile* is impossible", or "Nothing travels faster than light". That this is so shows that the first person idiom of "I am sitting here consciously writing these words" is not essential to the construction of a sentence which expresses what is both not conceivably false and yet not tautologically true. That there should be a *perpetuum mobile* invented tomorrow is not logically impossible; but there is not the slightest notion extant as to what such a device could be like. If something should be discovered to move faster than light the description of this fact would have to await the construction of a notational and conceptual framework *ab initio* before the event could be made at all intelligible to us. Such new sciences are possible. But that they are necessary before a *perpetuum mobile* or a velocity $> c$ can be countenanced makes it quite clear in what sense it is necessarily true that there can never be a *perpetuum mobile* or a velocity $> c$.

The sense then in which it is now necessarily true for me that I am sitting here consciously writing these words, or that a *perpetuum mobile* is impossible, must be strictly distinguished from the sense in which it is necessarily true that no triangle can be quadrilateral. The first two cannot be false because no conceivable evidence could show me that they are false. The last cannot be false because the very statement of its falsity is internally inconsistent. It is now inconceivable for me that I should be doing other than sitting here consciously writing these words, or that a *perpetuum mobile* should be built today; and it is inconceivable that there should be

---

[1] I.e., While I cannot negate the statement, I can entertain its negation.

a triangle which is quadrilateral. But the reasons for the inconceivability differ in each case.

Assessing the logical status of the principle: *if X is thinkable, then X is possible*, then, may best be done by entertaining its negation. That negation is of the form: There is an *X* such that *X is* thinkable and *X* is impossible. Now we will all agree that this describes an inconceivable state of affairs. But in what sense of inconceivable? Is $(\exists x)(Tx \cdot Ix)$ self-contradictory, or is it simply an hypothesis for which no conceivable evidence whatever could be forthcoming? Is it of the form $(\exists x)(Px \cdot \sim Px)$? Or is it simply like the statement: *It is not now the case that I am sitting here consciously writing these words?*[2] Is it necessarily true that everything thinkable/imaginable/picturable is possible, in that nothing could conceivably count as evidence against this, – or is it necessarily true in that the very idea of a counter-instance is itself self-contradictory?

IV

The former position is the more attractive because of the argument which opened Sect. II. In 'what is thinkable is possible' we seem to be arguing a matter of fact, – albeit a very abstract one. We do in fact think in the ways in which we do in fact think; we imagine in the ways we do in fact imagine; we picture as we do in fact picture. And in fact, what is logically impossible is never discovered to be either thinkable, imaginable, or picturable. Although an alternative to our ways of thinking, imagining, picturing cannot *ex hypothesi* be known to us, it does nonetheless seem venturesome to suppose "everything imaginable is possible" necessarily true in that its denial is self-contradictory. Although we have no idea of what it would be like to think differently, it does not seem demonstrably self-contradictory to suppose that we might have thought differently. To the logically equivalent way of stating the same principle – viz., what is impossible is unthinkable/unimaginable/unpicturable – the same argument must apply. Because although it is certainly true that if *X* is logically impossible, then *X* is unthinkable, it is again questionable whether the denial of this is logically inconsistent, or simply such that we can form no conception of evidence in support of it.

The statement then that *if X is thinkable X is possible* is necessarily true although it has a consistent negation. But it is still open to the objection that it appears to be an inference from a contingent statement to a logical statement. And I cannot see that this objection is unjust. What is or is not thinkable is contingent on the ways in which in fact we think. What is or is not logically possible is laid down in the rules concerning consistency and contradiction.

What has almost certainly happened here however, is that there has been a slide in the statement of this principle from one sense of "necessarily true" to the other, – from one sense of "conceivable" to another. Thus we know (1) that if *X* is thinkable, then it is not the case that *X* is inconceivable. And we know also (2) that if it is not the case that *X* is inconceivable, then *X* cannot be of the form $P \cdot \sim P$. (3) If *X* is not of the form $P \cdot \sim P$, then *X* is possible. While each of these

---

[2] Or like the statement: *M.I.T. built a perpetuum mobile today.*

uses of "inconceivable" is independently legitimate, there has been an illegitimate slide from (1) to (3), from one sense of "inconceivable" to another. Although it seemed that this term was functioning as a middle term of a hypothetical syllogism, we have here really two different concepts disguised by the one word "inconceivable". For "not inconceivable" in (1) meant that nothing could count as evidence in favour of its being inconceivable. But "not inconceivable" in (2) meant "not of the form $P \cdot \sim P$" And so the principle needs restating. It cannot simply read: If X is thinkable, then it is not the case that X is inconceivable, and if it is not the case that X is inconceivable, then X is possible. For the former conditional and the latter one *are* of different logical types, and hence cannot function in one argument as they have been supposed to do. And yet it is exactly this logical slide which is smuggled into most statements of the principle that *if X is thinkable, then X is possible.*

Chapter 15
On the Impossibility of Any Future Metaphysics

I

Consider 'Hume's dictum': that from a necessary proposition nothing contingent follows – and vice versa. The effect of this on speculative metaphysics is devastating, although few practicing metaphysicians realize how utterly their position has been devastated. I will spell this out by drawing the profound logical moral of Hume's dictum.

What the dictum comes to is this, that trans-type inferences, i.e., inferences across logical types, are impossible. But metaphysical arguments are necessarily trans-type. They proceed from statements concerning what *must be* the case to statements concerning what *is* the case, or vice versa. In short, they proceed from necessary propositions to contingent propositions – or vice versa. Indeed, what else *could* a metaphysical argument do? Any informative inference from contingent proposition to contingent proposition will be more the province of the scientist than the metaphysician. Observations constitute the appropriate appeal when determining whether some contingent conclusion can be reliably drawn from some contingent statement of fact. On the other hand, any inference from necessary proposition to necessary proposition will be more the province of the deductive logician than the metaphysician. Unpacking the semantical content of expressions constitutes the appropriate appeal here, and this is too limited an inquiry for the man who aspires to infer what *is,* from claims which could not be false, or who hopes to learn what *must be,* via inferences from what is the case.

Sooner or later, every metaphysical argument can be shown to be trans-type. Perhaps, then, the typical metaphysical inference is that encountered in the ontological argument, wherein a contingent existence claim (which it makes sense to deny) is inferred from a statement setting out which predicates are necessary constituents in our concept of God – a statement which, if true at all, could not but be true. Not every argument of speculative metaphysics is as transparently offensive as is the ontological argument. For example, Spinoza's version of the argument is

© Springer Nature B.V. 2020
N. R. Hanson, *What I Do Not Believe, and Other Essays*, Synthese Library 38,
https://doi.org/10.1007/978-94-024-1739-5_15

much more subtle. But every one of them *is* trans-type. Thus Kant, the destroyer of the ontological argument, allows his own transcendental deduction to proceed across types, even if very subtly. Were this not so, the argument would not be metaphysical at all, but strictly deductive, on the one hand, or strictly inductive on the other, alternatives Kant flatly rejects. That is, his argument (if not trans-type) would either turn on a purely formal entailment – and hence be non-falsifiable in the manner of a tautology – or it would constitute an informative scientific claim, and hence be falsifiable. But metaphysical arguments usually purport to be both informative, i.e., nontautological, and also apodeictically true, i.e., non-falsifiable. Hence they purport to be trans-type, and indeed *must* purport to be trans-type – *tertium non datur.*

Again, the possibilities in entailment are just these: (1) the entailment can be from what is necessary to what is necessary, (2) the entailment can be from what is contingent to what is contingent. But neither (1) nor (2) constitutes an informative inference. They are just 'semantical-unpackers'. If the move from what is contingent to what is contingent is genuinely informative, then the move cannot constitute an entailment. It will be more like an inductive inference, or a causal inference, and hence be non-necessary. So they will not be entailment-bound.

Traditionally, however, metaphysical arguments purport to be *both* informative (i.e., non-tautological) and also apodeictically true (i.e., nonfalsifiable). This means they must constitute informative entailments, to achieve which one's argument *must* proceed across types.

It may now be demonstrated that trans-type inferences are impossible. Suppose that P is necessary (1), and that Q is contingent (2), and that P entails Q (3). In Lewis' symbolism:

(1) $\sim \Diamond \sim P$	premise
(2) $\Diamond Q \cdot \Diamond \sim Q$	premise[1]
(3) $P \dashv Q$ (i.e., $\sim \Diamond (P \cdot \sim Q)$)	premise
(4) $P \dashv \sim Q$	premise
(5) $\Diamond Q$	(2)
(6) $\Diamond \sim Q$	(2)
(7) $\left[(P \dashv Q) \cdot \sim \Diamond \sim P \right] \dashv \sim \Diamond \sim Q$	Lewis and Langford 1959, 18:53[2]
(8) $\left[(P \dashv \sim Q) \cdot \sim \Diamond \sim P \right] \dashv \sim \Diamond Q$	
(9) $\sim \Diamond \sim Q$	(1, 3, 7)
(10) $\sim \Diamond \sim Q \cdot \Diamond \sim Q$	R.A. (6, 9)

---

[1] This rendering is necessary since '$\Diamond \sim Q$' is compatible with both 'Q is contingent' and 'Q is inconsistent'. (Cf. Aristotle, *Prior Analytics*, I.13: 32^a, 15ff; also Bochenski (1951, 56–57.)

[2] This doubling is necessary to guard against an inference from P to contingent Q when (2) is rendered as $\Diamond \sim Q$.

$(11) \sim \Diamond Q$ (1, 4, 8)
$(12) \sim \Diamond Q \cdot \Diamond Q$ R.A. (5, 11)

It is therefore not logically possible that a necessary proposition should entail a contingent proposition. Hence metaphysical argument is impossible.

II

This very formal proof is valid and completely general, more so than Lewis and Langford realized (cf. below, 247–248). But it will fail to carry conviction for most committed metaphysicians. They will feel that the conclusion is bought by arbitrary stipulations en route, each one constituting a *petitio principii*. This feeling is ground-less. But to convince the metaphysician of this one must construct an inference which is so obviously trans-type and yet so apparently a genuine entailment that he concedes that it is as carefully designed as any orthodox metaphysical argument could be – perhaps more so. Then we must unravel this very example itself, as a logical illustration of the kind of thing that can happen in every such putative trans-type inference.

Consider the following claim:

(a) "That $F_1F_2F_3F_4$ ... obtain is good inductive reason for expecting C."

entails

(b) "That F_4 obtains is a good inductive reason for expecting C."

Here, (a) can express a proposition which is necessary, and (b) can express one which is contingent. Moreover, (a) and (b) can express propositions between which a genuine entailment holds. By Hume's dictum, something is *prima facie* wrong here. But before popping the bubble, let us first blow it up to full size. Suppose that in (a), $F_1F_2F_3F_4$ take as their values the following statements of fact:

F_1: At present B's are regularly followed by C's.
F_2: In the past B's have always been followed by C's.
F_3: Our theories do, as a matter of fact, relate the concepts "being a B" and "being followed by a C". [This is to be understood as constituting no more than a contingent, factual refer-ence to what extant theories do actually do. No 'inference permit' is being smuggled in here.]
F_4: There are no known experimental or theoretical grounds for expecting otherwise.

All this constitutes good inductive reason for inferring to a C whenever a B occurs. This is not to say just that the truth of F_1–F_4 in fact constitutes sufficient condition for expecting C. Because, were F_1–F_4 known to obtain, it would be *irra-tional* to refuse to infer to C after witnessing another B. Indeed, when F_1–F_4 obtain we have before us the paradigm *meaning* of "having good inductive reason for infer-ring C whenever a B occurs". This would not be said in ordinary cases of providing sufficient conditions for inductive conclusions.

Thus, suppose that last April was showery, and that it was followed by a flowery May. And suppose that in all previous years showery Aprils have been followed by flowery Mays. Botanical research indicates abundant hydrogen oxide as a prerequisite for healthy plant metabolism. Moreover, nothing is known to suggest that May's flowers are completely independent of April's showers. Nor have we any expectation that 1960 will differ (e.g., nothing like a sharp increase in atmospheric radioactivity has been detected which might upset all natural regularities).

Should these constitute all the facts, then (should next April be showery) the *only rational thing* to infer would be that May will be flowery. (If this 'botanical' example seems too grossly Baconian, then any illustration from classical mechanics will serve as well, *provided* that F_3 is always understood to be a low-grade statement of fact, on a par (in this respect) with F_1, F_2, and F_4.)

In one sense of "is necessary," "That $F_1F_2F_3F_4$... obtain is good inductive reason for expecting C" expresses what is necessary. For values such as those prescribed, the denial of this is not simply false – it describes nothing intelligible at all. Not that such a denial is demonstrably self-contradictory; explicit definitions might fail to reduce it to inconsistency. Thus (a) is comparable to "No surface is at once everywhere red and green." The denial of this also leads to conceptual paralysis, but not because the idea involved is demonstrably inconsistent, which is, indeed, doubtful.

"That F_4 obtains is *a* good reason for expecting C" is clearly contingent, however. This is *a* good inductive reason for expecting May to be flowery *contingent upon* F_1, F_2, and F_3 actually obtaining. To expect C because of F_4 alone would be like expecting a Russian rocket to hit Mars tomorrow just because we have no ground for expecting otherwise. This inference is contingent upon the present stage of Soviet rocketry, and on the statistics of past successes and failures.

So, any one of F_1, F_2, F_3, or F_4 will be *a* good inductive reason for expecting C *contingent* on the remaining F's obtaining.

III

Hence, (a) can express a necessary statement, and (b) can express a contingent statement. Yet (a) appears to entail (b), just as a set of sufficient conditions for a contingent conclusion entails that any member of that set is a necessary condition for that same conclusion. However, *our* sufficient conditions are described in (a), which states what is necessary. This collides with Hume's dictum which is now established. Something *must* have gone wrong. But what? Unless we can say, the door is ajar for all manner of philosophical tomfoolery to rush in. The possibility of inferring across types is the lifeblood of speculative metaphysics, the secret of Cartesian rationalism, and the backbone of such things as the ontological argument. To get something out of nothing may seem a flippant way to characterize what most of the history of philosophy has tried to achieve. But the attempt to infer what *is* actually the case, from what reflection shows *must* be the case, is nothing if not the cooking of facts out of thoughts. Provide one valid example of such an inference and the future of clarity is in jeopardy. This is why any candidate for the title "trans-type inference" must be taken seriously, no matter how contrived it may appear at first. For, even to suggest that such an inferential move is possible is to sanction all pleas for a return to the capital-lettered, purple prose of the Grand Tradition in philosophical system building. It has been my intention to provide an inference which is as plausible a candidate for trans-typicality as is possible. In fact, this example has been framed to highlight those distinguishing characteristics of metaphysical argument which it is the purpose of the sequel to disentangle.

IV

But, attractive as it may be in some respects, this example is parasitic on subtle yet pernicious ambiguities in the expressions "is a good inductive reason" and "obtains".

On one interpretation of the phrase "good inductive reason", what (a) expresses *does* entail what (b) expresses. But *then* (b) is *necessary* and the inference is not across types. On the other interpretation, what (b) expresses is contingent, but then there may be *no* entailment at all. Consider: "F_4 is such that, when construed as a premise together with F_1, F_2, and F_3 (or some other appropriate set), it would be conceptually untenable to assert all these premises and deny that we have good inductive reason for expecting C." Thus interpreted, (b) expresses what is *necessary*. Truth can be determined by inspection and reflection, without recourse to observation. Here, then, we *do* have an entailment – but not across types. What (b) expresses is no longer contingent.

It is as if we argued

(a') "That P and $P \supset Q$ obtain is good deductive reason for concluding that Q"

entails

(b') "That $P \supset Q$ obtains is *a* good deductive reason for concluding that Q."

Similar considerations apply here. *If* there is an entailment between (a') and (b'), then (b') is necessary: otherwise it is just false. (I do not intend this as a covert restatement of Hume's dictum, to introduce which, here, would constitute a *petitio principii*.)

A second interpretation of (b) is this: "*There are* certain true contingent statements, $F_1F_2F_3$ (or some other appropriate set), such that, when conjoined with F_4 they give us good inductive reason for expecting C." This is contingent. It turns on the first two words; *there are* in fact true statements giving us good inductive reason for C, although there need not have been. Claims like F_1 F_2, F_3, and F_4, since they are all contingent, need not have been true at all, singly or in conjunction. But now there is no entailment. "That $F_1F_2F_3F_4...$ obtain is good inductive reason for expecting C" does not assert that F_1, F_2, F_3, and F_4 do in fact obtain. If it did, we would know clearly how that proposition might be false, and hence not necessary. All (a) claims is this: "*Allowing* that F_1-F_4 obtain we would then have good inductive reason for expecting *C*." Here, then, is the corresponding ambiguity in the term "obtains".

Again the deductive analogy:

"That *P* and *P* ⊃ *Q* obtain is good deductive reason for concluding that *Q*."

If this is understood to assert that *P* and *P* ⊃ *Q* do actually obtain, it is (insofar) falsifiable. They need not have obtained. And if, so understood, *this* is falsifiable, then so is (b'), "That *P* ⊃ *Q* obtains is *a* good deductive reason for concluding that *Q*." But now the conclusion *can* be false even when the premise is true. Hence any apparent entailment between (a') and (b') is not necessary, and hence not an entailment at all. Thus, on this interpretation what (a) expresses need not entail what (b) expresses. In short, if I claim "That *P* obtains entails *Q*", I have two choices. I can stress the word "obtains", which move will render the whole expression contingent. Or, I can stress "entails", in which case "obtains" loses all logical force whatever.

V

Only by running these two interpretations together can it appear that a necessary statement (a) entails a contingent one (b). Certainly (a) and (b) *can* express propositions between which an entailment *does* hold. But when that happens *both* (a) and (b) express necessary propositions; the inference is not across types. *Or*, (b) can express what is contingent and synthetic. But in that case (a) also expresses what is contingent, or (at least) what is synthetic. So, *were* there an entailment, it would not be trans-type. But there is no entailment since (a) could be false. Presumably, we would not wish to say that anything known to be false *entails* (b), or *entails* anything at all. It was part of Lewis' intention to exclude any such inferences (e.g., "That Indiana is an island *entails* that last May was flowery") from the class of genuine entailments. (The *fact that* a claim is false can have entailments. But a claim, known to be false, cannot itself have entailments; known falsehoods *per se* entail nothing.)

This much constitutes the resolution of the most plausible candidate I know of for trans-typicality in argument. This special illustration supports the general proof with which we began in a way which ought conclusively to demolish the very *idea* of metaphysical demonstration.

VI

However, consider now how an articulate and ingenious metaphysician might retort to all this. Until further notice, what follows is to be construed as a metaphysician's counterargument to what has preceded.

"Suppose it is argued that

(a″) 'Jones is a bachelor'

entails

(b″) 'Jones is male.'

Both (a″) and (b″) are contingent and (hence) synthetic. Both could be false. Yet the first entails the second. It might be argued that the original (a) and (b) are similarly related. While both are synthetic they are, despite earlier arguments to the contrary, entailment-bound.

"In much the way that the synthetic claim

'Jones is a bachelor' (a″)

entails the further synthetic claim

'Jones is male' (b″)

so does the synthetic claim

'That $F_1F_2F_3F_4...$ obtain, is good inductive reason for expecting C' (a)

entail the further synthetic claim

'That F_4 obtains is a good inductive reason for expecting C″ (b).

However, although undeniably synthetic, what (a) expresses is in an important sense necessary, as already outlined. It is not just that (a) sets out what are merely *in fact* necessary and sufficient reasons for expecting C. Its negation describes no intelligible state of affairs at all. We do not understand the *meaning* of its denial. Although synthetic logically (a) is necessary epistemically. *Logically, it could be false – its negation is not inconsistent. But epistemically it could not be false – its negation is unintelligible.* Insofar as Hume's dictum is construed as ruling out the possibility of inferring what is synthetic from what is analytic, the 'good reasons' example does not run counter to it. But insofar as Hume's dictum is construed as ruling out inferences from what is epistemically necessary to what is contingent, it may require revision, in light of a large class of respectable propositions, all logically synthetic, but whose denials describe nothing intelligible at all.

"The propositions in question [The metaphysician continues] are such as these:

'There are no perpetual motion machines.'
'Nothing travels faster than light.'
'A high speed microparticle's "state" cannot be precisely determined.'

These claims are synthetic. Whatever they entail will be synthetic, by a narrower version of Hume's argument. But whereas their consequences are likely to be contingent, and hence falsifiable, the claims themselves (i.e., those above) are necessary. To entertain their denials is conceptually paralyzing. We cannot say what it would be like for them to be false. But they are synthetic too: their denials do not reduce to inconsistency by L-transformations alone. Thus the inference from (a‴) 'Nothing travels faster than light'

to

(b‴) 'That positron's velocity is less than 3×10^{10} cm/sec'

is a possible candidate for epistemic trans-typicality. Both claims are transparently synthetic. But whereas the second reads like a mere statement of fact about positive electrons (falsehood seems as close as the changing of a number), the first statement could not intelligibly be denied. The reason? It is at the core of an enormous number of theoretically interlocked claims, all of which would fall apart semantically were this statement disproved. That is not to say that this couldn't happen: it is *logically* possible that it should happen. But the probability of its happening is 0.

"Hence [our metaphysician continues] Reichenbach was wrong in symbolizing the logically impossible as that which has zero probability. It is logically impossible that I should construct a quadrilateral triangle, but no probability-estimate of any kind attaches to this. It is logically *possible* that I should accelerate an electron faster than c, since to suppose this generates no contradiction. Nonetheless, as all our theories indicate, the probability of doing so is 0. And this is, in general, the analysis appropriate to *epistemically* necessary claims such as 'No surface is simultaneously red and green.'

"The denial of this is not demonstrably self-contradictory, but it has 0 probability of ever obtaining.

"Another such inference is that from

(a‴′) 'There are no perpetual motion machines'

to

(b‴′) 'The mechanism constructed yesterday by Professor Jones does, in fact, require more energy, than it produces.'

"Both claims are consistently deniable; (a‴′) entails (b‴′). But while the first compels assent since its negation is unintelligible, the second is no more than a local statement of fact, concerning what Jones *did* construct yesterday, and what are the *de facto* properties of his construction. Here is the counter-model to Hanson's earlier exposition, and it constitutes more the type of argument we metaphysicians recognize and employ.

"In short, there is another, 'super-Humean', variety of inference, a plausible candidate for epistemic trans-typicality. Thus (a) and (b) may both express what is synthetic, and hence what is possibly false. Nonetheless, perhaps (a) is necessary in that the probability of its *being* false is 0, while (b) is merely contingent in that the possibility of its being false is > 0. This might be relevant to understanding not only how laws of nature – themselves often epistemically necessary, or 'functionally *a priori*' – can entail transparently contingent observation-statements. It might also make clear what *kinds* of arguments metaphysicians since Plato have been concerned to articulate."

Thus for the counter-case of the metaphysician. Here his disquisition ends, becoming now itself a new object for scrutiny. At this point I am reminded of a conversation which ran: "Do you believe in baptism?" Answer: "Believe in it? I've seen it done!" Analogously, one might believe in the possibility of metaphysics on the grounds that they've seen it done. Our question now is just *what* is it they've seen done? The immediately foregoing might stand as a most attractive answer to that question.

We have been invited to entertain a *tertium quid*, a third possibility which avoids both the strictures of Hume's dictum and the emptiness of tautological demonstration. This consists in granting the force of Hume's ruling when construed as blocking inferences from analytic statements to contingent statements, or vice versa. That is, the metaphysician concedes that statements whose negations are inconsistent cannot entail statements whose negations are wholly consistent. This, presumably, has never been seriously in issue.

By distinguishing two senses of 'necessity', the metaphysical *tertium quid* arises. Granted, every analytic statement is necessary; its denial is demonstrably inconsistent so it could not but be true. But, it is urged, some *synthetic* statements are necessary too, necessary in that their denials, while consistent, are conceptually untenable, i.e., unintelligible.

That there are such statements is true. Indeed, this was how we characterized our first statement (a), consistently deniable, but not intelligibly deniable.

However, the further suggestion that a synthetic statement, necessary in this epistemic sense, could entail other synthetic statements *not* necessary in this epistemic sense – this suggestion is ruled out by an extension of the semantical content of our earlier formal proof.

It is probable that Lewis and Langford construe *necessity* as *analytic necessity*. Thus, "... a proposition is necessary if and only if, in point of fact, it does not have a contradictory... It is not easy to see what other reason could be assigned why some propositions are 'necessary truths'" (1959, 478).

There can be two different situations in which a proposition may 'not have a contradictory'. The first consists in the contradictory being inconsistent. The second consists in the contradictory being unintelligible. Lewis and Langford probably had only the first in mind. That part of the metaphysician's counter-claim which urges a second sense of necessity, as outlined above, is something to which one can unreservedly subscribe. But no trans-type inference is sanctioned by this subscription. The formal proof with which we began holds *irrespective* of what sense we attach to 'necessary'. This is not developed by Lewis and Langford. Thus their proof is more general than they realized. For if we distinguish necessity$_1$ from necessity$_2$– a proposition P being necessary$_1$ if $\sim P$ entails $Q \cdot \sim Q$, and being necessary$_2$ if $\sim P$ entails what is, although consistent, unintelligible – then, the general proof with which we began *apparently* demonstrated only that whenever P is necessary$_1$, its consequences must also be necessary$_1$ The *tertium quid* consisted in exploiting the notion of necessity$_2$. Our metaphysician urged that there are entailments between synthetic propositions, such that the first can be necessary$_2$, while the second is not. The proof, however, is as cogent when it is necessity$_2$ which is at stake. Hume's dictum can thus be generalized to claim that if a premise is necessary *in any sense* of that term, then its consequences will be necessary *in that same sense*.

It is left to the reader to expose the ambiguities which made the inferences concerning the velocity of light and *perpetua mobilia* seem trans-type in that special way which might help the cause of metaphysics. But it is assured in advance that something has gone astray; the *tertium quid* will sooner or later always collapse into a *tertium non datur.* Although this very conclusion is itself inductive, being based on special examples, and analyses such as those preceding, it is no less convincing for that. The probability that no metaphysician will ever produce an argument which is a genuine entailment and also moves across logical types is 1. So the probability of there ever being a valid metaphysical argument is 0.

References

Aristotle. 1995. In *The complete works of Aristotle*, ed. Jonathan Barnes. Princeton: Princeton University Press.
Bochenski, Joseph M. 1951. *Ancient formal logic*. Amsterdam: North-Holland Pub. Co.
Lewis, Clarence Irving, and Cooper Harold Langford. 1959. *Symbolic logic*. New York: Dover Publications.

Chapter 16
Good Inductive Reasons

I

"*F* is a good reason for *C*", if true at all, could not but have been true – even when *F* and *C* are contingent. Much in this claim is true. Much is misleading. Let us sort this out and, *en route*, discuss a certain pattern of inference, and "the problem of induction".

To begin with, the proposition above ought to be "*That F obtains* is a good reason for *C*". When *F* and *C* are contingent, 'good reason for *C*' would usually mean "good inductive reason for *C*".[1] This would be said only when (1) *C* can be inferred from *F*, and (2) *it is true that F.* It would be absurd to argue "It is false that *F*, so *F* is a good reason for concluding that *C*". So the original contention, qualified, comes to this: "*That F obtains* is a good reason for expecting *C*", if true, could not but be true. This appears to be sound. Goodness of reasons *seems* always to be settled by reflection alone. It is the legitimacy of an inference which is being judged when the expression '…is a good reason for…' is used thus. Further observations are not to the point; here we are but appraising the validity of a possible argument.

But consider now the inference:

F has always been followed by C. So that F obtains is a good reason for expecting C.

"*F* has always been followed by *C*" is clearly contingent. But according to considerations just explored, "that *F* obtains is a good reason for expecting *C*", if true at all, could not but be true.

The inference then is apparently from what is contingent to what, if true, could not but be true. This collides with the principle: *from what is factual nothing neces-*

---

[1] Cases like "That the table is brown is a good reason for claiming that the table is coloured" will not be considered here. These claims, *F* and *C*, although both contingent, do not oblige us to read 'good reason' as 'good inductive reason' – indeed, *cannot* oblige us to do so. For such values as this for *F* and *C*, the resulting statement is certainly such that if it is true at all, it could not but be true. But there is nothing perplexing, misleading (or even interesting) in such constructions.

© Springer Nature B.V. 2020
N. R. Hanson, *What I Do Not Believe, and Other Essays*, Synthese Library 38,
https://doi.org/10.1007/978-94-024-1739-5_16

sary follows. Call this "Hume's principle". Its function is to put every inference into logical equilibrium – from facts to facts, or from necessities to necessities. But supposed inferences from facts to necessities, or from necessities to facts, are suspect.

According to Hume's principle then, an inference like

April showers have always brought May flowers. So, that it is April and showery is a good reason for expecting May to be flowery,

is suspicious. It seems to be "trans-type". "April showers have always brought May flowers" is contingent. And "that it is April and showering is a *good reason* for expecting May flowers", if true at all, apparently could not but be true – so our original contention seems to dictate. Thus the inference is either suspect, or it is not what it appears to be. Or Hume's principle needs re-examination.

I will argue (1) that the inference is not suspect, but entirely legitimate, and (2) that it is in one respect *exactly* what it appears to be, namely, an inference from a statement of fact to a judgment as to the goodness of an argument. However, (3) in another respect the inference is *not* what it appears to be. Despite its being somehow correct to say that "That *F* obtains is a good reason for expecting *C*" is, if true, necessarily true – it is false to suppose that "April showers have always brought May flowers; so, that it is April and showery is a good reason for expecting May to be flowery" constitutes an inference from what is factual to what is necessary.

II

There are several reactions to these anticipated conclusions. One is to argue (against our original contention) that propositions of the form "That *F* obtains is a good reason for expecting *C*" are *always* enthymemic, and always contingent; only against a detailed contingent background could such a claim be true at all. This, however, should not be pressed too far. For suppose $\mathscr{F}$ to be the *composite* factual assertion:

(1) *F*'s have always been observed to be followed by *C*'s, and
(2) No known datum suggests that any *F* should *not* be followed by a *C*,

and

(3) An *F* obtains now.

(1), (2) and (3) conjointly are sometimes understood to constitute the force of "that $\mathscr{F}$ obtains Thus, for example, "That the diving glider's wings have just torn off...", "That his parachute is ablaze..." and "That Fangio's engine is seized..." would almost always be taken as composite in this way. "That the diving glider's wings have just torn off is a good reason for expecting a crash", – when the first part of this claim is construed as an elliptical reference to all the relevant evidence known about airborne gliders bereft of wings, – if true at all could not but be true. A man who agreed that all actually observed *F*'s have always been followed by *C*'s, i.e., that in gliding, wing-loss is always followed by a crash (Fact 1) and that no known datum counts against expecting any F to be followed by a C (i.e., that this was but an ordinary glider, with no auxiliary rockets or other crash-averting devices), (Fact 2), but who went on to deny that *this F's* obtaining is a good reason for expecting C, – such a man would have made a perfectly unintelligible remark. Thus, so long as "That $\mathcal{F}$ obtains…" is taken (as it often is) to be a compressed statement of all the relevant data there are, plus the further factual statement that no datum suggests anything to the contrary, – our original contention seems unexceptionable.

However, it remains arguable that "it is April and showery is a good reason for expecting May to be flowery" is always enthymemic, and always contingent. If true at all, it is only because a further premise is compressed within the claim. Clearly, if I treat "it is April and showery" as a premise and "May will be flowery" as a conclusion, any *entailment* between the two will require the further premise "Whenever there are April showers one may take it that there will be May flowers". That these two premises give good reason for expecting May flowers is logically true. But this only expands the enthymeme into *modus ponendo ponens*. When this *is* done, "that F obtains is a good reason for expecting C" becomes part of a transparently necessary argument. The factual claim, "F has always been followed by C" (or "April showers have always brought May flowers") thus becomes redundant. Only degenerately is this "that from which it is inferred" that F's obtaining is a good reason for expecting C. This returns Hume's principles to equilibrium, – by replacing the contingent "F has always been followed by C" with "Whenever F infer C". By converting the entire inference into "Whenever April is showery one may infer that May will be flowery; *so* that April is showery is a good reason for inferring May will be flowery", an obvious necessity results.

So what? The champion of "hidden-premise-validation" cannot mean that anyone who *does* infer from "April showers have always brought May flowers" to "that it is April and showery is a good reason for expecting May to be flowery" must *in psychological fact* have entertained the inference-permit "Whenever April is showery, one may take it that May will be flowery". This would be false. Rarely are "good reasons" claims made or authenticated by eliciting, contemplating, or whispering such permits. I *have* a good reason for expecting May flowers, when now in mid-April it is showering. Who would deny this just because I had not formed the inner speech pattern "Whenever F infer C"? I may never even have formulated such a hypothetical, – may never have read it or heard it. Psychiatrists may fail to fish up from my subconscious anything like "Whenever April showers infer May flowers". Yet I insist that its being a showery April *is* a good reason for expecting May flowers, – but deny that what justifies this is my having whispered the permit, or remembered past whispers. I did nothing but say what I said.

The "hidden-premise" move comes only to this: if one argued that F's obtaining is a good reason for expecting C and was then *challenged*, he would (ultimately) have to formulate "Whenever F infer C". Several things commend this view, – but "being true" is not one of them.

Agreed, if a man claims "That F obtains is a good reason for expecting C", yet just gapes when challenged, then we are likely to discount the claim. That F obtains may still *be* a good reason for expecting *C*, but such a man does not *have* good reason for expecting *C*; no more than had a parrot spoken. We do insist on the *possibility* of such substantiation, or as I would dub it "trans-substantiation". Thus the champion's thesis has become thoroughly subjunctive as well as being conditional:

> "Were Mr. X challenged to justify that *F* is a good reason for expecting *C*, then he would [if it *is* justifiable] formulate the inference-permit 'whenever *F* infer *C*". If unable to do this, his contention would not be justifiable. Settling the goodness of reasons turns on the elicitability of these permits".

This confounds two different questions:

(1) 'Is the *claim* "*F* is a good reason for *C*" justifiable?', and
(2) 'Is Mr. X justified in claiming "*F* is a good reason for *C*"?'

The "champion of the hidden premise" correctly argues that the answer to (2) may be "no" even when that to (1) is "yes". Moreover, the answer to (2) will rarely be "yes" unless the answer to (1) is also "yes". However, if it is argued that we are not entitled to answer "yes" to (2) except when X's defense consists in his actually eliciting the inference-permit which confirms (1), – placing it premise-wise into the argument, – this claim is just false. It is possible to answer "yes" to "Is X justified in claiming *F* is a good reason for *C*?" even where he would *not* defend his claim (if required to do so) by eliciting the permit necessary for an affirmative answer to "Is the claim '*F* is a good reason for *C*' justifiable?" There are other ways of showing the goodness of reasons. We regularly substantiate arguments, and the goodness of reasons, without any such recourse.

If, however, the champion's claim is only the colourless one to the effect that irrespective of how we *do* substantiate arguments it is always possible to elicit *some* inference-permit behind any legitimate "good reason" claim, – then this is absolutely true, and trivial. No one could deny this. Nonetheless, this tells us little about the concept of *showing the goodness of reasons* we actually have. It indicates nothing about what we *do* in meeting challenges against the goodness of our reasons. And this is as much a philosopher's responsibility as is describing what would be said if certain questions [rarely asked] *were* asked.

III

Consider a genuine challenge to: "April showers have always brought May flowers. So, that April is showery is a good reason for expecting May to be flowery". You question whether this is a *good* reason. You cannot want only to elicit from me the inference-permit "if April is showery, take it that May will be flowery". I just *have* taken it. Why force me to reformulate horizontally (as a premise) what I have just performed vertically (as an argument)? One *can* always re-phrase such claims. This is not in dispute. What you challenge though is whether this permit, and hence the inference itself, is allowable at all; whether the *facts* about April showers and May flowers really support such reasoning. Presumably, you are not demanding that I formulate a heretofore unexpressed inference-ticket, but are rather asking that I produce the factual cash-receipt with which I bought the ticket.

How could I satisfy you as to the goodness of my reasons in the present case? That I have lived through 35 showery Aprils and flowery Mays in several parts of the northern hemisphere would be relevant. So would national and international meteorological records. So would botanical data about the incidence of flowering in differing humidities. So would the general biological theory of plant metabolism. And so would the further fact that no piece of known evidence counts against a showery April being followed by a flowery May; that is, nothing climatologically unusual has occurred which would raise even a mild suspicion as to the expected sequence "April's showery – May flowery". If I cited all this, then if you are *compos mentis*, you ought to have been satisfied as to the goodness of my reasons for expecting May flowers. A questioner who, without being able to cite facts which makes this Spring different from all the others, yet remained unconvinced by such arguments, would either be an idiot or a philosopher, – or both. In either case, he would need therapy.

A philosopher's doubt might consist in denying that any finite recitation of facts could rationally justify an unrestricted use of the inference-permit "whenever April showers, infer May flowers". But what could be *more* rational than arguing "Since all observed F's have been C's, and since there is not only no known evidence for suspecting any F not to be a C, but also theories which link the very idea of being an F to the idea of being a C, – why then, whenever F infer that C"? This is not the universal categorical "all F's are C's" (extensionally interpreted), but the unrestricted permit "Since there are no grounds for doubting that F's are C's, the next time F obtains infer that C". It is not "April showers always bring May flowers", but "Since there are no grounds for doubting that April showers bring May flowers, the next time April is showery expect May to be flowery".

Such doubts mesh with the ancient question: Are universals categorical or hypothetical? one answer only being expected. But it all depends on *what* universals. That all, or 90% of, observed F's have been C's is categorical. But the inference-permit which this fact supports is hypothetical: "If F, then (there being no grounds for doing otherwise), infer that C". Or, "If F, infer with probability 0.9 that C". "All F's are C's" does not, by itself, disclose whether it is up for scrutiny *qua* inference-permit or *qua* the factual support for that permit. The "problem" of induction rests partly on this confusion, and partly on another connected with our apparently trans-type inference. Whenever anything like "all F's are C's" serves in an inference, it is often construed as an unrestricted categorical statement of fact. This must then, presumably, be justified by reference to some finite set of F's having been observed as C's. But this confuses the *permits* which structure inference, with the factual *support* for those permits.[2] To think that the inference from F to C is rational only

---

[2] In much of what follows I am indebted to Messrs. Toulmin, Garver, Maxwell and Feigl for stimulating discussion. Furthermore, Chapter III of Toulmin (1958) has been of particular help,

when "All F's are C's" is either an exhaustive description of all past and present F's, or else when a philosophical "justification" for inferring from a finite to an infinite set is appended, – this is to think that all permits, e.g., laws of nature, are never more than factual descriptions. This, plus the urge to view all inference as quasi-deductive, forces philosophers to "justify" moves from statements about facts to statements about arguments and possible arguments. Now although every permit and law may sometimes be construed as a description, indeed *must* sometimes be so construed, it is not necessary always so to construe them. When considering their roles in inferences it is often necessary *not* to construe them thus.

If I claim "All cases of showery Aprils are also cases of flowery Mays", you cannot know from this alone whether I am giving the history of all past Aprils or whether I am suggesting that, the facts being what they are, and there being no grounds for expecting otherwise, it may be inferred that Mays will be flowery when Aprils have been showery, – and this may be done until there is reason not to do so.

This distinction applies in complex statistical cases too:

Support = 84 % of all protons shot through C_{14} powder have been scattered through angles of less than 12°;

Permit = given 10,000 protons shot through C_{14}, infer that about 8400 of them will scatter through angles of less than 12°.

The second, not the first, is what figures in *modus ponens* reformulations of arguments attesting to the goodness of one's reasons. And when this presentation is effected, the permit is not a factual, categorical claim requiring further justification. What *would* require special justification is the suggestion that we ought not to *use* such a permit without tacking on philosophical appendages of the justificatory sort. *Why* not use the permit in a wholly unrestricted way, – the data being what they are? What philosophical impropriety would be involved in so doing? None, I submit.

Still, the reasonability of an inference may be established independently of any reference to the permit, – just by citing facts: "That April is showery is a good reason for expecting May to be flowery, *because* in every May following a wet April there has been flowering, dry Aprils have preceded only flowerless Mays, and because in fact nothing in experimental botany or theoretical biology gives any ground for expecting otherwise".

As against the "hidden premise" thesis, we often defend the reasonableness of our having inferred from facts to facts, not by formulating shadowy inference-permits, BUT BY CITING MORE FACTS. *Ex post facto* the validity of a parallel entailment can always be set out: *if F infer C; F,* – *so C*. Indeed, this *must* always be possible. But that in every case wherein we successfully defend F's being a good reason for C this submerged hypothetical *must* have been elicited, – this is factually false. Arguing further that when this hasn't been done, one's reasons for expecting C have not really been established as good ones, this is just philosophical legislation.

---

although Professor Toulmin may not altogether approve of the special uses to which I have put his arguments.

"Establishing the goodness of reasons" has its own logic. Delineating that logic consists in remarking that we often justify inferences from fact to fact by citing more facts; we show the good sense in inferring C from F not by formulating the submerged hypothetical "Whenever F infer C", but by citing such facts as that F_1 preceded C_1, F_2 preceded C_2, F_3 preceded C_3... Schematically:

PERMIT: Whenever F infer C

F has always been followed by C. So that F obtains is a good reason for expecting C.

$$\text{SUPPORT}: \begin{cases} F_1 & \text{was followed by} & C_1 \\ F_2 & \text{was followed by} & C_2 \\ F_3 & \text{was followed by} & C_3 \\ \dots \text{etc.} \end{cases}$$

And, more often than not, we justify such claims by citing the support, not the permit. Thus:

PERMIT: Whenever April is showery, infer that May will be flowery.

April showers have always brought May flowers. So that April is showery is a good reason for expecting May to be flowery.

$$\text{SUPPORT}: \begin{cases} \text{April 1958 was showery and May 1958 was flowery.} \\ \text{April 1957 was showery and May 1957 was flowery.} \\ \text{April 1956 was showery and May 1956 was flowery} \dots \\ \text{April 1888 was dry and May 1888 was flowerless.} \\ \text{Botanical experiments disclose that, without preparation in the form of} \\ \text{water intake, plants will not flower.} \\ \text{Biological theories of plant metabolism explain the necessary role of} \\ \quad \text{hydrogen oxide in the process of photosynthesis, such that in its} \\ \quad \text{absence this process cannot occur.} \\ \text{There is no known evidence which counts in favour of May's flowers} \\ \quad \text{being independent of April's showers.} \end{cases}$$

A man challenging the reasonability of this inference, could, when given the permit, retort "But I am asking *why?*", since the permit only restates horizontally the vertical argument which is being challenged. When given the support this retort is met, at least in most cases.

Now it might be charged that moving from such facts as that F's have always been followed by C's, to the claim that F's obtaining is a good reason for expecting

C, – that this is not an inference at all; not when one's only defence consists in citing more facts, namely the specific meteorological, botanical, and biological data which support the general claim that F has regularly preceded C. *Entailment* it may not be, granted. But *inference* it certainly is, as must be every case of drawing reasonable conclusions from evidence.

IV

Well, what about Hume's principle, and what about our ostensibly trans-type inference? We have seen that the *sentence*

'That F obtains is a good reason for expecting C'

can be used to express (at least) two quite different propositions. One of these, the one which held our attention initially, can be paraphrased thus:

'That $\mathcal{F}$ [i.e., $F_1 F_2 F_3... F_n$] obtains is a good reason for expecting C_n'

(where F_1 = all observed F's have always preceded C's; F_2 = experiments and theories connect the very idea of being an F with that of being followed by a C; F_3 = there is no known evidence leading anyone to believe that any F should not be followed by a C[3]; and F_n = we are here confronted with an F). That all this is a good reason for expecting C_n is necessarily true, – its negation is conceptually untenable. The onus is on him who denies this to provide an example wherein $\mathcal{F}$ obtains, yet fails to provide a good reason for expecting C_n. Our claim is that there is no such example, and cannot be one. But although this *sentence* figures in our April showers-May flowers inference, our original proposition does not. Paraphrasing that inference we have:

$F_1 F_2 F_3$. So that F_n obtains is a good reason for expecting C_n.

Whenever our first proposition is challenged, this is met by pointing out that such an array as $F_1 F_2 F_3... F_n$ is what is *meant* by having good reason for expecting C_n. But when what follows after the 'so' in our April showers-May flowers inference is challenged, two moves are open to the inferrer. One consists in formulating the inference-permit "Whenever F, infer C". This recasts the initial inference as a *modus ponendo ponens:* "Whenever F infer C; but F, – so C". This is formally valid. It could not but be true. And it validates the claim that F's obtaining is a good reason for expecting C.

---

[3] This is but one more, rather low grade, statement of fact. If "There *is* evidence that F's sometimes are not followed by C's" is merely factual, then so is "There is *no* known evidence that F's sometimes are not followed by C's". Nothing law-like is being smuggled in here. Since nuclear testing, meteoric dust, transits of Venus, and events of that ilk, have *not* in fact been spectacularly aberrant this year, then there is no evidence to support any expectation that this year will differ from any other *vis-à-vis* April showers and May flowers.

However, this "trans-substantiation" of "that F obtains is a good reason for expecting C" leaves Hume's principle in logical equilibrium. The left side of our original inference $(F_1\ F_2F_3...\ F_n)$ is completely swallowed up within the *modus ponens* restatement; it constitutes part of the hidden premise whenever $F_1\ F_2\ F_3\ ...\ F_n$ infer C. So, "That F_n obtains is a good reason for expecting C_n,", is only *apparently* inferred from "$F_1\ F_2\ F_3$ ", whenever one authenticates this by waving the inference-permit "whenever $F_1\ F_2\ F_3...\ F_n$ infer C_n". Hume's principle remains "horizontal" since here the factual regularities are converted into a premise of a valid argument.

Suppose, however, we consider "That F obtains is a good reason for C" merely as substantiated. Imagine, that is, that the claim is authenticated not by the inference-ticket "Whenever $F_1\ F_2\ F_3...\ F_n$ infer C, but by the factual cash which bought the ticket, "this F was followed by a C, that F by a C... F's have always been followed by C's, and nothing is known which leads us to expect otherwise.[4] Etc.".

Now the argument runs:

(1) this F, that F, those F's...*all* F's, have always been followed by C's (and there are no known grounds for expecting otherwise in any new case).
(2) F_n
(3) So F_n is a good reason for expecting C_n.

Whatever else may be said, this argument, $-$ (1) and (2), so (3), $-$ is not necessary, as a remarkable tortoise would readily have observed. Still, this *is* an inference. Its merits are settled by reflection on data we've got, not on expectations of new data we haven't got yet. And in this case "that F_n obtains is a good reason for expecting C_n is clearly contingent, $-$ contingent on $F_1\ F_2\ F_3...$

So, the April showers-May flowers inference is not a formal entailment, which is hardly news. One *might* deny that F is a good reason for expecting C, even while granting that F obtains, that F's have always been followed by C's, and that there are no known grounds for expecting otherwise in the present case. One *could* do this without contradiction. Nonetheless, so to deny F's credentials would not be rational. If, however, one can consistently deny F's goodness as a reason *despite* such factual support, then that F *is* a good reason for C can only be a factual claim; *the* claim that in some justificatory context F's status as a good reason for C was settled by appealing to further facts, $F_1\ F_2\ F_3$. Hence, it cannot be said that the inference, if true, could not but be true. That F is established as a good reason for C by reference to other facts $F_1\ F_2\ F_3$, $-$ this is itself a fact. *This* F need not have been a good reason for C. It just so happens that it is. Hume's principle is again left unmolested. When thus authenticated our inference is from facts to other facts (and other facts about facts). The inference is not across types.

Here then "That F obtains is a good reason for expecting C" is, if true, contingently true; it is not necessarily true as with our original value for the same sentence.

---

[4] Again, this is intended as but a low-order statement of fact, just like "There is nothing in the drawer", "There are no equations in the book", "There are no facts known which make tomorrow's sunrise doubtful".

When the goodness of a reason is authenticated not by a permit but by factual support, the result may be a statement which, although its credentials are settled solely by reflection on data we've got, is nonetheless consistently deniable. Only by supposing that the proposition expressed by "That F obtains is a good reason for expecting C" is *always* the kind of proposition which held our attention initially, can one be misled (as the author has been) into supposing that "April showers have always brought May flowers; so that it is a showery April is a good reason for expecting a flowery May" is a trans-type inference which throws Hume's principle out of logical equilibrium.

Still, we have taken a twist more tortuous than that taken by the tortoise. If we *do* meet challenges to establish a reason as a good one by citing further facts about, e.g., regularities in nature, then, while this may be inference it is not entailment, – nor can it be cooked into entailment.

Why try? Why not just admit that inferences from C's regularly having been followed by C's to "F_n is good reason for expecting C_n" are often completely justified by the factual observations *this F was followed by a C, that F was followed by a C, those F's were followed by C's,…, every experimental and theoretical consideration in fact leads us to expect F's to be followed by C's, and there is now no known evidence for expecting otherwise.* (Again, this last clause merely states a fact, on a par with those preceding.)

This *can* be converted into *modus ponens* by forcing the factual cash into the form: *whenever* $F_1F_2F_3… F_n$ *infer C.* But, as we've noted (1) this is rarely done, and (2) it is rarely necessary to do this in order to establish that F is a good reason for C. But whether or not the inference-permit *is* elicited, the move in question always constitutes an inference. When the permit is *not* elicited, however, the inference is not an entailment.

In this case the connective 'so' has unusual qualities. It remains possible to deny that F is a good reason for C, even while granting that F obtains, that F's have always been followed by C's, that every relevant theory seems to make "being followed by a C" an element within the very *idea* of "being an F", and that there are in fact no known grounds for expecting otherwise in the present case.

Yet, it would just be inconceivable that anyone should do this, namely, deny that F is a good reason for C. Were I asked to concede every fact relevant to precipitation in April and budding in May, and also to *deny* that April's being showery is a good reason for expecting May to be flowery, I would not know what to think. So, while it is logically possible that F is not in this case a good reason for C, it is conceptually untenable that it should not be.

If this *is* a non-entailment inference, then we may have here another assertion whose negation, while not demonstrably inconsistent, is nonetheless conceptually untenable. Other familiar candidates are:

No surface is everywhere red and green at t.
No two persons ever have the same pain.
No history-altering time-machine is possible.
Nothing counts as perpetual motion.
Nothing travels faster than light.

And now:

> When F's have always been followed by C's, when our theories relate the very concepts of "being an F" and "being followed by a C", when there are no known experimental or theoretical grounds for expecting otherwise, − then it is rational to infer a C whenever an F obtains.

The denial of this may not be logically inconsistent, i.e., Carnap's L-false. It is, however, inconceivably true, i.e., A-false.

If the assertion that *induction is a rational policy* is A-true, then it needs the same kind of justification as does any other A-truth; no less, − but no more. There is no *special* justificatory problem attaching to the claim that when $F_1 F_2 F_3$ obtain, it is rational to infer C_n when F_n obtains. This claim only poses afresh the general philosophical problem of showing what justifies any A-true-but-not-L-true proposition; any proposition, that is, whose negation, while not inconsistent, is nonetheless incredible.

To suppose there *is* some special justificatory problem about induction is to make a mistake which is the reciprocal of the one we almost made when, after considering our initial version of "*F* is a good reason for *C*", we supposed the April showers-May flowers inference to be trans-type. Although a justification is certainly needed when it is claimed that "That *F* obtains is a good reason for expecting *C*", − a justification often consisting in a recitation of further facts $F_1F_2F_3...$ − it is just an error to suppose that some analogous justification is required for "That $\mathscr{F}$ (i.e., $F_1F_2F_3...$ F_n) obtains is a good reason for expecting C_n", when $F_1 F_2 F_3... F_n$ constitutes all the relevant evidence known, plus the claim that this is all the relevant evidence known.

That "That *F* obtains is a good reason for expecting *C*" can express two logically different types of proposition, this is a logical fact which is as much missed by the man who thinks it ought always to be justified when it embodies the claim that induction is rational, as by the man who thinks it can figure in a trans-type inference.

It is A-true that F_n is a good reason for C_n when I can establish that F_n obtains, that F's have always been followed by C's, and that there is no known evidence in favour of any hypothesis to the contrary. I may not be able to show any inconsistency in the claim "*F* is *not* a good reason for *C*", even when backed as we have supposed. Because, the substantiation of the argument may consist only in reference to further facts. But even if I *do* recast the inference in *modus ponendo ponens* form, − if you ask me to justify *that*, I will do so by appealing directly to the facts that *F's* have always been followed by *C's*, and that there is no evidence known to warrant expecting anything different now. This *directly* substantiates the inference-permit "Whenever $F_1 F_2 F_3... F_n$ infer C_n", − not in that there is yet some higher principle which can reduce the negation of the inference-from-the-regularity-to-the-permit to inconsistency [this would require an infinite choir of ever-higher "trans-substantiations"], − but just in the sense that anyone who asks me to entertain both that

F's have always been followed by C's, that our theories both describe and explain this regular succession, and that no known datum supports any expectation to the contrary in any future case, – and also that F_n actually obtains,

AND ALSO

It is other-than-rational to infer C_n,

has given me nothing to entertain at all.

This does not reduce to the unilluminating legislation: "The inductive policy is just what we *mean* (or ought to mean) by proceeding rationally, – hence it's a jolly good way to proceed". The claim is rather that there is no thinkable alternative to so proceeding. The very idea of "justifying" what has no intelligible alternative[5], this idea survives only by successively posing as different problems.

Thus there *is* a problem of induction. But it is not the traditional problem philosophers have supposed it to be. It is the problem of showing how "The Inductive Principle" – in common with a large class of other philosophically perplexing statements – can at once be synthetic (in that its negation is not demonstrably self-contradictory), yet necessary (in that its negation describes no intelligible state of affairs at all).

Reference

Toulmin, Stephen. 1958. *The uses of argument*. Cambridge: Cambridge University Press.

---

[5] By analogy with our procedure when we are obliged to justify *which* of *several* intelligible alternatives ought to be adopted.

Chapter 17
A Budget of Cross-Type Inferences, or Invention Is the Mother of Necessity

17.1 Introduction

From what is contingent nothing necessary follows. And from what is necessary nothing contingent follows. Let us call this 'the Hume-Leibniz dictum'. These theses never occur full-blown in Hume or Leibniz, but the dictum is implicit within the philosophies history has come to associate with these thinkers. Thus Hume writes:

> Relations of ideas [are] either intuitively or demonstratively certain... [however,] matters of fact... are not ascertained in the same manner;... The contrary of every matter of fact is still possible; because *it can never imply a contradiction...* ([1748] 1999, 4.1–2, my italics)

And Leibniz says:

> There are also two kinds of truths, those of reasoning and those of fact. Truths of reasoning are necessary and their opposite is impossible; truths of fact are contingent and their opposite is possible. (1714, §33; 1951, 539)[1]

Stressing these differences in type between necessary and contingent propositions is proceeding toward a rejection, by Hume and by Leibniz, of the possibility of any genuine, nondegenerate, cross-type inference. For, if *P* states a possible matter of fact, so does ~*P*. But then neither *P* nor ~*P* implies what is logically false. This much cross-typicality is clearly barred by Hume.

What about inferences from contingencies to what is logically true, however? On this, Leibniz says:

I am indebted to Richard Smyth, Roger Buck, and Grover Maxwell for stimulation and criticism during the writing of this paper.

[1] Hanson quotes the translation given by Wiener, but deletes the (inconsistent) italicizing. – *MDL*

> According to the usage of logicians *the conclusion follows the weakest of the premises[conclusio sequitur partem debiliorem]* and cannot have more certainty than they. ([1704] 1916, 515; IV, xi, 14)

Hence, if Q follows from a set of premises of which one *(P)* is contingent, then Q cannot have more certainty than P. Therefore, anything entailed by P must also be contingent – so suggests Leibniz.

So Hume and Leibniz appear to show sympathy for the contention that from what is contingent nothing necessary follows.

Concerning the second half of the dictum, Hume and Leibniz are again far from explicit. Nonetheless, let Q be a "truth of fact"; then its "opposite is possible"; that is $\Diamond\sim Q$. Let P be a "truth of reason"; then its "opposite is impossible"; that is, $\sim \Diamond \sim P$. Suppose now that this P entails this Q; that is, $(\sim \Diamond \sim P) \dashv (\Diamond\sim Q)$. This runs counter to a theorem in Lewis and Langford (1959, 164, eq. 18.53)[2], to wit:

$$\left[(P \prec Q)\cdot(\sim \Diamond \sim P)\right] \prec (\sim \Diamond \sim Q)$$

Hume and Leibniz probably would have accepted this principle. Indeed, Hume's claim that "the contrary of [a] matter of fact… can never imply a contradiction" transforms (by contraposition) into the claim that "from what is necessary nothing contingent follows". But strict alignment with our texts invites us to call our dictum 'the Hume-Leibniz-Lewis-Langford dictum' (HLLL), which, to express it in yet different terms, will read:

> Any claim whose negation is consistent can have followed only from premises at least one of which has a consistent negation. And any claim entailed by premises whose negations are inconsistent must itself have an inconsistent negation.

In what philosophically significant sense a claim with an inconsistent negation can have followed from premises with consistent negations, this it will be our task to discover. Incidentally, Lewis and Langford, although denying that necessary propositions entail contingent propositions, consistently affirm that the former are *entailed by* the latter. This must be dealt with in the sequel.

Although never defended in full generality, this dictum has been valuable to modern philosophy. Qualitatively put, it states that no mere inspection of facts will secure certainty for a contingent conclusion. No amount of formal rigor will secure the factual truth of a conclusion unless contingent (and hence vulnerable) statements are fed into the premises. This logical blade has axed forests of wooden metaphysics to the ground. For, when it is seen that experience is insufficient to establish capital-lettered, necessary claims about the Cosmos – and when armchair reasoning is seen to be insufficient for determining what will be found in the world – much of traditional metaphysics collapses. Thus, the HLLL dictum cleared a new view of the nature of philosophical inquiry.

---

[2] Hanson uses a different logical notation from Lewis and Langford throughout this article. – *MDL*

Notwithstanding its methodological utility, however, the dictum's exposition must be sharpened. This is because we regularly encounter valid entailments that appear to move across types. If, in fact, they *do* so move, then Hume and Leibniz and Lewis and Langford must be put to the question. Moreover, no general condemnation of speculative metaphysics as consisting of cross-type inferences will have any validity unless the apparent cross-typicalities of logic and philosophy are reappraised. Of course, our venture here may fail entirely. After the following analyses it may still be necessary to concede that nondegenerate cross-type inferences do exist; if speculative metaphysics is suspect, it may have to be for reasons different from those set out here. But then the failure may be remarked in detail by others, and a rarely discussed perplexity will at least have been canvassed.

It is my objective to set out a "budget" of these putative cross-type inferences. Section 17.2 will delineate how from contingent propositions necessary propositions can apparently be inferred. After that it will be indicated how from necessary propositions contingent propositions can apparently be inferred (Sect. 17.3). Then an attempt will be made to resolve the perplexities proposed within each of these inferences (Sect. 17.4). Finally, a restatement of the HLLL dictum will be sought, such that it does not encroach upon certain valid inference-schemata (Sect. 17.5).

Before plunging into our budget, one or two cautions are in order. One might resist the HLLL dictum not because it is open to counter-instances, but because it is ineffective. P is necessary if its negation leads to contradictions; that this is so depends on having a technique for detecting inconsistencies within a potential infinitude of other consequences. But it is never ~P *simpliciter* from which we generate inconsistency; rather, it is ~P in conjunction with other unstated assumptions (e.g., rules of deduction; axioms – e.g., of reducibility, of infinity; principles – e.g., of mathematical induction, etc.). *If* contradiction follows from operations on ~P, then either (1) ~P is self-contradictory, *or* (2) the unstated assumptions contain contradictions, *or* (3) the "operator" is a poor logician. In principle, these other alternatives, (2) and (3), are never eliminable. But in practice they are. "Jones is a married bachelor" is a paradigm of what counts as a contradiction within a language – however true it is that this verdict involves contingent assumptions about the utility and permanence of certain explicit definitions and linguistic conventions within that language.

Our concern here is not with *establishing that* a given, particular claim is inconsistent. Difficulties such as those elaborated by Brouwer and Quine have bulked large within a literature quite different from what follows. Our present concern is with *what follows from* assuming a given claim to be inconsistent. Doubtless this assumption makes sense. Even if not one actual claim within a language L were ever demonstrated inconsistent in L, it would still make sense to elaborate the characteristics of a proposition *assumed* to be incontestably inconsistent within L. We can never demonstrate the existence of an ideal gas, but we know its properties intimately. So the esoteric difficulties which trouble finitistic logicians and which appear within recent discussions of analyticity will not deter us.

A final caution: Our guiding principle is to the effect that cross-type inferences are somehow semantically untenable. For, assume P to be an incontestably necessary claim (in that, within L, ~*P has already been discovered* to generate something of the form $F \bullet \sim F$). And assume that Q is incontestably contingent (in that, within L, ~Q generates nothing of the form $F \bullet \sim F$ – nor does Q either, of course); indeed, suppose that ~$Q \rightarrow (F \bullet \sim F)$ and $Q \rightarrow (F \bullet \sim F)$ themselves have been discovered to be inconsistent: nothing could better establish the consistency of Q, and of ~ Q, and hence the contingency of both. Then assume that P entails Q. But then,

$$(P \prec Q) \equiv (\sim Q \prec \sim P)$$

This is semantically absurd, since it states that an inference from what is necessary to what is contingent *is logically equivalent to* an inference from what is contingent to what is impossible – *reductio ad absurdum.* I should prefer to describe this as showing that, should an L-determinate P be taken to entail a contingent Q, this would have to be logically equivalent to a contingent ~Q entailing an L-determinate ~P – *reductio ad absurdum.* Two inferences could not possibly be more different. *Inferring from* what is contingent must be profoundly different from *inferring from* what is L-determinate. So, cross-type inferences in general are untenable, and the specific assumption that one might infer from a necessary to a contingent claim ends up in a contradiction. How, then, can we account for the candidates for inferential cross-typicality that follow?

17.2   Inferences from Contingency to Necessity

(1) "If it is contingently true that P and contingently true that Q, then the claim that P is contingently true and the claim that Q is contingently true are consistent claims." (Cf. Lewis and Langford 1959, 154, eq. 17.1.)[3]

The claim that $P \bullet Q$ is contingently true, this is itself contingent. $P \bullet Q$ need not have been true. But that the contingent truth of P is *consistent* with the contingent truth of Q, if this is true at all, it could not but have been true. Or else the claim is logically false. Settling whether the contingent truth of P is consistent with the contingent truth of Q, itself requires no contingent inquiry, only reflection. Whether the

---

[3] Comparisons with Lewis and Langford serve only as references, not as justifications, for our examples. 17.1 reads: "$(P \bullet Q) \rightarrow (P \circ Q)$." The example above is cast (at least) one metalanguage above this: $(P_{ct} \bullet Q_{ct}) \rightarrow (P_{ct} \circ Q_{ct})$. *This* reads: "That P is contingently true and that Q is contingently true jointly entail that the contingent truth of P is compatible with Q's being contingently true." References to Lewis and Langford are thus of analogical interest merely. A different notation in our example could have brought 17.1 directly into the argument: for instance, let

$R = P$ is contingently true
$S = Q$ is contingently true

Then example 1 would read: "If R and S obtain, then R and S are compatible" – which is a valid substitution instance of Lewis and Langford 17.1. The subscript notation set out in this footnote will be developed, however; it has advantages that outweigh its unfamiliarity.

contingent truth of $P \cdot Q$ entails anything of the form $F \cdot \sim F$ *is* a formal question. Yet we *do* justifiably infer the latter (i.e., the mutual consistency) from the former (i.e., the joint contingent truth). From observing two phenomena, we do infer that their descriptions are consistent. And from noting that the descriptions of those phenomena are contingently true, a higher-level consistency would also seem to follow. This higher-level consistency is what concerns us here, since, *prima facie*, our example constitutes an inference from what is contingent to what is necessary.[4]

(2) "That P is contingently true entails the claim that contingent P is logically possible." (Cf. Lewis and Langford 1959, 164, eq. 18.4.)

That P is logically possible just *is* the claim that P is self-consistent. This, again, is a formal claim – the claim that P entails nothing of the form $F \cdot \sim F$. This is settled by reflection alone. However, that P contingently obtains is itself a contingent claim. It makes sense to assert that $\sim P$ obtains. Which of the two ("P contingently obtains", or "$\sim P$ contingently obtains") states fact, requires contingent inquiry to settle. Still, inferring from "P is contingently true" to "P is logically possible" seems valid. Hence, another putative counter-instance to the HLLL dictum.

(3) "That $\sim P$ is contingently true entails the claim that $\sim P$ is logically possible." (Cf. Lewis and Langford 1959, 164, eq. 18.44.)

The analysis appropriate to example 2 (above) applies *mutatis mutandis* in this case.

(4) "If it is true both that P is contingently true *and* that Q is necessarily true, we can, from this conjunction, infer that Q is necessarily true." (Cf. Lewis and Langford 1959, 125, eq. 11.2.)

Again, that P is contingently true is itself a contingent claim. But that Q is necessarily true – this assertion is either necessarily true or necessarily false. It is certainly L-determinate. Its credentials are settled solely by reflection. Our example conjoins this contingent and this necessary claim; from this conjunction it follows that Q is necessarily true. Hence, a necessary conclusion from a contingent conjunction.

Nothing in classical or modern logic forbids conjoining necessary and contingent claims. Indeed, textbook discussions of the Square of Opposition usually

---

[4]This example may still contain an ambiguity that will affect future analyses. Two different things *might* have been intended by our example above:

(a) If $P \cdot Q$ is contingently true, then $P_{ct} \circ Q_{cb}$,

or

(b) Where P, Q are contingent, $P \cdot Q$ implies $P \circ Q$.

In (a), we infer that P' s contingent truth is consistent with the contingent truth of Q *from* "$P \cdot Q$ is contingently true".

In (b), we infer that P is consistent with Q *from* "$P \cdot Q$" – it being understood in this second case that both P and Q are contingent. Interpretation (a) is the one we use throughout.

include sample statements such as "All bachelors are male, *and* there are in fact bachelors"; that is, $(x)(B_x \supset M_x) \cdot (\exists x)(B_x)$. This can only be contingently true. It cannot be established by reflection alone; some appeal to the facts that support $(\exists x)$ (B_x) is required to establish the truth of the whole conjunction. Still, by detachment, from a conjunction one can infer to any constituent conjunct. Hence, from "All bachelors are male *and* there are bachelors" (contingently true) one can infer to "All bachelors are male" (necessarily true). The conclusion is incontestably necessary. Yet that from which it is inferred is contingent. Moreover, the inference is legitimate.

Note that, despite the ink poured over the Square of Opposition, no philosopher has noted that the *modified* A-form demanded by contemporary logicians (e.g., "All bachelors are male, *and* there are bachelors") sometimes conjoins a necessary with a contingent claim. This sanctions a cross-type inference. The ultimate consequences of this could conceivably prove more uncomfortable than ever did any existentially important inference from an *un*modified A-form (e.g., "All bachelors are male, so there exists a male bachelor").

Consider also the efforts of J. S. Mill to "justify" mathematics *via* experience, an enterprise far from dead even now. Suppose all mathematical propositions that are indisputably necessary to be compressed within the single (internally conjunctive) proposition Q. Suppose all of physical theory (θ) then to be represented as a "conjunction" of contingent elements P and the necessary elements Q just referred to. Then, from the contingent truth of θ (i.e., contingent-P-and-necessary-Q,) one could legitimately infer to Q. Attacks on Mill's argument might be weakened somewhat if the inference before us is valid. Mill's case would appear to be more than not obviously unsound; it seems to be sanctionable in a well-established form of inference.

(5) A degenerate analogon of this inference is produced thus: "P is contingently true; *so P is contingently true and Q* is necessarily true." Thus, "That there are bachelors is contingently true; *so* that there are bachelors is contingently true *and* that all bachelors are male is necessarily true."

A necessity can be conjoined with any contingent truth, the result being a contingent truth. Since one can infer P from P, one can also infer from P to P and (necessary) Q. But, when P implies a conjunction, it also implies either conjunct separately; i.e., if $P \to (R \cdot S)$, then $(P \to R) \cdot (P \to S)$. And, if $P \to (P \cdot Q)$, then $(P \to P) \cdot (P \to Q)$. Via this argument, then, contingent P implies necessary Q. Again, this is cross-type.

17.3   Inferences from Necessity to Contingency

(1) Consider the following abbreviations:

F_1:   At present B's are regularly followed by C's.
F_2:   In the past B's have always been followed by C's.
F_3:   All extant theories do, in fact, relate the concepts "being a B" and "being followed by a C".

(Here, this last constitutes no more than a contingent, factual reference to what presently accepted theories *do actually do*. This merely describes a property all theories concerning *B's* and *C's* now have. No "inference permit" is being smuggled in.)

F_4:   There are no known experimental or theoretical grounds for expecting anything other than a *C* to follow a *B*.

Entertain now the following inference:

(a) "That F_1, F_2, F_3, F_4 obtain is good inductive reason for expecting *C*."

    entails

(b) "That F_4 obtains is *a* good inductive reason for expecting *C*."

The inference from (a) to (b) appears to be cross-type; a necessary proposition entailing a contingent one. When F_1 to F_4 obtain, it would be *irrational* not to infer to *C* after witnessing another *B*. F_1 to F_4 obtaining constitutes the *meaning* of "having good inductive reason for expecting *C* whenever a *B* occurs." To understand what F_1, F_2, F_3, and F_4 mean and to understand what (a) asserts just *is* to realize that (a) could not but be true. Its semantic content guarantees its truth.

But (b) makes a contingent claim. For example, F_2's obtaining is *a* good inductive reason for expecting *C*, *contingent upon* F_1, F_3, and F_4 actually obtaining. To expect *C* because of F_2 alone would be like expecting one's brakes to hold just because they had held during their last 10,000 applications. (From only this we are more justified in expecting the brakes *not* to hold.) Only when further conditions $(F_1, F_3,$ and $F_4)$ are met will F_2 in this context constitute a good reason for expecting *C*.

Nonetheless, "That F_1, F_2, F_3, F_4 obtain is good inductive reason for expecting *C*" surely *entails* "That F_2 obtains is a good inductive reason for expecting *C*"? Thus our inference apparently runs across types: from what is necessary to what is contingent.

(2) Our final example requires some technical apparatus. "If contingent *P* entails contingent *Q*, then in fact either *Q* obtains or *P* does not." That is, if contingent *P strictly implies* contingent *Q*, then *P materially implies Q*. (Cf. Lewis and Langford 1959, 137, eq. 14.1.) In symbols:

$$\left[\left(P_{ct} \prec Q_{ct}\right) \prec \left(P_{ct} \supset Q_{ct}\right)\right]^{n}$$

As before, the subscript '*c*' signifies "contingent"; '*t*' signifies "true"; '*n*' will signify "necessary". Note again that these denote *actual components within* our examples of inference, not merely metalinguistic comments on the inferences; the corresponding superscripts on the other hand will serve as such comments. 'P_{ct}' thus stands for "the proposition *P* is contingently true"; 'Q_{nt}' would stand for "the proposition *Q* is necessarily true".

The formula above then reads:

That *P* is contingently true entails that *Q* is contingently true; therefore, if *P* is contingently true, *Q* is also contingently true.

(The final superscript 'n' is thus an aside to the effect that the entire inference is necessarily valid.)

Now let P and Q take contingent values; for instance, let P = "Jones is a bachelor", and Q = "Jones is male". Then $P_{ct} \rightarrow C_{ct}$ reads: "That *Jones is a bachelor* is contingently true entails that *Jones is male* is contingently true" – and all this is necessarily true.

The material implication, however, would seem to require a different analysis. All that '$P_{ct} \supset Q_{ct}$' means is that *in fact* $\sim P_{ct} \vee Q_{ct}$. (Cf. Whitehead and Russell 1950, 1:94, eq. $_*1.01$.) Now $\sim P_{ct} \vee Q_{ct}$ is clearly contingent. A disjunction of contingent claims is itself contingent (with the obvious exceptions of $P \vee \sim P$ and its derivatives, $P \vee \sim P \cdot Q$, $P \vee \sim P \cdot Q \cdot R$, etc.). All we are being informed of by '$P \supset Q$' is that either Jones is not a bachelor or Jones is male. *In fact*, one or the other (or both) of these contingencies obtain. We are *not* here being modally constrained, as we would be were we informed that it *must* be the case that $\sim P \vee Q$ – that any Jones *must* either be male or else must be other than a bachelor. Of the possibility of some Jones being neither male nor other than a bachelor (e.g., a female bachelor) we are told only that this never in fact happens – not that it could not happen.

This sanctions the following inference:

$$\left[\left(P_{ct} \prec Q_{ct} \right)^n \prec \left(\sim P_{ct} \vee Q_{ct} \right)^c \right]^n$$

And this echoes example 2 above; "If the contingent truth of P entails the contingent truth of Q, then *in fact* either it is contingently true that Q or it is contingently true that $\sim P$". Since it is necessary that every bachelor be male, then in fact every person consulted during the recent census was either male or not a bachelor (or both). Again, that every consulted person *must* be either male or not a bachelor goes beyond what '$P \supset Q$' actually claims; '$P \rightarrow Q$' would be required to convey the idea that there could be no exception. So the claim '$\sim P \vee Q$' must even be considered provisionally to have a consistent negation.[5]

Lewis and Langford write (1959, 137) that "whenever a strict implication can be asserted, the corresponding material implication can also be asserted. The converse

---

[5] This conclusion will seem inconsistent with Lewis and Langford, 18.7: $(P \rightarrow Q) = [\sim \Diamond \sim (P \supset Q)]$. No simple and satisfying resolution of this tension occurs to me. Since 18.7 entails $(P \rightarrow Q) \rightarrow (P \supset Q)$ and since this is equivalent to $(P \rightarrow Q) \rightarrow (\sim P \vee Q)$, the analysis we have tendered for the latter still seems to hold. Since $\sim P_{ct}$ and Q_{ct} are both contingent, their disjunction must also be contingent, at least in every case where $Q \neq P$.

In general, it would appear that in an inference such as $[(P_{ct} \rightarrow Q_{ct})^n \rightarrow (\sim P_{ct} \vee Q_{ct})^c]^n]$ – e.g., "Since every bachelor cannot but be male, then in fact every person consulted during the current census will either be male or not a bachelor" – we are moving from an intensional analysis of a term to its extensional analysis. The *vel* in the apodosis invites this interpretation. 18.7 in Lewis and Langford encourages one to overlook this shift, but, inasmuch as a very natural interpretation of $\sim P_{ct} \vee Q_{ct}$ is extensional, a very natural and wholly legitimate way of interpreting the inference before us would be as cross-type.

does not hold; ... the assertion of a strict implication is a stronger statement than the assertion of the corresponding material implication." The justification of this contention thus rests on assuming the validity of the cross-type inference just explored.

Any two randomly chosen propositions are such that either one materially implies the other or vice versa. Not so with entailment. Randomly chosen propositions *might* be logically independent. There must be a *deductive* connection between two assertions before they can conspire in an entailment. "P entails Q" means "It is impossible that P should obtain but not Q" "P materially implies Q", however, means only "It never *in fact* happens that P obtains when ~Q also obtains". Thus, it never happens that bachelors weigh more than two tons. Hence, being a bachelor materially implies weighing less than two tons. But being a bachelor does not *entail* weighing less than two tons; "weighing less than two tons" is no part of the semantic content of "being a bachelor". One cannot deduce that Jones weighs less than two tons simply from his being a bachelor.

Thus, if Q necessarily follows from P (that is, $P \rightarrow Q$), then, in fact, you will never discover P in the absence of Q (that is, $P \supset Q$).

That one can infer from a deductive claim to a factual claim is at the heart of the system of strict implication.

All of these putative cross-type inferences can be set out schematically as follows:

THE SCHEMATA OF CROSS-TYPE INFERENCES[6]

Contingent $\rightarrow$ Necessary

(1) $[(P_{ct} \bullet Q_{ct})^c \dashv (P_{ct} \circ Q_{ct})^n]^n$

(2) $[(P_{ct})^c \dashv (\Diamond P_c)^n]^n$

(3) $[(\sim P_{ct})^c \dashv (\Diamond \sim P_c)^n]^n$

(4) $[(P_{ct} \bullet Q_{ct})^c \dashv (Q_{nt})^n]^n$

(5) $[(P_{ct})^c \dashv (Q_{nt})^n]^n$

because $P_{ct} \dashv (P_{ct} \bullet Q_{nt})$

Necessary $\rightarrow$ Contingent

(1) $[(F_1F_2F_3F_4 \rightarrow Rc)^n \dashv (F_4 \rightarrow Rc)^c]^n$

that is, "$\rightarrow R$" = "... is a good inductive reason for c..."

(2) $[(P_c \rightarrow Q_c)^n \dashv (P_c \supset Q_c)^c]^n$

---

[6]Again, remember that superscripts serve merely as comments upon the propositions preceding them. They are never part of the inference, but (as it were) logician's asides characterizing the previously stated inferences.

$(P_{ct} \vee \sim P_{ct})^n$ thus reads: "Either P is contingently true, or it is not the case that P is contingently true"; and the superscript 'n' marks this disjunction as necessary – although this characterization is itself no part of what the disjunction under consideration asserts.

17.4   On Resolving Apparently Cross-Type Inferences

Our budget of cross-type inferences now spreads before us. Should they go unchallenged, they could conceivably bolster mystery-mongering metaphysics – and blunt the instruments of analysis. More important, they might even perplex us concerning the very nature of deduction.

Let us reconsider each inference in turn. It may be possible to draw the intellectual sting from them. Perhaps we can relieve the conceptual pain caused by these metaphysical progeny of the Ontological Argument, that most notorious of all cross-type inferences.[7]

(1) "If it is contingently true that P and contingently true that Q, then the claim that P is contingently true and the claim that Q is contingently true are consistent."

The guarantee of the validity of this inference is not logical, perhaps not even conceptual, in character. The guarantee is "ontological", although this is not to be misunderstood. We are being told that *the world* is such that a contingent proposition (P_{ct}) and its negation $(\sim P_{ct})$ cannot both be true at once. Or that Q (in Q_{ct} above) cannot take as its value $\sim P$. This is a reasonable claim. But I am not convinced that its guarantee is purely logical. The implication, that is, seems not to be strict. Perhaps the connection is at most one of material implication: *in fact*, there is no as-yet-discovered case where a contingent proposition (P_{ct}) and a contingent proposition (Q_{ct}) are simultaneously true, but where the claim that P_{ct} and the claim that Q_{ct} fail to be consistent. But, could one *deduce* their consistency just from the fact that P_{ct} and Q_{ct} could simultaneously obtain? What is the nature of the *reason* for denying this, as most of us would? Why *not* suppose that the world could contain "contradictory" states of affairs? Why *not* suppose that an otherwise well-formed language could conceivably contain contingent descriptions of the form "P_{ct} and $\sim P_{ct}$"?[8] There *are* good reasons why we should not suppose this. But might it not be argued that these reasons go beyond any narrow theory of deduction? Such reasons are usually systematic, broadly conceptual, or even "metaphysical" in nature. We cannot, I submit, *establish* that nature does not at once embrace inconsistent facts simply by explicit definitions, linguistic conventions, and *modus ponendo ponens*. Presumably, it might be just barely possible, and barely intelligible, to dispute this issue. Hegel may be a case in point; his reaction to inconsistency was apparently distinguishable from that of most philosophers. This is not intended as a convincing reference. The onus must, of course, rest on the Hegelian to make *his* position intelligible. But, rather than simply lay down the issue against the possibility of his

---

[7] "G necessarily has the properties α, β, γ, δ, and ε (for 'existence'); therefore *in fact* G has the property ε."

[8] "*The Mona Lisa is smiling* is contingently true, and *it is not the case that the Mona Lisa is smiling* is contingently true" might be a *prima facie* candidate for this status. Special semantical questions arise here which it is not our purpose to explore. But the sentential form of such an utterance is at least similar enough to the possibility mooted above to warrant some patience with what follows.

doing so, I should prefer to allow claims as to what is or what is not ontologically possible to rest ultimately upon considerations more "philosophical" than any narrow appeal to what deductive theory can or cannot sanction.

So, in fact, concurrently true contingent propositions may never yet have been discovered to be inconsistent. But, so far as I can see, this no more allows one to *deduce* their consistency from their truth than would the historical fact that no bachelor has ever exceeded two tons allow us to deduce that Jones weighs less than two tons just from his being a bachelor. Jones's sex *can* be deduced from his being a bachelor. That he weighs less than two tons can be inferred from his being a bachelor. But one must distinguish deductive inferences like "that Jones is a bachelor entails that he is male" from inferences of a different kind, e.g., "that P and Q are each contingently true 'entails' that the contingent truth of P and the contingent truth of Q are consistent". This latter seems to me not to constitute genuine entailment. Unless special and arbitrary semantic conventions are adopted, can it really be represented as any more than a material implication? It seems to be no stronger an inference than that from lightning to thunder – and no weaker than that from sunspots to a wheat failure.

Now, my suggestion that P_{ct} and Q_{ct} may only *materially* imply the consistency of P_{ct} and Q_{ct} is open to the following counter-move:

If $[(P_{ct} \cdot Q_{ct})^c \supset (P_{ct} \circ Q_{ct})^n]^n$ is taken as a premise and $(P_{ct} \cdot Q_{ct})^c$ as another premise, then these two entail $P_{ct} \circ Q_{ct}$, which is necessary. But both premises are contingent; $P_{ct} \cdot Q_{ct}$ is certainly contingent. And $(P_{ct} \cdot Q_{ct})^c \supset (P_{ct} \circ Q_{ct})^n$ is no more than $\sim(P_{ct} \cdot Q_{ct})^c \vee (P_{ct} \circ Q_{ct})^n$ – which, in turn, is equivalent to $\sim[(P_{ct} \cdot Q_{ct})^c \cdot \sim(P_{ct} \circ Q_{ct})^n]$. Since an internally consistent conjunction containing a contingency must itself be contingent, it follows that both premises are contingent. Yet the conclusion, $P_{ct} \circ Q_{ct}$, is necessary (either necessarily true or necessarily false; i.e., it is L-determinate). Hence, we have here yet another inference that looks cross-type, namely:

$$\left\{ (P_{ct} \cdot Q_{ct})^c \cdot \left[(P_{ct} \cdot Q_{ct}) \supset (P_{ct} \cdot Q_{ct}) \right]^c \right\}^c \prec (P_{ct} \cdot Q_{ct})^n$$

Precisely what is involved in designating as "contingent" the material implication $(P_{ct} \cdot Q_{ct}) \supset (P_{ct} \circ Q_{ct})$, that is, $\sim[(P_{ct} \cdot Q_{ct}) \cdot \sim(P_{ct} \circ Q_{ct})]$, will be explored below in example 4. This is but a degenerate sense of contingency. Its effect on the general philosophical utility of the HLLL dictum is negligible, as we hope to disclose.

(2) "That P is contingently true entails the claim that contingent P is logically possible."

In another place[9] the perplexities of this apparently cross-type inference are resolved somewhat as follows. The claim that P_{ct} obtains is a contingent claim. The claim that P_{ct} need not have obtained; it could have been false. However, that P is *contingent* is not a contingent claim. If true at all, it could not but have been true. For this just *is* the claim that the negation of P (within some language L) is consistent.

---

[9] Cf. Hanson, 'It's Actual, So It's Possible' (1959); reprinted in this volume.

This we settle by reflection, not observation. Such is the teaching of Hume and Leibniz.

~P will be consistent whether P is contingently true or contingently false. So the assertion "P is contingently true" is composite. It consists of (1) a necessary component, viz., "P is contingent (true or false); that is, P and ~P are consistent"; and further, of (2) a contingent component, viz., "contingent P is in fact true." That P *is logically possible* is inferred only from the necessary component (1). That P is logically possible just *is* the claim that P is self-consistent; i.e., entails nothing of the form $F \cdot $ ~F. (Cf. Lewis and Langford 1959, 159, eq. 18.1)

Hence, that P is logically possible is the claim that P is self-consistent. And that P is contingent (true or false) is the claim that P and ~P are each self-consistent. The inference is thus only apparently cross-type. Its logical bones look like this: "P is self-consistent, therefore P is self-consistent." This is not problematic.

Even thus boiled down, however, our example does not constitute an equivalence. The conclusion, "P is logically possible", follows not only from P's being contingently true; it follows also from the claim that P is necessarily true. Hence, from "P is logically possible" one cannot infer that P is contingently true; P might be necessary. But from P's being contingently true (or false) it does follow that P is logically possible, i.e., self-consistent.

Remarking the composite character of the claim "P is contingently true" does not settle all present pretensions to cross-typicality. Let it be agreed that "P is contingent" is a necessary claim (true or false), and that "contingent P is true" is contingent (true or false). Still, there remains a sense in which the conclusion, viz., "P is logically possible", can be inferred from the conjunction of the first two, that is, "P_c is contingent *and* P_c is true". This entire conjunction is itself contingent; its truth cannot be determined by reflection. Nonetheless, that P is logically possible follows, by detachment, from the conjunction itself – or at least from one of the conjuncts. Thus (P is contingent *and* P is true)^c $\dashv$ (P is logically possible)^n. But this has now become an instance of a general case, to be examined below under example 4.

It might be argued that, although we have resolved these first two examples in different ways, they ought to have had the same analysis. Concerning example 1, we suggested that, at most, it constituted a material implication, not an entailment. Example 2 was felt to comprise a composite premise – from one component of which the conclusion followed necessarily. However, in (*ibid.* 162, eq. 18.3), Lewis and Langford set out $(P \circ Q) = \Diamond (P \cdot Q)$; this was Lewis's definition of consistency in his *Survey of Symbolic Logic* (1918). Since example 2 has the form $P \dashv \Diamond P$, substitution will give us $(P \cdot Q) \dashv \Diamond (P \cdot Q)$, which, by 18.3, gives us example 1, $(P \cdot Q) \dashv \Diamond (P \circ Q)$.

Hence, 1 and 2 are really the same example. Either of our two resolutions should apply to each case. That is, it might be said that both, as they stand, constitute at most a material implication; *in fact*, the protases have never been found to hold where the apodoses have not also held. Or, if it is insisted that these are genuine entailments (as opposed to being mere material implications), the respective protases must be construed as being composite. Just as the claim "P is contingently true" is composite, so is the claim "P and Q are contingently true". These assert (a) contingency, and (b) truth; and (a) and (b) are not the same. In both examples the apodosis follows *only* from (a) the contingency (respectively) of P and Q (in example 1) and of P (in example 2). To say of a proposition that it is contingent is to make a *necessary* claim about it (i.e., either necessarily true or necessarily false). Therefore, if we insist that these two examples constitute genuine entailments, then the apodoses can follow only from this necessary component (a) in the protases. $P \circ Q$ is the same as $\Diamond P \bullet Q$. And, in our examples, this follows directly from the claim that $P \bullet Q$ is contingent, that is, from $\Diamond(P \bullet Q) \bullet \Diamond \sim(P \bullet Q)$. So, in examples 1 and 2, if the protases are *not* composite, the inference can be no stronger than material implication. But, if the inference *is* stronger, i.e., is an entailment, then the protases must be composite.

(3) "That $\sim P$ is contingently true entails the claim that it is logically possible that $\sim P$."

All the considerations just explored in 2 above apply in this case.

(4) "If it is true both that P_{ct} and that Q_{nt}, we can from this conjunction infer that Q_{nt}."

To repeat: The conjunctive premise is contingent. Its truth cannot be assessed by reflection alone. The inference is valid; it embodies only the rule of detachment. How is this to be squared with the HLLL dictum? The consequences of leaving it unchallenged could be serious. For, if it stands, no *general* philosophical objection to inferring necessary conclusions from contingent premises can be tenable; no *general* attack on a kind of argumentation exemplary in speculative metaphysics can succeed.

Notice this: The characterization "… is a necessarily true proposition" covers (at least) two different kinds of reference. This is well known. That all bachelors are male is necessarily true within our language L, in that its semantic content renders its negation inconsistent, *à la* Hume and Leibniz. However, that no surface is simultaneously pink and purple all over will also be necessarily true within the same L – but not for the same reason as the bachelor example. It is not so much that the negation of this color-proposition is easily reducible to inconsistency; this may not even be possible. Still, this proposition's negation is conceptually untenable – unintelligible. That there is a surface at once all purple and all pink is only an *apparent* proposition. It gives us nothing whatever to entertain. No coherent conception answers such a description. Another example of this "weaker" necessity might be "There are no *perpetua mobilia*". The grammatical negation of this, within the appropriate L, will express an apparently consistent claim. But it is pretty certain that no one now has any clear, scientifically articulable idea of what a *perpetuum mobile* (First Type) would be like. Hence, since "There are no *perpetua mobilia*" is a claim that now lacks an intelligible negation (in that no one can give semantic sense to the physics of this negation), this justifies calling it a necessary claim. It is invulnerable, even though formally deniable.

Could the class of contingent statements also be subdivided? Can it be said that (1) "No bachelor exceeds two tons" is contingent in *just* the way (2) "No bachelor exceeds two tons *and* no bachelor is female" is contingent? Both are contingent in that establishing their truth requires more than reflection. Indeed, one way of negating (2) might even be said to be consistent, but surely not in the way in which that of (1) is.

The denial of "$P_{ct} \cdot Q_{nt}$", moreover, has some extraordinarily peculiar properties, properties never encountered in "ordinary" contingent statements. Presumably, one could deny that $P_{ct} \cdot Q_{nt}$ either because it is false that P_{ct} or because it is false that Q_{nt} – or both. If "$P_{ct} \cdot Q_{nt}$" is denied because it is false that P_{ct}, the situation is not novel. Were the facts different, "$P_{ct} \cdot Q_{nt}$" would then express a contingent truth. But when "$P_{ct} \cdot Q_{nt}$" is rejected not because expresses what is factually false, but because "Q_{nt}" expresses what is logically false, then the situation becomes bizarre. For in this case one could apparently convert a contingently false conjunction (viz., $P_{ct} \cdot Q_{nt}$) into a contingently true one not by altering facts or by denying contingent claims, but by negating an inconsistency. In a long conjunction of genuinely contingent propositions, some of which are false, one could not know which to negate without making observations. When it is only an inconsistent conjunct to these that needs alteration, however, no observation is required. Symbol inspection will readily determine the miscreant proposition. But when two genuine contingent propositions are conjoined to form an "ordinary" factually false conjunction, symbol inspection is never sufficient for determining which of the two is false.

Hence, just as there are two distinguishable senses of "… is necessary", so also (although in a different way) there may be two distinguishable senses of "… is contingent." "$P_{ct} \cdot Q_{ct}$" *is not contingent* in the *same sense as is* "$P_{ct} \cdot Q_{nt}$".

Moreover, there is some *conceptual impropriety* in referring to a conjunction of a contingently true proposition and a contradiction as contingently false. To say that "Today is Friday *and* some bachelors are married" is *contingently* false, sounds queer, to say the least. The impropriety consists in its being clear that the falsity within such a conjunction lies not in the contingent component *per se*, but rather in the fact that a contradiction has been conjoined to that component, forming thereby an unacceptable conjunction. So, let us provisionally distinguish a hard sense of contingency from this merely "logical" sense. Distinguish the sense in which it is contingently false that the earth has two moons from the sense in which it is "contingently false" that the earth has one moon *and* some bachelors are married.

Finally, that a conjunction of 30 propositions might be adjudged *contingently* false because one of the conjuncts was *inconsistent* would not, as a matter of linguistic practice, ever be entertained seriously.

It might be countered that this analysis is unnecessary since there is *no* sense in which a conjunction of a contingently true claim and a contradiction is contingently false. Perhaps such a conjunction generates *only* an impossible (i.e., self-inconsistent) proposition. There are serious objections to this line. There is no way of characterizing $P_{ct} \cdot Q_{nt}$ as true other than by calling it "contingently true". It cannot be necessarily true. Thus, analogously, "All bachelors are male, *and* there are in fact bachelors" can at most be contingently true. Its truth cannot be established by reflection alone. But then, if $P_{ct} \cdot Q_{nt}$ is contingently true, there must be some sense in marking $\sim (P_{ct} \cdot Q_{nt})$ as contingently false. If P is contingent, $\sim P$ is also contingent; and $P_{ct} \cdot Q_{nt}$ is a legitimate substitution for P.

If it counts as a logical principle that

(a) A contingent conjunction will be false when either conjunct is false, and that
(b) If R is contingently true, then $\sim R$ is contingently false, and that
(c) If $R = (P_{ct} \cdot Q_{nt})$ and $\sim R = \sim(P_{ct}^{\cdot} \cdot Q_{nt})$, where it remains true that P_{ct},

then, granting all this, it is certainly reasonable to suggest that, when Q_{nt} is false, then $P_{ct} \cdot Q_{nt}$ is contingently false. Just to lay it down that, here, $P_{ct} \cdot Q_{nt}$ must be characterized as impossible, would be to knife through a Gordian knot of concepts. A natural extension of established logical principles has resulted in a case for calling $P_{ct} \cdot Q_{nt}$ "contingently false" whenever Q_{nt} is false. Furthermore, we have given *reasons* for resisting this natural extension. But simply to *rule* that, here, $P_{ct} \cdot Q_{nt}$ is impossible is not giving reasons. Hence, it constitutes no proper analysis of the case before us.

This distinction between types of contingencies, if it is a sound one, is not merely a scholastic proliferation of minute differences, lacking in any philosophical motive. Rather, it provides just what is needed to mark the notion of contingency involved in our first set of examples as degenerate and conceptually harmless.

Nonetheless, since the entailment before us is formally sound, one cannot avoid the responsibility of modifying the HLLL dictum. It would appear to be an analytic philosopher's obligation to block all cross-type inferences save those which are logically degenerate and, hence, conceptually harmless. It may be theoretically quite difficult to express this exception in a general way. But the attempt will have to be undertaken.

(5) "P is contingently true; *so* P is contingently true and Q is necessarily true." That is, $P_{ct} \to (P_{ct} \cdot Q_{nt})$.

This ultimately sanctions "P is contingently true; *so* Q is necessarily true", that is, $P_{ct} \to Q_{nt}$. But this is only one of the "paradoxes" of strict implication. Since a necessary proposition (Q_{nt}) is implied by any proposition whatever, it is certainly implied by *any* contingent proposition (including P_{ct}). Such an inference *would* be cross-type. But it would also be obviously degenerate, even by one of the leading conceptual principles of strict implication. P_{ct} and Q_{nt} may be thoroughly independent propositions, between which no one would ever suppose any deductive connection to obtain. The *raison d'être* of Lewis and Langford's system is to give sense to the idea of independent propositions, between which no entailment will hold – even though some material implication may hold. Yet, even within this system, Q_{nt} will be entailed by any P whatever; *ergo*, $P_{ct} \to Q_{nt}$. Perhaps this reflects more on the LL system of strict implication than on the general status of cross-type inferences.

17.5   Concerning Inferences from Necessity to Contingency

Logicians are more likely to feel that this variety of inference is clearly invalid – a feeling not generally held with respect to our earlier examples. In addition to the argument immediately preceding Sect. 17.2 of this paper, the following one might also be convincing:

(1) P is necessary	$\sim \lozenge \sim P$
(2) Q is contingent	$\lozenge Q \cdot \lozenge \sim Q$
(3) P entails Q	$P \dashv Q$ that is, $\sim \lozenge (P \cdot \sim Q)$
(4) Necessities entail only necessities	$[(P \dashv Q) \cdot (\sim \lozenge \sim P)] \dashv \sim \lozenge \sim Q$ [10]
(5) $\sim Q$ is possible	$\lozenge \sim Q$ (from 2)
(6) Q is necessary	$\sim \lozenge \sim Q$ (from 1, 3, 4)
(7) $\sim Q$ is both possible *and* impossible	$(\lozenge \sim Q) \cdot (\sim \lozenge \sim Q)$ (from 5, 6)

Reductio ad Absurdum
In the light of this demonstration it becomes difficult to construct plausible candidates for the title "Inference from Necessity to Contingency". Nonetheless, our two examples are probably about as strong as could be found.

(1) "(a) 'That F_1, F_2, F_3, F_4 obtain is good inductive reason for expecting C.'

entails

(b) 'That F_4 obtains is *a* good inductive reason for expecting C.'"

Since this putative cross-type inference has been dealt with elsewhere[11], the problem will not be engaged in detail.

Although an attractive candidate for cross-typicality, this is parasitic on ambiguities in "is a good inductive reason". On one interpretation of this expression, what (a) expresses *does* entail what (b) expresses. But then (b) is also necessary; i.e., the inference does not cross types at all. On another interpretation, what (b) expresses *is* contingent, but then there is no entailment at all between (a) and (b).

One way of expanding (b) so that its semantic content spreads before us, is this:

> F_4, when construed as a premise together with $F_1, F_2,$ and F_3 (or some other suitable set), is such that conceptual paralysis would result from asserting all these premises (F_1 to F_4) and at the same time denying that we have good inductive reason for expecting C. This means that any one of these premises, for instance, F_4, necessarily *must* be *a* good reason for C; it constitutes part of the *meaning* of "… having good inductive reason for expecting C".

---

[10] Lewis and Langford (1959, 164, eq. 18.53). This step may beg the very question at issue. But since it would seem to be part of the apparatus of *any* modal logic, I can see no reason not to use it here.

[11] Cf. Hanson, 'On the Impossibility of Any Future Metaphysics' (1960); 'Good Inductive Reasons' (1961); both reprinted in this volume.

Now the inference is horizontal: from a necessary claim to another necessary claim. No more is said here than would have been said had we argued:

(a') "That P and $P \rightarrow Q$ obtain is good deductive reason for Q"

entails

(b') "That $P \rightarrow Q$ obtains is *a* good deductive reason for Q

Here too, *if* (b') is entailed by (a'), (b') must be necessary. Otherwise, (b') is just contingently false; $P \rightarrow Q$, by itself, is not *in fact* a good deductive reason for Q.

Another, alternative, way of expanding (b) is this:

> There are in fact certain contingently true claims (F_1, F_2, F_3) such that, in conjunction with F_4, they actually do give us good inductive reason for expecting Q.

This could hardly be more than contingent. There *need not have been* true statements F_1, F_2, F_3, giving us good inductive reason for C (there need not have been any statements at all) – but *in fact* there are.

But now there is no entailment whatever between (a) and (b). (a) remains necessary; it sets out the semantic conditions for having good inductive reason for expecting an event. However, now (b) states the fact [wholly independent of (a)] that in our world *there are* conditions F_1 to F_3 which, when conjoined with F_4, *do in practice* warrant our inference to C. This does not follow from (a) at all; (b) might remain true even were (a) demonstrated (e.g., by a logician) to be self-contradictory.

In general, then, what begins apparently as a startling cross-type inference ends as an understandable confusion at best, and a disguised ambiguity at worst. Deductive theory is done no harm by this example, and speculative metaphysics is certainly done no good.

(2) Concerning now the inference,

$$\left[\left(P_{ct} \prec Q_{ct} \right)^n \prec \left(P_{ct} \supset Q_{ct} \right)^n \right]^n$$

this is not so devastating as at first it appears.

One *could* just say, e.g., that Lewis and Langford's system must be ill founded, since it sanctions this cross-type inference. But this might be a *petitio principii*; Lewis and Langford have been useful in all our appeals thus far. Why jettison the system when the very rules we have been relying upon turn up something uncomfortable? There is a less *ad hoc* line of attack.

We have observed that $\sim P_{ct} \vee Q_{ct}$ will itself be contingent; its negation, $\sim (\sim P_{ct} \vee Q_{ct})$, would at least be consistent. Thus, "That it is contingently true that 'The moon is made of limburger cheese' materially implies that it is contingently true that 'The earth is flat'," *means* only that *either it is not the case that "The moon is made of limburger cheese" is contingently true, or else "The earth is flat" is contingently true* – and this disjunction is certainly contingent. The negation of this is also in principle contingent. But who would ever recast this as a *strict* implication, e.g., "That the earth is flat is *deducible* from the moon's being made of limburger cheese"? There is no such strict implication. "$P \dashv Q$" means that $\sim \Diamond (P \cdot \sim Q)$. But it *is* possible, however, that the moon should be made of limburger cheese while the earth is other-than-flat. The entailment relation cannot hold here, therefore; although a material implication can.

So this peculiar inference would never even come up for attention save when $P \dashv Q$ does actually hold – i.e., when Q *is* deducible from P. But then our problem is one we have already faced: If $P \dashv Q$ is necessary, then, of course, $P \dashv Q$ *does actually obtain*, whatever that may mean (i.e., $P \supset Q$ is factually true).

This is either an exploitation (for "philosophical" purposes) of some degenerate sense of contingency *or* an elaborate way of remarking one of the paradoxes of material implication.

That is, since $P \supset Q$ *must* be true when its values are chosen so that $P \dashv Q$ might just as well have been asserted, why then naturally $P \supset Q$ follows from $P \dashv Q$. In this case, however, it is difficult to construe "$P \supset Q$" as being contingent at all. A related point is that, whenever $P \supset Q$ is true, it follows (as one of the paradoxes of material implication) from any proposition whatever. *Eo ipso*, it follows from $P \dashv Q$. Nothing of special philosophical interest emerges.

Suppose now that someone reasoned: "That the moon is made of limburger cheese (M), is deducible from the earth's being round (E)." He argues, in effect, that $E \dashv M$. This is just the claim that $\sim \Diamond (E \cdot \sim M)$. A critic might try to reveal the error by deducing $E \supset M$ from $E \dashv M$. Since this is the same as deducing $\sim E \vee M$, that is, "Either the earth is not round, or the moon is made of limburger cheese", and since this latter disjunction is actually *false*, the critic might succeed thereby in demonstrating the untenability of the original argument, viz., $E \dashv M$.

Would one have been endorsing a genuine, nondegenerate cross-type inference [namely, $(E \dashv M) \dashv (E \supset M)$] in trying thus to expose the original error?

No. No more than would a logic teacher be endorsing a fallacy by writing "If Jones is Welsh then he is human; he is human, therefore he is Welsh" *in order to demonstrate*, by this unacceptable conclusion, the unsoundness of the argument. We often *follow through* on an inference to reveal its shortcomings. This does not signal acceptance of that inference. Similarly, although we may follow through with $(P_c \dashv Q_c) \dashv (P_c \supset Q_c)$ to shake confidence in $P_c \dashv Q_c$, this does not rob us of the capacity to criticize the *general* status of the inference.

Both these inferences remind one of the ancient inference-schemata: "P is necessary, therefore *in fact P*" and "P is impossible, therefore *in fact ~P*". Logicians of many ages and of many persuasions have been seduced into thinking that, in virtue of such inference-patterns as these, logical necessities might be thought to be factually efficacious. Latter-day Aristotelians, intensional logicians, and Idealist thinkers have felt it somehow to be informative about our world to be told that, for instance, $P \vee \sim P$ obtains in it.

Thus: "It is impossible to square the circle; hence *in fact* no one in America, during 1960, found a quadrature of the circle." But the apodosis of this claim cannot be *merely* contingent, inasmuch as the very idea of an American in 1960 actually succeeding in squaring the circle is not even an intelligible idea. In the remote past the notion of a constructive quadrature of the circle could never have been more than a radically confused idea. Just putting the expression "in fact" before the apodosis will not infuse the latter with factual content. The world becomes no richer episte-mologically just because some logician moves through it announcing "In *fact* all triangles have three vertices", "*In fact* all bachelors are male", "As *a matter of fact* in our world, either a proposition or its negation is true", etc. A claim that excludes no possibility whatever cannot convey anything factual. It cannot express anything logically novel. At most, its message may be psychologically new, as when we learn, e.g., that transfinite addition is possible.

A claim like $P \vee \sim P$ can appear to be informative only when one conflates (1) the claim that it is contingent whether or not it is P, or $\sim P$, that does in fact now obtain, with (2) the logical fact that $P \vee \sim P$ exhausts all possibilities for anything whatever obtaining. Only by confounding the observations necessary to determine whether it is P, or $\sim P$, that *does* obtain, with the distinct fact that no third possibility beyond these two can exist, only so can $P \vee \sim P$ appear to have any factual content.

One is reminded of the story told of G. E. Moore, who, happy with his wife's having given birth, was stopped by a well-wisher and asked: "Is the child a boy or a girl?" To this, Moore answered, "Yes!" His friend was asking for information about a contingent matter. But, without further punctuation, his question begged for Moore's "tautological" interpretation. Only by conflating Moore's answer with the obvious intent of the question could one construe this as the expression of a neces-sary fact. Only by confounding the reflection that babies are either male or female with the further fact that it requires observation to determine which, could one suppose that here is a description of fact which is invulnerable. In this case we have a transparent howler before us. Treating "$P \vee \sim P$" as if it were both necessary *and* informative may not be a transparent howler, but it is at least translucent.

Certain "realistic" discussions of the nature of logic make their case largely by forcing these distinct considerations onto the same track. Such discussions have not yet realized, apparently, that *invention is the mother of necessity.*

17.6   Restating the HLLL Dictum

In the history of philosophy some cross-type inferences have been harmful. The subhistory of speculative metaphysics is replete with examples. It was the HLLL intention to design a criterion against which these miracles of argument (which, since they can accomplish virtually anything, actually accomplish nothing) might be exposed for what they are. However, there are inferences, in theoretical logic and in everyday reasoning, which apparently *do* proceed across types. It has been our purpose to provide some understanding of these. Our verdict is that they are all conceptually harmless. They are harmless either because they rest *not* on logical principles alone, but on wider metaphysical, systematic assumptions, *or* because they trade on degenerate inferences, of which $P \rightarrow P$ might be taken as an example;

the "paradoxes" of material and strict implication provide others. From such as these, speculative metaphysicians gain nothing, and deductive logicians lose nothing.

Restating the HLLL dictum can be achieved, therefore, as follows:

From what is contingent nothing necessary follows, save only in harmless and degenerate cases. These exceptions exist only by there being a weak, merely logical sense of "contingent" quite distinct from what the designers of our "dictum" ever had in mind.

And from what is necessary nothing contingent follows, save again in those harmlessly degenerate cases in which "contingent" takes on a semantic force totally unlike what is at issue in the HLLL dictum.

Any *unquestionable* example of a contingent statement remains such that it neither entails nor is entailed by any *unquestionable* example of a necessary statement.

References

Lewis, Clarence Irving. 1918. *A survey of symbolic logic*. Berkeley: University of California Press.

Lewis, Clarence Irving, and Cooper Harold Langford. 1959. *Symbolic logic*. 2nd ed. New York: Dover Publications.

Hanson, Norwood Russell. 1959. It's actual, so it's possible. *Philosophical Studies: An International Journal for Philosophy in the Analytic Tradition*. 10 (5): 69–80.

———. 1960. On the impossibility of any future metaphysics. *Philosophical Studies: An International Journal for Philosophy in the Analytic Tradition*. 11 (6): 86–96.

———. 1961. Good inductive reasons. *The Philosophical Quarterly*. 11 (43): 123–134.

Hume, David. [1748] 1999. *An enquiry concerning human understanding*, ed. T. Beauchamp. Oxford: Oxford University Press.

Leibniz, Gottfried Wilhelm. 1714. *The monadology*. In *Leibniz selections*. 1951. Ed: P.P. Wiener. New York: Scribner.

———. [1704] 1916. *New essays concerning human understanding*. Trans. Alfred G. Langley. Chicago: The Open Court Pub. Co.

Whitehead, Alfred North, and Bertrand Russell. 1950. *Principia mathematica*. 2nd ed. Cambridge: Cambridge University Press.

Chapter 18
The Irrelevance of History of Science to Philosophy of Science

I

There is but one question before us: can a philosopher utilize historical facts without collapsing into the "genetic fallacy"? If he can, will his analyses be improved?

Failure to answer this question has vitiated many discussions concerned with the role of historical facts within philosophy of science, as well as the role of logical analysis within history of science. Some philosophers have set their sight on *Weltphilosophie*, noting that every historian has one. Explicitly or implicitly it controls his selection of salient subjects, his alignment of data, his conception of the over-all objectives of the scientific enterprise, and his evaluations of the heroes and villains within the history of science. That the historian's interpretation is shaped by covert cosmic commitments is clear in the writings of Waddington, Bernal, and Needham. It is apparent also in the works of Whewell, Meyerson, and Poincaré. Moreover, unspoken and unspectacular *Weltphilosophien* provide the intellectual reticulum in terms of which we must view even our most honored "objective" historians of science – Tannery, Duhem, Sarton, and Koyré. As has been suggested recently by Professor R. Cohen at the Xth International Congress of History of Science, to be human at all is to fit the elements of one's outlook into uncriticized philosophical patterns; this is no less true for eminent historians of science than for the rest of us.

Those who stress the silent operation of a *Weltphilosophie* in the studies of historians of science then suggest that without philosophical awareness and acuity the reader must remain at the mercy of the historian's unspoken assumptions. "[The physicist's] dominant faculties, the doctrines prevalent around him, the tradition of his predecessors, the habits he has acquired, the education he has received will serve him as guides, and all these influences will be rediscovered in the form taken by the

---

Presented in a symposium on "The Mutual Relevance of the History and the Philosophy of Science" at the fifty-ninth annual meeting of the American Philosophical Association, Eastern Division, December 28, 1962.

© Springer Nature B.V. 2020

N. R. Hanson, *What I Do Not Believe, and Other Essays*, Synthese Library 38, https://doi.org/10.1007/978-94-024-1739-5_18

theory he conceives." (Duhem 1893, 377)[1] Presumably, it requires a logician critically to understand what is significant in the great expositions of the evolution of modern science. The conclusion is that *history of science without philosophy of science is blind.*

A more modern plea for the use of philosophy of science within history of science has concerned not the capital-lettered "isms" that direct the historian's work, but rather the conceptual details making up that work; not the philosophical architecture, but the conceptual bricks and beams constituting the structure of particular histories of science. We have all encountered naive accounts of scientific discovery with their crude uses of terms like *law, cause, explain, predict, observe, verify, refute, deduce*; even conceptions like *science* and *discovery* themselves are accorded the same unsatisfactory treatment. One often reads histories of science for illumination about the genesis of such "philosophical lubricating terms". The light cannot go on when these terms trip off tongues untried and untrained, indeed tied into knots by factual overconfidence. One almost imagines that some historians[2] cloak their conceptual confusions in clouds of data-clusters: dates, editions, acquaintances, and genealogies. But in the best histories this is not so. Mach's analyses of Newton's laws, Duhem's studies of *force* and theory, Koyré's treatise on Galileo's law of freely falling bodies, and Rosen's examination of the interrelations of *circulus* and *orbis* in the work of Copernicus – these are cases of philosophical acuity leading to the conceptual clarity that makes great history of science. But some will mark these as the brilliant exceptions that prove the dull rule. Thus the indisputable suggestion that closer attention to the logical structure and deductive consequences of key "philosophical" concepts within the history of science would help most historians immeasurably; *ergo* philosophical insight is what historians of science need more of!

Recently (at the Xth International Congress) I stressed a third kind of interpenetration between history of science and philosophy of science. This had little to do with the *Weltphilosophie* of historians, still less with the concept-spectra exploited in their expositions, and more to do with the understanding of their *arguments*; the prime target was the arguments of scientists that historians purport to illuminate. My suggestion was that we should take our scholarly spotlights away from the architecture, and away from the bricks and beams. We ought to play them more on the structures, interrelations, and, indeed, the engineering connections that have made science *the* intellectual concern of our time. Let me say a little more about the centrality of *argumentation* within the endeavors of both historians of science and philosophers of science.

Logicians are concerned with *arguments,* logicians of science with scientific arguments. Their enquiries presuppose answers to worries about the conceptual "stuff" of arguments: unless you know *what is being* argued you cannot determine the argument's soundness. Unless you understand the historical force of concepts in seventeenth-century science, what *force* and *gravitation* and *mass* meant to Kepler,

---

[1] Hanson does not give the translator for this quotation. –*MDL*

[2] For example, Berry (1946), Whittaker (1960), and Nordenskjöld (1946).

Galileo, Huygens, and Newton, you cannot determine the soundness of those particular classical arguments in which such concepts so gloriously figured.

The "higher-level perplexities" involved in understanding the historians' *Weltphilosophie* themselves assume detailed knowledge of the actual arguments which, for the historian, constituted a scientific advance. But few have stressed the dependence of their broader interests in science on the analytic assessment of scientific *arguments*. To parody the diverse approaches: philosophers of science like Buchdahl, Hesse, and Toulmin often focus on understanding the evolution of scientific *concepts* as constituting the intersection of philosophy of science and history of science. Philosophers like Mandelbaum, Meyersohn, and Cohen have their sights on the ultimate *direction* and objectives of historians of science. Doubtless, both kinds of concern are important in any critical understanding of the literature of history of science. But to me it is in the detailed analysis of the detailed arguments of scientists *and* historians where philosophy can most help, and *be helped.*

Keynes (1952) shows us the way. He argues that no scientific statement is ever *probable* in itself, but probable *only* on the assumption of given evidence. To say of a proposition that it is probable or has a probability of 0.9 is, for Keynes, like saying that it is "equal to" or "greater than" or "divisible by". Such relational characterizations make no sense whatever when only one of the relata has been designated. No; Keynes perceives that the probability relation is an inferential connection between scientific premises or initial conditions, and observable consequences – a connection the assessment of which must always be deductive in form. The analysis of arguments in these terms is an enterprise for which the logician of science should have received some rigorous training. Assuming an advanced familiarity with a scientific subject matter, then, the logician of science should be capable of assessing the formal cogency of arguments of, e.g., "steady-state" cosmologists as against "big-bang" theorists: he should be able (in principle) to determine which claims of reasoning are the "best made", which conclusions are most likely *on the evidence given,* which assumptions *en route* are most and least vulnerable. He ought to be able coolly to reconsider the experimental evidence available to microphysicists in 1931 and determine therefrom who had the best arguments – those who quickly opted for the existence of anti-particles (like the positron), or those (like Bohr and Rutherford) who sought to reinterpret the shocking cloud-chamber tracks of Anderson, Blackett, and Occhialini, and the perplexing "negative-energy" equations of Dirac, in terms of more familiar ideas well known in the 1920s. The philosopher of science should place into logical counterpoise the explanations of Asaph Hall in 1896 and Einstein in 1916 – explanations of the disturbing secular advance of the perihelion of Mercury – in order to see which investigator, on the evidence before him, reasoned most relentlessly toward his conclusion. This does not always mean that rigorous and precise determinations of the probability of past scientific arguments are within the easy grasp of the logically trained historian. There will always be difficult "twilight" cases: most of the really important cases may lie in this region. But midnight is still very different from noon. The *very* probable and *highly* unlikely are always separable on *logical* grounds, given a carefully formulated set of initial conditions on which to base one's inferences. This, even though

any assignment of a probability like 0.92 to Fizeau and one like 0.71 to latter-day
Newtonians may be practically beyond achievement.

> When we argue that Darwin gives valid grounds for our accepting his theory of natural
> selection, we do not simply mean that we are psychologically inclined to agree with him; it
> is certain that we also intend to convey our belief that we are acting rationally in regarding
> his theory as probable. We believe that there is some real objective relation between
> Darwin's evidence and his conclusions, which is independent of the mere fact of our belief,
> and which is just as real and objective, though of a different degree, as that which would
> exist if the argument were as demonstrative as a syllogism. We are claiming, in fact, to
> cognize correctly a logical connection between one set of propositions which we call our
> evidence and which we suppose ourselves to know, and another set which we call our con-
> clusions, and to which we attach more or less weight according to the grounds supplied by
> the first. (Keynes 1952, 5)

Keynes goes on:

> It would be as absurd to deny that an opinion *was* probable, when at a later stage, certain
> objections have come to light, as to deny, when we have reached our destination, that it was
> ever three miles distant; and the opinion still *is* probable in relation to the old hypotheses,
> just as the destination is still three miles distant from our starting point. (7)

In other words, for Keynes the probability relation that obtains between a conclusion and its premises is so "objective" that one can characterize it in a time-
independent way at any future date. One can determine the probability obtaining
between some conclusions advanced by the young Darwin and the evidence or data
from which they were drawn; no matter what mature findings may have been made
by Darwin later in his research, that original estimate of the original conclusions'
probability, on the basis of the original evidence, remains fixed for all time.

Two things are immediately clear from this. (1) Such logical evaluations of historically significant arguments are not "subjective": they do not depend on the logicians' or the historians' prejudices or choice of heroes – no more than would a
mathematician's evaluation of the soundness of a purported proof of Fermat's last
theorem have to rest on extra-formal considerations. Given a premise set, a putative
conclusion either does or does not follow. If it does, it does so necessarily. If it does
not, then the assertion that it does is inconsistent. Similarly, given a set of physical
premises and initial conditions, a physical consequence will either have a probability P on these premises necessarily or the assertion that its probability *is* P will be
demonstrably self-contradictory. Mapping this out as a path for philosophers
through the jungles of history of science, I am expressing a thesis counter to that of
Professor Cohen: "… while philosophers of science think *about* deductively-
formulated theories, they had better do so inductively" (Xth International Congress).
My claim is that *while philosophers of science think about inductively formulated
theories, they had better do so deductively.* Or, at least, the justification for a philosophical analysis had better never consist in any gross appeal to the facts.

(2) Our assessments of which argument at time t was the *best* argument (given
the data available) will not always award the guerdon to the argument that is ultimately correct. This point is of the utmost importance to any historian or philosopher of science: that scientific advance and rigorous logic do not always walk arm

in arm is an exciting disclosure, but it should always be spelled out in logical detail, not painted poetically in words (as historical scholars are sometimes wont to do). Consider Galileo's (correct) contention that the instantaneous velocity of a falling body is functionally related to the duration of its fall rather than to the distance it falls: Duhem's proof that this was based on a formally fallacious argument was a triumph of analytical scholarship. Similarly, today's universal recognition that Archimedes was *not* a Copernicus of antiquity, save in a *very* special and debatable way – since, on evidence then available, the arguments of Hipparchus and Apollonius were logically much preferable – this is the kind of conceptual vortex that quickens history of science into genuine intellectual excitement. As a probe for the testing of such vortices, I submit that the logical analysis of arguments within the history of science is no less rewarding than is concept-genealogy (as with Toulmin) or the recognition of the pervasive *Weltphilosophien* (referred to by Cohen). Indeed, for the understanding of the turbulent ripples in the flow of western science – as when we face the counter claims of Gold vs. Gamow, Anderson vs. Rutherford, Hall vs. Einstein, Adams vs. Airy, Young and Fresnel vs. the later Newtonians, Lavoisier vs. Priestley, Kepler vs. Brahe, Copernicus vs. Müller, etc., attention to the logical cogency of rival arguments is of maximal scholarly value. Even when final decisions elude the investigator, such a confrontation of historically important arguments can strip the history of science to its logical bones. At such moments, logical analysis of the historically significant arguments (on the evidence then available) might even be *identified* with history of science at its best. The giants were of this analytical cast: Tannery, Duhem, and Koyré.

Here then is the "hot" junction box which connects the conceptual circuitry in history of science with that of philosophy of science. Professionally, the logician and the historian will often be concerned exclusively with the rational wiring within that box – the scientific argument itself – and not just with the intricate intellectual geometry leading to it and away from it, nor with the lights that may go on in the world of science, and the illumination afforded by historians of science, as a consequence of that circuitry and that junction box being designed as they are. The historian of science and the logician are both concerned with the structure of scientific ideas. These concerns fuse into one when the scientific *argumentation* of the past takes the spotlight.

But all this has fallen into my earlier aphoristic mold: that history of science without philosophy of science is blind. I must now undertake to show that philosophy of science without history of science is empty.

II

The foregoing looks parochially professional. It seems to survey the ways philosophers of science tell historians of science how to do their jobs better. Historians will quickly retort that, since I have only characterized what they do in their everyday work anyhow, it is only a constellation of platitudes. This reaction must now be qualified with a few "plongitudes" – a few plunges into the troubled waters which separate history of science and philosophy of science.

The maelstrom within these waters is the elusive, yet pervasive "genetic fallacy", around which I have rowed for these seven pages. When experimental psychologists, social anthropologists, and "kultur vultures" (e.g., Lévy-Bruhl, John Dewey, Talcott Parsons, and George Mead) sail glibly into strictly logical discussions concerned with the semantical content of technical concepts and the logical structure of formal arguments – how satisfying it is sometimes to hear the bold retort: "That is merely a matter of fact!" There can be no doubt about it, within the history of philosophy illumination has been lost and scattered through clouds of conceptually irrelevant historical detail. A simple question is asked concerning whether a given conclusion follows from a given argument and, too often, the air becomes charged with quotations from Plato and Aristotle "who thought it did" and Spinoza and Gassendi "who thought it did not". Indeed, *the* standard "goof off" amongst professional philosophers is to serve up a tray of facts when what is really needed is the sharp scalpel of analysis. We are all sometimes guilty of this – when weary or disinterested, or rushed. Some of us are always doing this. But we all know that this is a poor excuse for philosophy, just as covering fences with colorful thickets is a poor excuse for town and country planning.

Some of our greatest philosophers of science: Schlick, Carnap, Reichenbach and Popper, have been sensitive to the ways in which scholars sometimes dull the scalpels of philosophy by burying them in the historical gravel. Conceptual clarity is primarily the result of unfettered logical analysis: allusions to actual occurrences in the history of science were at most illustrations (for Schlick, Carnap, Reichenbach, and Popper) of arguments which commended themselves on rational grounds alone. What does it matter that von Neumann, Jeffreys, or Clerk Maxwell invoked the probability calculus this way or that? They were concerned to explain and predict the workings of physical nature: they rarely faced the logical structure of probability arguments *per se*. What does it matter that Mach, Newton, and some of the Schoolmen thought of laws of nature as statistical summaries of observed data and, as such, generable (or "deducible") from the facts? Again, these men were Natural Philosophers, not philosophers of science.

That X is done universally does not in itself make X the universally correct thing to do. That all past and present scientists do X or say that X – or are said by historians to have done or to have said that X – does not in itself make X the correct thing to do or to say. Philosophy of science is, like all philosophies, not simply a rehearsal and recitation of what *is* done and said; it is also an analysis and an appraisal of the *rationale* and logical justification of scientists doing and saying what they do. Just as a child has not defended his misbehavior by the claim: "Johnny does it too" or "Johnny did it too", so also – it can be argued – the real business of philosophy of science is in no way furthered or illuminated by pronouncements like "Heisenberg says it too" or "Newton did it too".

This much seems completely to have sundered history of science from philosophy of science – and let no man join what reason reveals as sundered. That will constitute my leading conclusion regarding the logical relevance of history of science to philosophy of science. The former has *no* logical relevance whatever: should anyone ever attempt to buy off the validity of an argument by reciting facts out of the history of science, he deserves the scolding inevitably to ensue.

Nonetheless, when stressing that history of science and philosophy of science had a common concern in the structure and function of scientific *arguments*, some softening of this rigid logical proscription was heralded. Let no man completely sunder disciplines that are intimately connected through their common concern with ideas, concepts, reasoning, and the argumentation of scientists.

As we have been told, philosophy has no subject matter. But philosophy *of science* has, namely, science. It is all very well for a philosopher of science to argue *"If there were* a discipline in which conservation principles P_1 and P_2 held and within which laws L_1 and L_2 were adhered to, then from initial conditions I_1 and I_2 conclusions C_1 and C_2 would strictly (i.e., deductively) follow". If what the philosopher of science says in such a context is true at all, it is necessarily true. If it is not true, then it is logically false. And no facts about the theoretical constitution of present or past scientific events can have any logical bearing on the appropriate appraisal of the philosopher's analysis. But it still remains that the philosopher of science may be discussing no genuine state of affairs at all! This is the reaction of most practicing microphysicists when they read Reichenbach's *Philosophic Foundations of Quantum Mechanics* (1944), and it is a standard response of historians of science when they confront works in the philosophy of science, especially within the tradition built up in the wake of George Sarton. Historians see in the works of such "formalistic" philosophers of science as Carnap the "fallacy of misplaced abstraction". Without some concrete treatment of the *de facto* development and present state of modern science, philosophy of science strikes many as unilluminating. But those philosophers of science who shy away from "historicism" find the facts within history of science equally unilluminating. To the historian such philosophy of science is often unilluminating because it does not enlighten one about any *thing'*, nothing in the scientific record book is treated in such symbolic studies. To the philosopher, histories of science are often unilluminating because, as a result of their chaotic diffuseness, they never reflect monochromatically: only spectra of concepts and arguments result. For the historian formal philosophical analyses are often empty. For the philosopher the historian's factual compendia seem blind. This suggests a loose analogy within the development of theoretical hydrodynamics and aerodynamics.

The rigorous mathematical explorations of Euler and Bernoulli were models of logical precision even though they dealt with a highly fluid and unstructured subject matter. The relationships between velocity and pressure, between boundary layers and turbulence, between flow direction and "lift", are beautifully mapped within the elegant algebra of these accomplished mathematicians and their inspired followers. Alas: the elegance was illusory. The algebra and the elegant analysis were all based on the assumption of an *ideal fluid* – one utterly lacking in resistance and viscosity. The result was that practical hydrodynamicists, ship designers, civil engineers, plumbers, and aeronautical enthusiasts could not use one line of what the Euler-Bernoulli theoretical tradition had produced. There *are* no ideal fluids! Oil, water, and air all offer considerable resistance and have pronounced viscosity (thank heavens). A more "pragmatic" discipline was quickly fabricated. It was called "hydraulics". This was no elegant, axiomatically generated calculus in the Euclidean manner. It was, rather, a chaotic collection of recipes, hints, descriptions, and techniques – a plumber's tool box. But without knowledge of this kind we should never have understood the phenomenon of heavier-than-air flight, much less actually built aircraft. A considerable interplay between practical aerodynamics and classical hydrodynamics has at last been effected, despite the fact that some of the standard problems within aerodynamic theory have been completely beyond any general mathematical treatment. (Many perplexities arise through the required use of partial, nonlinear differential equations of the second order in time – for which no *general* mathematical solution can grind out past or future "state-descriptions" of phenomena, comparable to what is encountered with the linear differential equations of the first order as encountered within Newtonian mechanics.) Here again is an intellectual contest: classical hydrodynamic theory, following Euler and Bernoulli, generates the sharpest possible answers to a cluster of beautifully formulated hypothetical questions. The only difficulty is that these answers cannot help in practical hydrodynamics, wherein there has never been an ideal fluid with properties and behavior like those so magnificently described in the eighteenth and nineteenth centuries. Against this there are *practical* hydrodynamics and hydromechanics – the sophisticated recipe compendium originally called "hydraulics". Within this discipline one knows that every element corresponds to some observed phenomenon. But it is difficult to interrelate these observed phenomena, to see any rhyme or reason in the connections they do manifest, or even to formulate physically cogent questions, much less provide logically satisfactory answers. When practical aerodynamics was just getting off the ground, classical hydrodynamics was viewed as a mathematical toy. It was an elegant but empty discipline. Its perennial preamble was: *"If there were* an inviscid, nonresisting, and irrotational fluid, it would be observed to do the following things...."* Against this presentation the *facts* concerning what kind of fluids there really are must remain wholly irrelevant. The formal hydrodynamicist seems thus only to be working through the structure of an argument; plumbers and plummets are thus beside the logical point.

   Still, a position with which we can all be sympathetic was adopted by practical hydrodynamical engineers and aerodynamicists. They had to learn about fluid media *de novo*, without any help from the lofty ivory towers of the theoreticians. How much more valuable the work within the Euler-Bernoulli tradition would have been had these thinkers immersed themselves somewhat in a study of *what there is*! Their analyses would not have been made one whit more rigorous – or less rigorous – by so complicating their premises. But the results would have looked more like military strategy than like chess, more like physics than like pure algebra. By analogy, the analyses of the philosopher of science pick up nothing in rigor or elegance, nor do they lose, when the rubric: *"If there were* a science in which..."* is dropped and the premises become instead: "Within experimental hydrodynamics it is observed that...." But the illumination afforded by uncompromising philosophical analysis beginning with the sciences *as they really are* can be as rewarding as the efforts of Euler and Bernoulli would have been had their immense powers been turned on the *de facto* subject matter of hydrodynamics and not upon the properties of non-existent ideal fluids. Surely, the arguments of philosophers can only gain in stature when directed at the conceptual perplexities and the perceptual complexities actually known to occur at the frontiers of science.

"Purified" logical and philosophical studies can be found throughout the literature, studies concerned with deciding between statistical hypotheses, the construction of models, the nature of theoretical terms, the verification and falsification of theories, the axiomatic rewriting of classical mechanics, etc.[3] These intrinsically valuable exercises can only increase in their timely value when the author indicates that the *occasion* for his unflinching analysis is some flesh-and-blood perplexity possessed by physicists[4] or some complexity encountered within chemical theory[5] or some beastly ambiguity badgering biologists.[6] That his "springboard" problems are *real* problems is, of course, no guarantee that the philosopher's analysis will be cogent, sound, and valid. The latter must be assessed in terms invulnerable to any form of the "genetic fallacy". Nonetheless, if a critic's appraisal of the philosopher's analysis is justified, it will remain justifiable whether or not the philosopher has chosen to begin with a *de facto* scientific problem rather than with some sundry suppositions about hypothetical sciences from which rigorous inferences are guaranteed.

There are at least two ways of "cheating" in our examinations of western science. One way is to begin with the data and problems as they actually obtain or did obtain, but then, because of difficulties in generating analyses from such intricate and recalcitrant beginnings, to befog the result with clouds of facts. Ask about consistency, validity, or redundancy, conceptual connections, or the design of an hypothesis – and your answer comes back studded with quotes and dates.

The other "cheat" way is to secure a rigorous analysis and argument at any cost, even to the extent of adjusting the starting point so that it corresponds to no actual scientific problem.

The "hard" way – the *only* way – is to begin with an accurate description and delineation of some experimental or theoretical perplexity, one with which no historian of science could quarrel. This then would be subjected to a philosophical analysis characterized by a rigor that any logician might respect. As an ideal this may be unattainable. But it does possess maximum heuristic value. And in putting the matter thus we can at last demarcate the relationship between history of science and philosophy of science.

---

[3] Cf. Braithwaite (1953); Carnap (1962); Reichenbach (1944); Bunge (1959); Hutten (1956); McKinsey et al. (1953); and Popper's articles on probability.

[4] Cf. Sciven (1954).

[5] Denbigh (1953).

[6] Pirie (1952).

III

The logical relevance of history of science to philosophy of science is nil. Staring at novel facts has never made old arguments invalid, new arguments valid (or vice versa). Fresnel, Fizeau, and Foucault *did* prove that light was undulatory, despite the "granular" discoveries of Hertz and Einstein (photoelectric effect), Compton, and Raman. These later investigators did not show the "three F's" to be wrong, but disclosed only that light is more complex than they had imagined. Similarly, Aristotle's analysis of Eudoxos' astronomy, the critiques Buridan and Oresme leveled at Aristotle's theory of motion, Gassendi's remarks about Osiander's preface to *De Revolutionibus Orbium Coelestium*, Berkeley's examination of infinitesimals and the idea of absolute space, Peirce's analysis of the discovery of Kepler's Laws, Duhem's account of some of Galileo's demonstrations, Mach's demolition of the classical concept of *mass,* Schlick, Feigl, and Grünbaum on relativity, Reichenbach and Feyerabend on microphysics... the internal validity of such philosophical studies of the sciences depends only on questions of logic and conceptual analysis. Historical data just cannot function legitimately in appraisals of the philosophical and logical acceptability of these great works.

But already a patent artificiality is clear from this anxious attempt to avoid the genetic fallacy. Schlick may not contradict himself, and his arguments may be philosophically illuminating and conceptually enriching. But if he just doesn't have the facts about Special Relativity, its genesis, or its present state, Schlick's considerable insights must be adjudged somewhat sterile within the literature of philosophy of science. Had Duhem never read *Il Saggiatore,* had Peirce never opened *De Motibus Stellae Martis*, and had Berkeley never perused Newton's *Principia*, their works would strike us rather as do the crackpot's "proof" of a sixth dimension, the attempts of Soviet politicians to abrogate the Uncertainty relations, Lindsay's "disproof" of Einstein, and the Paduan philosopher's rejection of Galileo.

For a work in philosophy of science to be shot down by philosophers, it must at least get off the ground. This is done only via a runway of facts concerning the history and present state of the science with which the investigator is concerned. Such facts are not germane to the sophisticated professional appraisal of the intellectual flight and logical maneuvers demonstrated thereafter. But the philosopher of science who does not know intimately the history of the scientific problem with which he is exercised is not even airborne. His analytical skill may be admirable, but it does not take us anywhere.

So, history of science and philosophy of science are not logically related: to claim that they are would be either to underestimate or to misunderstand the genetic fallacy. But the risk of inferring that there is thus no connection at all between the two is the risk that philosophers of science may not know what they are talking about, a verdict none of us can accept silently.

References

Berry, A.J. 1946. *Modern chemistry; Some sketches of its historical development*. Cambridge: Cambridge University Press.

Braithwaite, R.B. 1953. *Scientific explanation: A study of the function of theory, probability and law in science*. Cambridge: Cambridge University Press.

Bunge, Mario. 1959. *Metascientific queries*. Springfield: C.C. Thomas.

Carnap, Rudolf. 1962. *Logical foundations of probability*. Chicago: University of Chicago Press.

Denbigh, K.G. 1953. Thermodynamics and the subjective sense of time. *The British Journal for the Philosophy of Science* 4 (15): 183–191.

Duhem, Pierre. 1893. L'école anglaise et les théories physiques. *Revue des questions scientifiques* 34: 345–378.

Hutten, Ernest H. 1956. *The language of modern physics: An introduction to the philosophy of science*. London: Allen & Unwin.

Keynes, John Maynard. 1952. *A treatise on probability*. London: Macmillan.

McKinsey, J.C.C., A.C. Sugar, and Patrick Suppes. 1953. Axiomatic foundations of classical particle mechanics. *Indiana University Mathematics Journal* 2 (2): 253–272.

Nordenskjöld, Erik. 1946. *The history of biology, a survey*. Trans. Lennard Bucknall Eyre. New York: A. A. Knopf.

Pirie, N.W. 1952. Concepts out of context: The pied pipers of science. *The British Journal for the Philosophy of Science* 2 (8): 269–280.

Reichenbach, Hans. 1944. *Philosophic foundations of quantum mechanics*. Berkeley: University of California Press.

Scriven, Michael. 1954. The age of the universe. *The British Journal for the Philosophy of Science*. 5 (19): 181–190.

Whittaker, E.T. 1960. *A history of the theories of aether and electricity*. New York: Harper.

Chapter 19
The Idea of a Logic of Discovery

I

Is there such a thing as a 'Logic of Discovery'? Do we even have a consistent *idea* of such a thing? The approved answer to this seems to be "No". Thus Popper argues "The initial stage, the act of conceiving or inventing a theory, seems to me neither to call for logical analysis nor to be susceptible of it" (1959, 31). Again, "… there is no such thing as a logical method of having new ideas, or a logical reconstruction of this process". (32) Reichenbach writes that philosophy of science "… cannot be concerned with [reasons for suggesting hypotheses], but only with [reasons for accepting hypotheses]" (1938, 382). Braithwaite elaborates: "The solution of these historical problems involves the individual psychology of thinking and the sociology of thought. None of these questions are our business here" (1953, 20–21).

Against this negative chorus, the 'Ays' have certainly not had it. Aristotle (*Prior Analytics*, II.25), and Peirce (*Collected Papers*, I.188)[1] hinted that in science there may be more problems for the logician than just analyzing the completed arguments supporting already-invented hypotheses. But contemporary philosophers are today unreceptive to this. Let us try once again to discuss the distinction F. C. S. Schiller (1921) made between the 'Logic of Proof' and the 'Logic of Discovery'. We may indeed be forced, with the majority, to conclude 'Nay'. But only after giving Aristotle and Peirce a sympathetic hearing. Is there *anything* in the *idea of a "logic of discovery"* which could merit the attention of a tough-minded, analytic logician?

It is unclear what a logic of discovery is really a logic of. Schiller intended nothing more than "a logic of inductive inference". Doubtless his colleagues were so busy sectioning syllogisms, that they usually ignored inferences which mattered in science. All the attention philosophers now give to inductive reasoning, probability,

---

[1] Neither of the references given here by Hanson are especially relevant to the subject at hand. In a related article (Hanson 1965), he refers to *Posterior Analytics*, II.19, a source which better backs his claims, along with the two listed here. –*MDL*

© Springer Nature B.V. 2020
N. R. Hanson, *What I Do Not Believe, and Other Essays*, Synthese Library 38,
https://doi.org/10.1007/978-94-024-1739-5_19

and the principles of theory-construction, would have pleased Schiller. But, for Peirce, the work of Popper, Reichenbach and Braithwaite would seem less like a *Logic of Discovery* than like a *Logic of the Finished Research Report*. Contemporary logicians of science have described how one sets out reasons in support of an hypothesis *once it is proposed*. They have said almost nothing about the conceptual context within which such an hypothesis is initially proposed. (In this Mario Bunge and Leonard Nash are distinguished exceptions.) Both Aristotle and Peirce insisted that the proposal of an hypothesis can at least be a reasonable affair. One can have good reasons, or bad ones, for suggesting one kind of hypothesis initially, rather than some other kind. These reasons may differ in type from those which lead one to accept an hypothesis once suggested. (This is not to deny that one's reasons for proposing an hypothesis initially may sometimes be identical with reasons for later accepting it.)

One thing must be stressed. When Popper, Reichenbach, and Braithwaite urge that there is no logical analysis appropriate to the psychological complex which attends the conceiving of a new idea, they are saying nothing which Aristotle or Peirce would reject. The latter did not think themselves to be writing manuals to help scientists make discoveries. There could be no such manual. ("There is no science which will enable a man to bethink himself of that which will suit his purpose" (Mill 1973, 285; Book III, ch. 1)) Apparently they felt that there is a *conceptual* inquiry, one properly called "a logic of discovery", where *logic* is used in its broad, traditional sense, which is *not* to be confounded with the psychology and sociology appropriate to understanding how some investigator stumbled on to an improbable idea in unusual circumstances. There *are* factual discussions such as these latter. Historians like Sarton and Clagett and Professor Nash have undertaken such circumstantial inquiries. Others, e.g., Hadamard and Poincaré, have dealt with the psychology of discovery. Mario Bunge, in *Intuition and Science* (1962), is also a contributor in this area. But these are not logical discussions. They may not even turn on conceptual distinctions. Aristotle and Peirce thought they were doing something other than psychology, sociology, or history of discovery; they purported to be concerned with a *logic* of discovery: theirs was a philosophical inquiry about the formal structure of reasoning which constitutes scientific innovation and discovery.

This suggests caution for those who reject wholesale any notion of a logic of discovery on the grounds that such an inquiry can *only* be psychology, sociology, or history! That Aristotle and Peirce deny just this has made no impression. Perhaps Aristotle and Peirce were wrong. Perhaps there is *no* room for logic or analysis between the psychological dawning of a discovery and the final justification of that discovery *via* successful predictions. But this should come as the conclusion of a discussion, not as its preamble. If Peirce is correct, nothing written by Popper, Reichenbach or Braithwaite cuts against him. Indeed, these authors do not really discuss what Peirce wishes to discuss. For Peirce renewed again what Aristotle called "abduction"; Peirce referred to it as "Retroduction", which here we will designate *RD*. This is to be contrasted with "Hypothetico-Deduction", which we'll dub *HD*. The point is that although Peirce's cursory analysis of Retroduction is not at all adequate, *some* RD account might yet make sense of the *idea* of a Logic of

Discovery – and this is something which the HD account not only does *not* do, it usually insists that it cannot be done, that no sense can be attached to the idea of a Logic of Discovery.

Let us begin this somewhat uphill argument by distinguishing

(1) Reasons for accepting an hypothesis H, from
(2) Reasons for suggesting H in the first place.

This distinction is surely in the spirit of Peirce's thesis. Despite his arguments, most philosophers seem to wish to deny any *logical* difference between these two. This must be faced. But let us shape the distinction carefully before denting it with criticism.

What would be our reasons for accepting an H? These will be those we might have for thinking H *true*. But the reasons for suggesting H originally, or for formulating H in one way rather than in another, these may not be those reasons one requires before thinking H true. They are, rather, those which make H seem a *plausible type of conjecture*. Now, no one will deny *some* differences between what is required to show H true, and what is required for deciding that H constitutes a plausible kind of conjecture. The question is: are these logical in nature, or more properly called "psychological" or "sociological"? More generally, are there *fundamental* differences between the HD and RD accounts, or are they just "psychological" in nature?

Or, one might urge (as does Herbert Feigl) that the difference is just one of refinement, degree, and intensity. Feigl argues that considerations which settle whether H constitutes a plausible conjecture are of the *same type* as those which settle whether H is true. But since the initial proposal of an hypothesis is a groping affair, involving guesswork amongst sparse data, there *is* a distinction to be drawn; but this, Feigl urges, concerns two ends of a spectrum ranging all the way from inadequate and badly selected data, to that which is abundant, well-diversified, and buttressed by a battery of established theories. The issue therefore remains: is the difference between reasons for accepting H and reasons for suggesting it originally, one of logical type, or one of degree, or of psychology, or of sociology?

Already a refinement is necessary if our original distinction is even to survive. The distinction just drawn must be re-set in the following, more guarded, language. Distinguish now

(1′) reasons for accepting a particular, minutely-specified hypothesis H, from
(2′) reasons for suggesting that, whatever specific claim the successful H will make, it will nonetheless be an hypothesis of one *kind* rather than another.

Neither Aristotle, nor Peirce, nor (if you will excuse the conjunction) myself in a host of earlier papers, sought this distinction on these grounds. The earlier notion was that it was some particular, minutely-specified H which was being looked at in two ways: (1) what would count for the acceptance of that H, and (2) what would count in favour of suggesting that same H initially.

This way of putting it is objectionable. One must therefore object to Aristotle's account, to Peirce's, and to my own earlier, as wholly inadequate: they are vague, ambiguous and even incoherent. The issue is, whether (*before* having hit on an hypothesis which succeeds in its predictions) one can have good reasons for anticipating that the hypothesis will be one of some particular *kind*.

II

The insight of the HD analysis consists in distinguishing the rational activity of the natural scientist from that of the mathematician, a distinction which Popper, Reichenbach, Braithwaite, Bergmann, and Carnap perhaps draw better and more finely than did earlier inductive logicians like Hume, Mill, Jevons, Venn, and Johnson. The mathematician argues "typically" when he entertains certain premises solely to "unpack" them. His concern is neither with their contingent truth or falsity, nor with that of the conclusions unpackable therefrom. It is the *unpacking relationship* itself which alone interests the formal scientist. The natural scientist, however, cares not only about consistency within a universe of discourse; he is concerned also with the contingent truth of claims about the universe in which we live. That a statement follows from *some* premise cluster may be a necessary condition for its descriptive utility. But it is not sufficient. False conclusions can follow validly from contingently false premises, or from logically false ones.

If each premise is contingently true, and if the deduction is valid, the conclusion will have "about" the same probability as its premises. But problems seldom come to the scientist thus. Rarely is he given a list of claims and charged to draw up another list of their consequences. Usually he encounters some anomaly, and desires an explanation. It cannot follow from any *obvious* premise cluster, else it would not be anomalous. So, one proceeds to cluster *some* established truths with hypotheses to see whether they may not jointly entail the anomaly. But *now* estimate the probability: the anomaly's description is assumed to be correct. The available premises obtain. From the joint probability of the anomaly plus these obvious premises one now estimates the probability of an hypothesis which, when conjoined with the premises, entails the anomaly.

The HD account is concerned not only with *conclusion deducing*, but with *hypothesis testing*. Hypotheses are tested by linking them with already confirmed statements to form a premise cluster. From this cluster, observational consequences are generated. If these are confirmed, the hypothesis is to that extent confirmed. This is the meaning of "having good reasons for accepting H". But if further consequences turn out false, the probability of the hypothesis diminishes. And that is the meaning of "not having good reasons for accepting H".

Much scientific reasoning and argumentation displays this HD pattern. Whenever the extension of a partially confirmed theory is in question, one generates further observational consequences of the theory and checks them against the facts. Indeed, detecting flaws in apparatus, and deviations in measuring instruments – as well as the theoretical discovery of "unexpected" phenomena – consists largely in deductively decomposing the premise clusters of theoretical science. This sets out the "logical expectations" of a given theory, and hence highlights any deviation from these expectations. The very identification of an event as "anomalous" depends on this HD elaboration of familiar premise clusters.

The HD theorist attends thus to the scientist's inferences from contingent premise clusters to observationally vulnerable conclusions. The RD account focuses rather on the explanation of anomalies. RD enthusiasts think scientific argumentation to consist first in the recognition of anomalies, and then in the hunt for some premise cluster which, if confirmed, would explain the anomaly. This premise cluster will contain initial conditions and an hypothesis, the form of which "reveals itself" psychologically by its initial absence from the cluster. Thus, that the law of Universal Gravitation had an inverse square *form* seemed clear to the young Newton from the logical gap left in the cluster of known mechanical laws when he assumed that such laws were sufficient to explain *all* mechanical phenomena – the tides, hydrodynamics, ballistics, celestial motions, etc. A further hypothesis was needed. But although *it* was not discovered until 1687, Newton perceived its form "lurking" in the very statement of his problem in 1665. Even then he could have given good reasons for anticipating that the ultimately-successful hypothesis would be of the inverse square *kind.*[2] So while the HD account pictures the scientist with a ready-made theory and a store of initial conditions in hand, generating from these testable observation statements, the RD account pictures him as possessing only the initial conditions and an upsetting anomaly, by reflections upon which he seeks an hypothesis, or a kind of hypothesis, to explain the anomaly and to found a new theory. Again, the HD account focuses on *hypothesis testing*; the RD account is concerned with *anomaly explaining.*

Some signal events in history have involved reasoning of this kind. The discovery of Neptune by the Inverse Problem of Perturbations, and of the neutrino, are characterizable thus. Just as the discovery of Pluto, and of the antiproton, seem much better described in HD terms. Here one runs out the consequences of an accepted theory and tests them. In the RD case, some facts surprisingly fail to confirm the consequences of an accepted theory; one then argues from these to some new hypothesis, or hypothesis-kind, which may resolve the anomaly.

HD and RD proponents both recognize that their formal criteria for success in argument are *precisely the same.* In this Peirce was no less acute than Popper. Thus, imagine that one scientist argues from premises A, B, C and hypothesis H, to conclusion D (which, although originally unexpected, ultimately is confirmed in fact). Another encounters the anomalous fact that D, and conjoins this with A, B, and C so as to "corner" an hypothesis H which, when bracketed with A, B, and C, will "explain" D. Both scientists have been arguing; both have been using their heads. Differently. But the criterion for their having succeeded with their different tasks will be simply this: that D follows from A, B, C, and H. If either the first or the second scientist was mistaken in thinking D to be entailed by A, B, C, and H, then his reasoning fails.

---

[2] That is, given Kepler's $T^2 = r^3$, and Huygens's $F = r/T^3$, it follows that the 'F' of gravitation must equal $1/r^2$.

But if the *logical* criteria for success or failure of reasoning in either case are the *same,* then whatever distinguishes these two scientific arguments must be nonlogical, and therefore (so the position develops) must be *merely psychological.* This is the strong form of the thesis that, though the aspects of scientific thinking distinguished by the HD and RD accounts may be interesting to psychologists, they contain nothing of importance for philosophers and logicians. It is the thesis that only psychological considerations distinguish 1' and 2' as set out earlier. My first objective is to attack that conclusion.

Consider a logic teacher presenting a problem to his class. One orthodox assignment might be this: "Here are three premises A, B, and C. From these alone generate the theorem, D." The teacher is here charging his students to find what follows from premises written "at the top of the page". This is related to the traveler's puzzlement when he asks, "here I am, river to the left, mountains to the right, canyon ahead; *where do I go from here?"*

Contrast with this the different assignment a logic teacher might give: "Here is a theorem D. Find any three premises A, B, and C from which D is generable". Here, he gives his students D written, as it were, "at the bottom of a page". He asks them to work back from this to three premises which, if written at the top of the page, will be that from which D follows. Analogously, the traveler's question would be *"would I be able to return here from over there? or there?"*

These two queries of the traveler will be answered, and appraised, by the same geographical criterion; "is there a geographical route connecting point A with point B?" Whether one is at A asking if he can get from there to B, or asking while at B whether he could return from some other point A *back* to B – the ultimate geographical issue is only whether some traversible route connects A and B.

Similarly, the criteria for assessing the logic students' answers are the same whether the teacher asks his question in terms of premise unpacking, or in terms of premise hunting. "Is there a logical route connecting A, B, C with D?" Whether one is at D and looking for some A, B, C, H from which he could get back to D, or whether one begins at A, B, C, H and asks whether he can make it to D – that these are different is not relevant in strict logic. The question of the existence of a route, logical or geographical, is independent of whether the route is traversed from one end to the other, or from the other end to the one: from A, B, C, H, to D or from D to A, B, C, H.

It is often supposed that when considering the *form* of an argument one should think of it as if it were *mathematical.* It is imagined that the ways logicians and mathematicians argue illuminates the issue of logical form. This is false. Mathematicians no less than other reasonable men argue sometimes from premises to conclusions, and sometimes from an anomaly to its explanation *independently* of any *general* mathematical question of whether some logical route connects the beginning point of the argument with its terminus. The actual arguments of pure mathematicians are just like ours. They have an arrow built into them; they progress from a starting point to a finish line.

The *logical form* of an argument, however, does not progress at all. It is static, time-independent, problem-neutral – above the battles of natural science and formal science alike. Hence, if deducing is what logicians and mathematicians *do* when arguing from premises to conclusions, then the word "deductive" cannot distinguish the formal characteristics of one kind of argument as against others, i.e., probabilistic, analogical, etc. If deduction is what someone does during the *de facto* business of reasoning, then alternative ways of proceeding with one's reasoning might be different and might have different names, e.g., "hypothetico-deduction", "retroduction", etc. This may be so even though from a *strictly* formal standpoint nothing may distinguish such procedures.

Just as arguing from premises at the top of a page down to a conclusion differs from working from a conclusion "up" to premises at the top, even when the logical form of each will be identical to that of the other – so also, arguing from initial conditions plus hypothesis, A, B, C, H, down to an observation statement D is different from working "up" from an anomaly D to some H which, when conjoined with initial conditions A, B, C will entail, and hence explain, the anomaly. This, although the logical structure of each procedure is the same as that of the other. The only question here is "does some logical route connect A, B, C, and H with D?"

The HD account centres on hypothesis testing. It stresses the generating of observation statements D from premises A, B, C, and H. When the D's square with the facts, H is insofar confirmed. The typical description gives A, B, C as known, H as conjectured, while D_1, D_2, D_3 ... have yet to be "unpacked" from this premise cluster. The analogy between what the mathematician does during some of his problem solving and what the scientist is taken to do by the HD philosopher is instructive. The natural scientist does not know in advance *what* observation statements D_1, D_2, D_3 maybe generable from A, B, C, and H. This is what makes this HD procedure an indirect test of H (after it has been formulated and conjoined with A, B, and C). In both mathematics and natural science, arguments often exfoliate deductively; they proceed from the "top of the page" down to the D-statements. This does not identify the two procedures, however. The formal scientist is not concerned with the empirical truth of A, B, C, or H or of the conclusions drawn therefrom. That a conclusion D is validly generable from premises A, B, C, H, contingent truth or falsity aside, this will be his one concern. A natural scientist proceeding in the HD manner, however, will begin with initial conditions A, B, and C established as true. The status of H remains unknown. After D is deduced from this set and discovered to describe the facts, H may be said to have become "probabilified". The natural scientist's concern is to determine whether a given H can thus be raised to the same degree of acceptability as the initial conditions A, B, and C. This he settles by enlarging and diversifying the set of observation statements D_1, D_2, D_3,...the regular confirmation of which will systematically raise H's probability. This distinguishes the epistemic context within which the mathematician and natural scientist work. Still, *vis-à-vis* the *direction* of argument, the mathematician *and* the natural scientist will both on occasion argue from the top of the page down, and this is traditionally described as "deducing". This is often the thrust of Sherlock Holmes' comment: "Simple deduction, my dear Watson."

When wearing his RD cap, the natural scientist begins his inquiry in puzzlement. This is the normal context of discovery. After unpacking a well-established theory, replete with hypothesis H, into the expected observation statements D, he discovers that nature is not described by some of these latter. His normal expectations (and those of the theory) are thus thwarted. Hence he is puzzled. He has no reason to doubt initial conditions A, B, and C; their independent verification is what made them initial conditions. But he is astonished to note that the apparently *orthodox* hypothesis H does not, when conjoined with A, B, C, generate descriptions of the facts. Thus the question: "Given the anomaly D, and initial conditions A, B, C – from the hypothesis H' (i.e., any hypothesis other than H) does D follow when H' is bracketed with A, B, and C?"

Consider these two schemata:

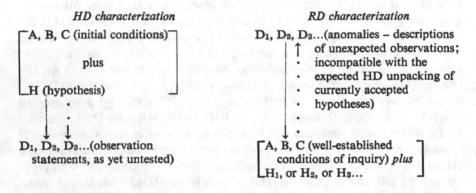

HD characterization	*RD characterization*

⌈A, B, C (initial conditions)⌉

plus

⌊H (hypothesis)⌋

D₁, D₂, D₃...(observation
statements, as yet untested)

D_1, D_2, D_3...(anomalies – descriptions
of unexpected observations;
incompatible with the
expected HD unpacking of
currently accepted
hypotheses)

⌈A, B, C (well-established
conditions of inquiry) *plus*
H_1, or H_2, or H_3...⌋

Notice that the solid arrows represent the *actual* order of the scientist's argument. The "beginning" in the one case is H plus A, B, C, which set is then unpacked into the heretofore-unformulated D_1, D_2, D_3. In the other case, the occasion for the inquiry is the anomaly D: the rational moves from that point are towards a premise cluster A, B, C, H which can "explain" the anomaly. The dotted arrow, however, represents the *logical* order of the progressions. It points the same way in both cases – towards D_1, D_2, D_3; hence the logical criteria for appraising the validity of arguments of either form above are identical. Here then are two argument-schemata which, *vis-à-vis* logical structure, are the *same* argument, but, *vis-à-vis* their *de facto* development within the problem-solving context, are clearly different and not merely psychologically so! The HD "starts from" initial conditions and an hypothesis and terminates in low-level observationally testable statements. The other "begins with" statements of actual observations – ones unexpected on an HD basis – and terminates in a statement of initial conditions A, B, C, and some heretofore-unformulated hypothesis H.

Consider again the claim that this difference can be no more than psychological since both argument and schemata are identical in logical form. This cannot be correct: the same conceptual probe leads to quite different reactions.

Thus consider the premise set, A, B, C, *and* the claim "John is a bachelor." If these four premises are consistent, everything, D, which follows from them will also be mutually consistent, e.g. "John is unmarried", "John is male", "John is an adult", etc. But begin now from the low-level claim, D: "John is male". This *can* be shown to follow from A, B, C, *and* "John is a bachelor". But it also follows from A, B, C, *and* "John is a married uncle". These two premise sets, however, are not consistent with each other. Since conceptually different answers result from this probe, the two characterizations must therefore be conceptually different and not merely psychologically so.

Here it might be objected, "Yes, 'John is a bachelor' and 'John is a married uncle' *are* inconsistent, and any premise sets in which they are imbedded will also be inconsistent. But these two premise sets are not inconsistent with respect to what is required in order to generate the single conclusion 'John is male'. With respect to that conclusion, being married or unmarried is irrelevant. Indeed this must be so, by the principle that if p entails q (and q is not necessary) then $\sim p$ cannot also entail q. The only analysis is this: that when $(r \bullet p)$ obtains and $p \to q$, then q will follow – and it will follow also from $(\sim r \bullet p)$ and $p \to q$. Hence a single anomaly D (i.e., q) can follow from two mutually incompatible premise sets only when the incompatibility plays no immediate role in the deduction."

This is an extremely potent objection. But in my opinion it leaves the conceptual issue unscathed. It remains that A, B, C, H and R, if consistent, will entail only compatible conclusions D_1, D_2, D_3 ..., etc. But an anomaly, D_3, might be explained not only by *different* premise sets – by A B C H R and M N L O P – but also by *incompatible* premise sets – by A B C H R and by A B C H ~ R – where R and ~ R (like 'bachelor' and 'married uncle') are admittedly redundant to the derivation. Redundant or not, the conceptual distinction persists and rules out the "mere psychology" interpretation. When A scores a genuine logical point against B, it does not help B to disclose the point to be 'redundant' or 'trivial' or 'insignificant'. Only if the point is logically *invalid* can B squirm free.

Moreover, in distinguishing premise sets *as embedded in scientific theories*, no premises are wholly redundant in the degenerate logical sense. For, although R and ~R may be redundant for this *one* accounting of D, they will not be redundant in general (as would a tautology) in the business of distinguishing the whole theories in which these arguments occur. Thus, in the wave theory of light, R may signify that a light ray *decelerates* on entering a denser medium, whereas in the particulate theory ~R will signify that the light *accelerates* on entering a denser medium. But neither R nor ~R will be needed immediately in the deduction of D – e.g., the proposition that sines of the angles of incidence and refraction stand in a certain ratio to each other. Nonetheless, *explaining* this latter phenomenon will involve reference ultimately not just to the premise set A, B, C and H (which may be identical in both the wave theory and the particle theory): the explanations will sooner or later involve A, B, C, H, and R, on the one hand and A, B, C, H, and ~ R, on the other. So the conceptual difference remains, and is not trivialized by the redundancy move just noted.

III

My objective has been to argue that there is far more scope for the exercise of reason and analysis within the "context of discovery" than most philosophers of science have been willing to grant. This is certainly not meant to deny that scientific discovery may depend *essentially* on "intuition", "insight", "the inspired hunch", and "sheer genius". The IBM Corporation will never invent a mechanical Nobel Prize winner! However, as an area of inquiry which, in my opinion, has received far too little examination by logicians and philosophers, the Context of Discovery should be recognized as having logical credentials of its own and should not simply be relegated to being a kind of "puzzling out in reverse" of what will end up as the finished Research Report. Unless philosophers, logicians and historians turn their considerable analytical powers in this direction, we shall have to continue to suffer inaccurate and "merely poetic" descriptions of such things as *argument by analogy*, the *simplicity* criterion, the role of *symmetry* in the development of theories, the function of *aesthetic elegance* in the design of research programs and the feeling for *explanatory fertility* which so often guides a scientist's experimentation and observation. These latter are not simply to be adjudged the outcome of the scientist's genetic constitution and up-bringing: rather, they constitute the kinds of conceptual consideration which can make a man's research, indeed the total conception of his discipline, what it is for him during periods of profound perplexity. Appeals to analogy, symmetry, simplicity and elegance have a *rational* function within the Discoverers' attack on the unknown. These designate influences which shape the arguments of discoverers. If we simply leave all this to the *obiter dicta* of psychologists and the synoptic overview of historians, it is unlikely that we shall ever see anything worth pursuing in the idea of a Logic of Discovery.

References

Braithwaite, R.B. 1953. *Scientific explanation: A study of the function of theory, probability and law in science*. Cambridge: Cambridge University Press.

Bunge, Mario. 1962. *Intuition and science*. Englewood Cliffs: Prentice-Hall.

Hanson, Norwood Russell. 1965. Notes toward a logic of discovery. In *Perspectives on Peirce*, ed. R.J. Bernstein, 42–65. New Haven: Yale University Press.

Mill, John Stuart. 1973. *A system of logic, ratiocinative and inductive*. In *Collected works of John Stuart Mill, VII*, ed. J.M. Robson. Toronto: University of Toronto Press.

Popper, Karl R. 1959. *The logic of scientific discovery*. London: Hutchinson.

Reichenbach, Hans. 1938. *Experience and prediction; An analysis of the foundations and the structure of knowledge*. Chicago: The University of Chicago Press.

Schiller, F.C.S. 1921. Hypothesis. In *Studies in the history and method of science*, ed. Charles Singer, vol. 2, 414–446. Oxford: Clarendon.

Part V
Religion

Chapter 20
The Agnostic's Dilemma

An agnostic maintains himself in a state of perfect doubt concerning God's existence, a position I regard as unsound. The agnostic achieves his equipoise of dubiety only by shifting his ground where logic requires him to stand fast.

Is religious belief reasonable? This question pivots on reactions to the claim 'God exists'. This claim could be false. Its denial is consistent, hence the claim is synthetic. Otherwise it would be as uninformative to be told that God exists as it is to hear that bachelors are male.

Distinguishing theists from atheists, and these from agnostics, depends on there being alternative answers to the question "Does God exist?". The theist answers "Yes". The atheist answers "No". The agnostic doesn't know, or cannot decide.

There is a fund of subtle literature concerning this existence claim. Sometimes it is construed as synthetic but necessarily true. But this would make atheism impossible, which it is not. This point also cuts against 'God exists' being analytic. Again, some think the claim to be factual, yet established beyond all reasonable doubt. This makes atheism unreasonable, which it is not.

Many theologians hold the claim 'God exists' not to be central to the core of religious belief at all. In different ways, Niebuhr, Tillich, and Braithwaite have argued that the role of belief within human life remains fundamental whatever our decisions about the logical or factual status of the claim 'God exists'. Apparently it matters little to the reasonableness of one's religious beliefs whether or not he believes in God: indeed, it might remain reasonable for one to persist as a believer even after further thought has led him to deny God's existence.

This apologia has gained in popularity what it has lost in rationality. Clearly, a rational man will not continue to believe in what he has grounds for supposing does not exist. Nor will he maintain belief in that chain of claims which hang on a proposition he no longer thinks is true.

Hence, in this paper, 'God exists' is a synthetic claim; it could be false. Moreover, the claim could be contingently confirmed, as some theists say it already is. What have theists, atheists, and agnostics been arguing about, if not whether this existence

© Springer Nature B.V. 2020   305
N. R. Hanson, *What I Do Not Believe, and Other Essays*, Synthese Library 38,
https://doi.org/10.1007/978-94-024-1739-5_20

claim is, or can be, factually established? Logically, the claim belongs in the center of our discussion. Historically, that is where it always has been. Despite the hocus-pocus of theologians, the claim is also central within the lives of genuinely religious people. Surely most streetlevel believers would be affected in their religion by the disclosure that the New Testament was a forgery, or by a demonstration that God could not exist – assuming such a disclosure or demonstration to be possible.

Many theists will not be moved by these considerations. They will insist that 'God exists' is not the sort of claim that could be amenable to scientific observation, or even to logical scrutiny. Both reason and the senses fail when issues which turn on faith arise. This, of course, is a flight from reason. If neither logic nor experience can be allowed to affect our attitudes towards God's existence, then no argument and no ordinary experience can affect the theist's belief. However, it then becomes a university's function to stress that religious belief, so construed, is not reasonable. Nor is it connected with ordinary experience – since, if the latter cannot count against such belief, then neither can it count for it. A university must help young adults to distinguish positions for which there are good grounds from other positions for which the grounds are not so good. When the theist lets his appeal collapse into faith alone, he concedes that his position rests on no rational grounds at all.

The agnostic, however, cannot adopt any such theistic device. He must grant, without qualification, that 'God exists' is contingent. He feels, nonetheless, that there are no compelling factual grounds for deciding the issue one way or the other. After the atheist has exposed as inadequate all known arguments for God's existence, someone will ask, "But can you prove God does not exist?". Instead of realizing he has already done this, the atheist often hedges. This the agnostic mistakenly makes the basis for his universal dubiety.

If the argument between theists and atheists could have been settled by reflection, this would long since have been done. The theist's appeal to faith cannot settle any argument. So the agnostic adopts the only alternative, viz., that the argument concerns a matter of fact – whether or not God does in fact exist. But he remains in an equipoise of noncommitment by proclaiming that neither theist nor atheist has factual grounds for supposing the other's position to be refuted. How in detail does the agnostic argue this point?

Consider some logical preliminaries: entertain the claim 'All A's are B's'. If this ranges over a potential infinitude, then it can never be completely established by any finite number of observations of A's being B's. 'All bats are viviparous' receives each day a higher probability – but it is always less than 1, since the claim ranges over all past, present, and future bats, anywhere and everywhere.

This claim is easily disconfirmed, however. Discovering one oviparous bat would do it. Consider now the different claim: 'There exists an A which is a B'. This can never be disconfirmed. Being told that some bat is oviparous cannot be disconfirmed by appealing to everything now known about bats, as well as to all extant bats. The 'anywhere-everywhere' and 'past-present-future' conditions operate here too. However, we can confirm this claim by discovering one oviparous bat.

So, 'All A's are B's' can be disconfirmed, but never completely established. 'There exists an A which is B' can be established, but never disestablished.

'There is a God' has never been factually established. Any account of phenomena which at first seems to require God's existence is always explicable via some alternative account requiring no supernatural reference. Since appealing to God constitutes an end to further inquiry, the alternative accounts have been the more attractive; indeed, the history of science is a history of finding accounts of phenomena alternative to just appealing to God's existence.

Thus there is not one clearcut natural happening, nor any constellation of such happenings, which establishes God's existence – not as witnessing a bat laying an egg would establish 'There is an oviparous bat'.

In principle, God's existence could be established with the same clarity and directness one would expect in a verification of the claim 'some bats are oviparous'. Suppose that tomorrow morning, after breakfast, all of us are knocked to our knees by an earshattering thunderclap. Trees drop their leaves. The earth heaves. The sky blazes with light, and the clouds pull apart, revealing an immense and radiant Zeus-like figure. He frowns. He points at me and exclaims, for all to hear.

"Enough of your logic-chopping and word-watching matters of theology. Be assured henceforth that I most assuredly exist". Nor is this a private transaction between the heavens and myself. Everyone in the world experienced this, and heard what was said to me.

Do not dismiss this example as a playful contrivance. The conceptual point is that were this to happen, I should be entirely convinced that God exists. The subtleties with which the learned devout discuss this existence claim would seem, after such an experience, like a discussion of color in a home for the blind. That God exists would have been confirmed for me, and everyone else, in a manner as direct as that involved in any noncontroversial factual claim. Only, there is no good reason for supposing anything remotely like this ever to have happened, biblical mythology notwithstanding.

In short, not only is 'God exists' a factual claim – one can even specify what it would be like to confirm it. If the hypothetical description offered above is not rich or subtle enough, the reader can make the appropriate adjustments. But if no description, however rich and subtle, could be relevant to confirming the claim, then it could never be reasonable to believe in God's existence. Nor would it then be reasonable to base one's life on such a claim.

What about disconfirming 'God exists'? Here the agnostic should face the logical music – but he doesn't. What he does do, and as an agnostic must do, is as follows:

The agnostic treats 'God exists' as he should, as a factual claim the supporting evidence for which is insufficient for verification. However, he treats the denial of that claim quite differently. Now the agnostic chooses the logical point we sharpened above. No finite set of experiences which fail to support claims like 'Oviparous bats exist' and 'God exists' can by itself conclusively disconfirm such claims. Perhaps we have not been looking in the right places, or at the right things. We do not even know what it would be like to disconfirm such claims, since we cannot have all the possibly relevant experiences. But we do know what it would be like to establish that 'God exists'. Variations of the alarming encounter with the thundering God described above would confirm this claim.

The logical criterion invoked when the agnostic argues that 'there is a God' cannot be falsified applies to all existence claims. Hence, he has no grounds for denying that there is a Loch Ness Monster, or a five-headed Welshman, or a unicorn in New College garden. But there are excellent grounds for denying such claims. They consist in there being no reason whatever for supposing that these claims are true. And there being no reason for thinking a claim true is itself good reason for thinking it false. We know what it would be like to fish up the Loch Ness monster, or to encounter a five-headed Welshman, or to trap the New College unicorn. It just happens that there are no such things. We have the best factual grounds for saying this. Believers will feel that 'God exists' is better off than these other claims. They might even think it confirmed. But if they think this they must also grant that the evidence could go in the opposite direction. For if certain evidence can confirm a claim, other possible evidence must be such that, had it obtained, it would have disconfirmed that claim.

Precisely here the agnostic slips. While he grants that some possible evidence could confirm that God exists, but that it hasn't yet, he insists that no possible evidence could disconfirm this claim. The agnostic shifts logical ground when he supposes that evidence against the 'God exists' claim never could be good enough. Yet he must do this to remain agnostic. Otherwise, he could never achieve his 'perfect indecision' concerning whether God exists. For usually, when evidence is not good enough for us to conclude that X exists, we infer directly that X does not exist. Thus, the evidence fails to convince us that there is a Loch Ness monster, or a five-headed Welshman, or a New College unicorn; and since this is so, we conclude directly that such beings do not exist. These are the grounds usually offered for saying of something that it does not exist, namely, the evidence does not establish that it does.

The agnostic dons the mantle of rationality in the theist vs. atheist dispute. He seeks to appear as one whose reasonableness lifts him above the battle. But he can maintain this attitude only by being unreasonable, i.e., by shifting ground in his argument. If the agnostic insists that we could never disconfirm God's existence, then he must grant that we could never confirm the claim either. But if he feels we could confirm the claim, then he must grant that we could disconfirm it, too. To play the logician's game when saying that 'there are no oviparous bats' cannot be established, one must play the same game with 'there is an oviparous bat'. Even were a bat to lay an egg before such a person's very eyes, he would have to grant that, in strict logic, 'there exists an oviparous bat' was no more confirmed than its denial. But this is absurd. To see such a thing is to have been made able to claim that there is an oviparous bat. By this same criterion we assert today that 'there are no oviparous bats'. We take this to be confirmed in just that sense appropriate within any factual context.

The agnostic's position is therefore impossible. He begins by assessing 'God exists' as a fact-gatherer. He ends by appraising the claim's denial not as a fact-gatherer but as a logician. But consistency demands he either be a fact-gatherer on both counts or play logician on both counts. If the former, he must grant that there is ample factual reason for denying that God exists, namely, that the evidence in favor of his existence is just not good enough. If the latter, however – if he could make logical mileage out of "it is not the case that God exists" by arguing that it can never be established – then he must treat 'God exists' the same way. He must say not only that the present evidence is not good enough, but that it never could be good enough.

In either case, the conclusion goes against the claim that God exists. The moment the agnostic chooses consistency he becomes an atheist. For, as either fact-gatherer or logician, he will discover that there are no good grounds for claiming that God exists. The alternative is for him to give up trying to be consistent and reasonable, and assert that God exists in faith. But then he will have to doff the mantle of rationality which so attracted him when he adopted his original position.

The drift of this argument is not new: it is not reasonable to believe in the existence of God. Reflective people may have other grounds for believing in God's existence, but these hinge not on any conception of 'having good reasons' familiar in science, logic, or philosophy. The point is that the agnostic, despite his pretensions, is not more reasonable than the atheist or the theist. The next step for him is easy: if he chooses to use his head, he will become an atheist. If he chooses to react to his glands, he will become a theist. Either he will grant that there is no good reason for believing in the existence of God, or he will choose to believe in the existence of God on the basis of no good reason.

Chapter 21
What I Don't Believe

There is no good reason for belief in the existence of God

It may disarm the gentle reader to learn that I am not a trained theologian. Although once religious, and a serious student of 'The Arguments' for the existence of God, the writer has had no explicit academic preparation for an essay like this one. These pages, then, must be construed as the good natured testament of a reluctant disbeliever – whose studies in logic, analytical philosophy and philosophy of science have ground the lenses through which he looks at life and death, perhaps never again with the innocence once possible.

Disarmament being thus effected, I will now loft my 'Belief Missile' (not too 'ballistically', I trust). THERE IS NO GOOD REASON FOR BELIEF IN THE EXISTENCE OF GOD.

Within our culture and civilization the very sight of such words often hurts the reader's eye. This may be even more the case for readers of this journal. Yet this is the Belief-Thesis to which my own untrained reflections, however simple minded, have driven me. The structure of these cerebrations will now be set out for the now-perhaps-not-so-gentle-reader.

Most of the thousand-odd animated conversations that have erupted from my 'advocatus diaboli' pronouncement five sentences above (at cocktail parties, student gatherings, panel meetings… etc.), have ignited some inevitable religious enthusiast to flame-throw my thus-announced belief back at me with the hot retort:

'Well, can *you* prove that God does *not* exist?'

– already signalling thereby that nothing purporting to be such a proof will ever constitute a proof *for him*! And when, as experience and puzzlement have taught me always to do, I decline even to gesture at such a proof such an enthusiast smiles triumphantly 'round at the relieved believers there in attendance, as if he were Saint George and the dragon of disbelief were now dead from his dialectical dart.

© Springer Nature B.V. 2020
N. R. Hanson, *What I Do Not Believe, and Other Essays*, Synthese Library 38,
https://doi.org/10.1007/978-94-024-1739-5_21

These many 'Saint Georges' who argue thus miss the heart of the conceptual issue, in my opinion.

There is no proof (in George's sense) that green goblins do not exist on the far side of the moon. There is no proof that a blue Brontosaurus does not exist in Brazil. There is no proof that the Loch Ness monster does not exist. But the non-existence of such proofs does not give us the slightest reason for supposing that goblins, Brontosaura or monsters *do* exist! In this sense of 'proof' (St. George's), there can never be a proof that Shangri-la does not exist. Or that the Abominable Snowman does not exist! Or that Flying Saucers do not exist! Nonetheless, anyone who inferred from the absence of *this* kind of proof that there *was* good reason therefore to suppose that Shangri-la, the Abominable Snowman and Flying Saucers *do* exist, such a one would have badly confused two radically different senses of the expression 'proved the existence of *X*'. One proves the existence of a prime number above 1,000 by calculating. But to prove the existence of some as-yet-undetected organism one must do more than calculate; one must look and see. Proof of the *non-existence* of some specified mathematical entity will also proceed by calculation. But a proof of the non-existence of a living organism is no more a task for simple calculation, computation or cerebration than would have been the discovery of the organism itself. Just as these constitute different kinds of proofs for positive existence, so also they correspond to different kinds of proofs for non-existence. Our 'cocktail party Saint George' insists on being provided with a proof of that first kind (a formal, deductive demonstration), when all that is appropriate is a proof of the second kind (an inductive description of gathered evidence). That is, George demands that he be shown by a formal demonstration how his claim 'God exists' is somehow self-contradictory, redundant, or otherwise unacceptable. When his demand is not met (as it cannot be), he concludes mistakenly that there is no proof of the non-existence of God – where by 'proof' he means only a claim-sequence structured *à la* the Euclidean model. What is too often overlooked is that a 'proof of *X*'s non-existence' usually derives from the fact that there is no good reason for supposing that *X does* exist. Since there is no good reason whatever for supposing that green goblins do exist, *that* fact is normally what is *meant* by reference to the 'proof' that green goblins do not exist. Proving that George is not at home is usually accomplished by showing that there's no good reason for supposing that he *is* at home – i.e. no one has seen him, no noise or stirring is anywhere apparent, the beds are all empty and the house is dark.

By the same logic, the lack of any conclusive, formal (deductive) proof that God does not exist provides no reason whatever for supposing that God *does* exist. On the contrary, a 'proof' that God does *not* exist might very well be felt to be the name of that enterprise which consists in reviewing all the preferred reasons which purport to show that God does exist, and then demonstrating that none of these reasons are *good* reasons.

Thus, just as we prove that a bike has not been stolen by revealing that there is no good reason for supposing that it *has been* (e.g., by opening the garage door and disclosing the bike itself), so also it could be argued that a proof that God does *not* exist turns on no more than the demonstration that there is no good reason whatever for supposing that he *does* exist.

This is just as we should expect things to be where the claim 'God exists' is construed as a factual claim, synthetic in form. And it *should* be so construed. Why? It is factual because it purports to inform us as to 'what is the case'; alternative states of affairs are logically possible, but (as this claim asserts it) these alternatives do not actually obtain. Further to remark that 'God exists' is 'synthetic' is only to note that the negation of that claim is not itself demonstrably inconsistent, nor does it entail anything self-contradictory. In short, atheism is a logically consistent position. (Which agnosticism is not, as we shall show soon.) If there is anything wrong with atheism, it is wrong with respect to the facts of existence. If atheism is erroneous it is so in the manner of claims like 'Boston is south of Miami' and 'Eagles outweigh elephants'. These are errors as to the facts; if theism or atheism is in error, it is with respect to *the facts.*

Or, to put the matter another way, if it could be *logically* demonstrated that God exists – if 'God exists' were logically true or conceptually necessary, like 'bicycles have two wheels' – then the defeat of atheism would be analogous to the disclosure of mathematical error. Nothing more than rigorous deduction from 'self-evident' truths would be required to demonstrate to all rational men that God necessarily exists in just the sense that a prime number greater than 1,000 necessarily exists. But no religious enthusiast, and no sober theologian, has ever produced such a proof. This judgment surely includes Augustine, Anselm, Descartes and Malcolm. Therefore the proof, if there is one, probably rests on other grounds – grounds rather like what Galileo provided when he proved that Jupiter had moons, and proved that Saturn had rings. Galileo proved these things to his skeptical scholastic opponents by having them look heavenward through his telescope – by contriving to let them have the experiences which he had had. He could not dialectically drive them into the concession that such things did exist, argument alone would not force their affirmation – since the existence of Jovian satellites and Saturnian circles are matters of fact and not simply issues concerning logic or mathematics. So similarly, and for much these same reasons of propositional analysis, a proof that God exists must turn on matters of fact also, albeit matters of fact of a kind quite different from any so far noted.

My claim here, then, is only the traditional one that factual-synthetic claims cannot be proven *a priori*. Why is this so? Well, since what *are* the facts is discerned only by discovering which of *many* logically possible alternative states of affairs *do* actually obtain, reflection alone can never establish which facts *must* obtain (since sundry alternatives must always be consistently conceivable). Thus both 'an organism X exists' and 'an organism X does not exist' are factual claims – as against e.g., 'a number X exists' and 'a number X does not exist', which are either logically-true-and-logically-false, or logically-false-and-logically-true. Claims about the existence of non-formal entities are expressed in synthetic propositions the negations of which are equally synthetic, equally consistent, and equally factual (i.e. factually true or factually false). Concerning which of the two factual claims 'organism X does exist' and 'organism X does not exist' actually states what *are* the facts, *a priori* reflection must be forever insufficient. [This much is little more than an unspectacular restatement of the familiar analytical doctrine which, in other places, I have dubbed 'The Hume-Leibniz Dictum'; see e.g., 'A Budget of Cross-Type Inferences', in this volume, page 261].

My contention has always been, is now, and will be in this article, that the claim 'God exists' is a factual-synthetic claim in just this sense. It asserts the existence of a non-formal entity, in language the negation of which is consistent. Moreover, it is a factual-synthetic claim for which there are, in my view, no good reasons whatsoever for concluding that it is true; there is no evidence open to all, no objective data which compel skeptics against their expectations, to concede that such a claim expresses what are indubitably the facts.

'God exists' is thus a factual claim – it purports to inform us concerning what is the case. 'God exists' is thus also a synthetic statement – its negation (i.e. 'It is not the case that God exists') is a perfectly consistent proposition. It entails nothing of the form Q-and-not-Q. That is, atheism is a logically possible position; formal reflection alone is insufficient to reveal atheism to be a self-contradictory thesis. Theists and atheists, therefore, are at odds concerning the facts. The former is convinced of the truth of an assertion of the form 'X exists'. The latter is convinced of either the counter-claim 'X does not exist' (which is also factual and synthetic), or the ostensibly quite different kind of remark 'There are no good reasons for asserting, X exists' (which seems somehow more analytical than factual – more a reflection of criteria concerning the goodness of reasons, than a straightforward description of what is, or is not, the case).

It has been already indicated how *this* atheist (the writer) would undertake to support the claim 'It is not the case that God exists' – to wit, by considering all the reasons that have been offered for supposing that God *does* exist, and then undertaking to show *seriatim* that none of these are *good* reasons. That is, it may very well be that *some* factual-synthetic claims derive as much from reflection on the goodness of reasons offered in support of these claims, as from any direct confrontation with the denotata of those claims. This is precisely what many theists urge – that 'God exists' is a factual conclusion drawn from reflection and analysis of the world in which we live. My argument will be the mirror-image of that. I will urge that God does not exist precisely because the reasons theists advance for supposing that he does exist are all poor reasons (singly and *en bloc*) for that conclusion. This is just the way in which one supports claims like 'It is not the case that green goblins exist' and 'It is not the case that a Loch Ness monster exists'. One simply evaluates the reasons certain believers offer in support of the claim that such entities *do* exist, and then shows them to be poor reasons. Or, at the very least, a critic may contend that the reasons offered by believers are quite compatible with alternative explanations, and hence they are equally in support of the negations of the proffered claims.

All of which is a concededly long-winded way to say that 'God exists' *could* in principle be established for all factually – it just happens not to be, certainly not for everyone! Suppose, however, that on next Tuesday morning, just after our breakfast, all of us in this one world are knocked to our knees by a percussive and ear-shattering thunderclap. Snow swirls; leaves drop from trees; the earth heaves and buckles; buildings topple and towers tumble; the sky is ablaze with an eerie, silvery light.

Just then, as all the people of this world look up, the heavens open – the clouds pull
apart-revealing an unbelievably immense and radiant Zeus-like figure, towering up
above us like a hundred Everests. He frowns darkly as lightning plays across the
features of his Michaelangeloid face. He then points down – *at me*! – and exclaims,
for every man, woman and child to hear:

> I have had quite enough of your too-clever logic-chopping and word-watching in matters of
> theology. Be assured, N. R. Hanson, that I do most certainly exist.

Nor is this to be conceived of as a private transaction between the ultimate
Divinity and myself – for everyone in the world witnessed, 'knew by acquaintance',
what had transpired between the heavens and myself, and all men heard what was
entoned to me from on high. TV cameras and audio-tapes also recorded this event
for all posterity.

Please do not dismiss this example as a playful, irreverent Disney-oid contrivance. The conceptual point here is that *if* such a remarkable event were to transpire,
I for one should certainly be convinced that God does exist. That matter of fact
would have been settled once and for all time. Indeed, every single witness of this
singular happening, i.e., everyone living on that Tuesday morning, would be equally
convinced of this factual state of affairs – just as completely as they might be convinced of the existence of rain during a deluge, or death and destruction during a
war. The Frisco quake, the immolation of the Hindenburg, the destruction of
Hiroshima are settled in our minds as established facts on much more restricted
evidence than our example hypothesizes. The intricate subtleties with which the
learned devout now discuss this existence claim would seem, after such a universal
experience as was just imagined, to resemble a discussion of color in a home for the
blind. Such a discussion would be hollow and thin were the blind suddenly to be
given their sight in one dramatic instant. (Indeed, let *that* happening be an additional
event in our story.) So similarly, several centuries of logic-chopping about God's
existence, and his properties – those also would evaporate within a searing universal
experience such as was suggested above. That God exists would, through this
encounter, have been confirmed for me and for everyone else in a manner every bit
as direct as that involved in any non-controversial factual claim. Just as everyone can
experience that fire is hot, and that water is wet, so also *ex hypothesi* everyone would
then know that God exists. These would be the experienced and incontrovertible
facts. [But remember, such facts still cannot be 'proved' by a mathematical and logical demonstration. It would still be possible for logicians to entertain a philosophical
doubt concerning whether fire need be experienced as hot, that water is felt as wet,
and that God exists. Only, the possibility of such a consistent series of philosopher's
doubts in no way provides a good reason for believing that fire is not hot, that water
is not wet and – granted the events delineated above – that God does not exist.]

Thus, not only is 'God exists' a factual and synthetic claim, one can even specify
in detail what it would be like to confirm it. Of course, the excessively-Hollywood
description set out earlier will not seem to be rich enough, or subtle enough, or serious enough, or awe-inspiring enough to match the God-conception of most serious
theologians today and yesterday. But who would deny that, were such a happening

to take place, this would constitute a factual proof of God's existence? Would it not even be *relevant* to such a proof? Or, if the pagan thunderings of our hypothetical description are just too offensive to satisfy the serious devout, then any alternative factual description may be substituted instead. Surely the believer can make whatever adjustments may be appropriate in order to set out 'the event' through which God's existence was established *for him* – or *would* universally establish it for all men. Perhaps a more persuasive factual hypothesis would involve re-telling some particularly persuasive Biblical story – now imagined to be enacted again? Or perhaps some marvelous tale of a mystic in Madagascar, now modeled in actuality. Or perhaps 'the experience' will consist in looking at long lines of limping ladies leaving Lourdes, their crutches and lameness left behind? What factual data *are* relevant? Some must be (if 'God exists' is to have any semantical content). Whatever these data are for you, substitute them for the factual account imagined: but make sure that your conclusion 'God exists' is as forcefully entailed by your data as the same conclusion would have been entailed, for me, by mine!

Whatever the once-gentle reader requires to render 'the experience' sufficient to establish the fact that God exists – for *him*, at any rate – let it now be substituted for the childish technicolor account offered earlier.

One thing, however, is absolutely certain. To repeat: the reader *must* be able to cite some actual happening, some genuine experience, some *de facto* description of events which would be relevant to his conviction that God exists. Because if no such factual description, however rich and subtle, and theologically sophisticated, could even be germane to further confirming the claim 'God exists' for a believer, then it could never be reasonable for him to believe in God's existence. This ultimate fact would be totally unsupported in experience of any kind. Were this so, it would not be at all reasonable to base one's entire life on such a claim. Who would adjudge it reasonable now to shape one's life in the image of Apollo, or Wodin, or Ra? No one, surely. Why? Because no one's experiences today are any different because of commitments to such divinities, no different from what they would be without any such reference at all.

There is no good reason now for supposing that anything even remotely resembling our hypothetical example ever took place anywhere in this world – Biblical mythology to the contrary notwithstanding. The imaginative excesses of Old Testament and New Testament literature – although once construed literally by believers (e.g. remember Bishop Ussher's dating of Genesis as having obtained in 4004 B.C.) – can hardly serve as more than symbolic signals for the serious believer today. (The Fundamentalist, e.g. the Missouri Synod Lutheran, admirable though he may be for the intensity, energy and enthusiasm of his faith, has no compelling *reasons* for his belief. *His* arguments could hardly convince other believers, much less other non-believers. Reasons, remember, are more than just glandular activities; they are more than personal postures assumed because of emotional inclinations – however fervent. Reasons for beliefs are the objective, discursive embodiment of those considerations which, were anyone at all to be convinced of their cogency, the beliefs thereby inferred would be held by everyone who appreciated the reasons.)

Now in most cases where experiences available to all are insufficient to establish some factual claim, further supporting data and reasons may be offered by way of arguments which *circumstantially* strengthen the likelihood that the claim is true – or some reinterpretation of the available experiences is suggested as giving a deeper insight into 'what is the case'. Our present knowledge of Lunar and Martian topography has accumulated thus. But 'the arguments for the existence of God', from Aristotle through Augustine, Anselm, Thomas, Descartes, Mercier and Niebuhr, have suffered seriously from the logician's scalpel. Not a single one of the 'classical' arguments for the existence of God is conclusive, several are not even promising of conviction – something which has been demonstrated far more firmly by critical theologians than by atheists. Paul Tillich has often conceded that, were the logical strength of existence arguments to be the sole consideration, atheists would long since have carried the day – just as disbelievers in goblins, witches, devils, angels and flying saucers have long since carried the day. Arguments for these latter beings are weak and vague – so disbelief is the preferable course. Arguments for God's existence are also weak and vague, as every critic since Gaunilon has amply demonstrated – so, again, disbelief is preferable. The *factual* evidence in support of the existence of such fanciful entities is also wholly inadequate – and the ancillary arguments which are intended to patch up such weaknesses in the evidence are barely beyond contempt. The situation is not different within Christian theology: not one of 'the great arguments' is sufficient to force the logically alert atheist into a position of devout belief. This is in contrast to the standard situation within mathematics and logic, where 'doubting Thomases' are quite often forced into initially unattractive positions by the sheer rigor of an opponent's logic! But the 'First Cause', 'Ontological' and '...from Design' arguments have no rigor. Their conclusions, i.e. 'Therefore God exists', can be resisted by the atheist for any of a number of traditional logical reasons – notably *Petitio principii* and *ignoratio elenchi*. Either said conclusion is covertly packed into the premises, or it does not follow from the premises *at all*.

But if the arguments are insufficient even for critical theologians, and if the publicly-available evidence is insufficient for everyone, what *does* establish it as a fact that God exists for anyone who initially doubts just that? Is this a matter of fact the reasons in support of which could quiet the logical doubts of all rational skeptics?

Well, the hypothetical Greco-Roman example delineated earlier would do it for me! I mean that quite seriously. This indicates that I can at least specify circumstances which, were they to obtain, would completely convince me that God does exist. But nothing like this has ever happened to me, nor is it likely to happen in the foreseeable future.

At this point the now-hostile reader will snarl: 'God does not have to provide evidence that will convince the likes of *you* that he exists! he has better, more elevating things to do. Your overweening arrogance, N. R. Hanson, is surpassed only by your crude opacity; that you have even a primitive conception of God's nature, and the vaguest inkling of manifestations of his presence in our universe, is in no way apparent from your Neanderthaloid noises...' etc.

Such inevitable emotional reactions are understandable, even if somewhat painful. My only objective has been to explore some conceptual and logical credentials of the claim 'God exists' so that I can judge whether, for *me,* this could express a belief that I could ever honestly hold. It has never been any part of my intention to 'make Christians' heads roll' or to engage in *ad hominem* dialogue of any kind. Nor have I ever urged a believer to abandon his beliefs! (Although many energetic believers have urged me to abandon mine.) Beliefs are private matters and all of us may believe what we please. My only concern is that whatever beliefs one does hold should be recognized for what they are. Either they rest upon rigorous arguments and tight logical inferences from unquestionable pre-misses; or, they rest on incontrovertible experiences of what are the facts of this world of ours; or they rest on something else – e.g., emotions of fear, love, insecurity – or on other psychological needs which all of us humans have in abundance, and meet in a variety of different ways. My sole aim here in this paper has been to indicate why pure reason and factual experience have thus far been wholly insufficient to make 'God exists' a creditable belief for *me!*

For me to move in thought, however, from the inconclusiveness of arguments and evidence in support of the claim 'God exists' to the conclusion that, therefore, it is not the case that God exists – this continues to strike many of my devout friends as unjustified, and even irrational. It echoes the kind of inference which seems to me so unwarranted in 'the arguments'. Many serious scholars recommend to me that I have no more going for me, *vis-à-vis* inferential rigor, than would support a thoroughgoing agnosticism.

What is an agnostic? An agnostic ostensibly maintains himself in a balanced state of perfect dubiety concerning God's existence. For him, reason and experience are equally insufficient to establish God's existence. But, by the same token, he muses, reason and experience are insufficient to establish God's *non*-existence. Therefore, the agnostic remains in equipoise concerning the entire issue – neither assenting to the God-belief with any conviction, nor dissenting from it in the spirit of the atheist.

In my view, the agnostic achieves his equipoise of dubiety only by logically inadmissible devices; he shifts his ground where consistency requires him to stand fast.

The agnostic grants, of course, that 'God exists' is a factual, contingent, synthetic claim. Its denial he acknowledges not to be self-contradictory. In a logically different universe of existence, perhaps, there need not have been a God at all; perhaps that is the one thing on which theists, atheists and agnostics can all agree. It is a factual statement about this universe, and this state of existence, that there is a God. But, for the agnostic, this is precisely the contention which the believer has never established, not to the agnostic's satisfaction at least. Therefore, the agnostic does not surrender himself over in a posture of acquiescent belief. He feels strongly, nonetheless, that there are no compelling factual grounds for deciding the issue the other way either. He nods neither pro nor con. Indeed, after the atheist (following a centuries' long tradition of intra-theological criticism) exposes all of the standard arguments for God's existence as question-begging, inconsistent or non sequiturs – and truly not a one of them inferentially guarantees the conclusion that God exists –

after such an exposé the agnostic will often join the theist in the noted counter-query: 'But can *you* prove that God does not exist?'

Now the atheist, instead of realizing that he has already done just this – he has done this by reviewing and stressing the fact that there are no good reasons for believing that God *does* exist – instead of basking in that achievement, the atheist often flounders and hedges with respect to his position. From this uncertain stammering the agnostic mistakenly concludes that the atheist and the theist are equally at sea, and that *this* therefore sanctions the agnostic's universal dubiety.

Again, could the issue between theists and atheists but have been settled by deduction alone – by pure cerebration without any appeal to experience – this would have been done eons ago. But, again, matters of fact cannot ever be resolved by reflection alone. Experience of some kind is required. And this is where the believer's eggs hit the skeptic's fan! Whose experience? What *kind* of experience? Everyday sense experience fails to tip the scales even slightly toward the theist's position. Nor does the precise and refined experience of the scientific laboratory, all the pious pronouncements of retired laboratory researchers to the contrary notwithstanding. Ultimately some kind of extra-rational faith, some kind of mystical communion between the individual believer and his God, is introduced in such a way as to nettle both the atheist and the agnostic – two searchers for truth who at least insist on *generality* for all solutions to human perplexities. There is nothing generalizable about personal faith. It is particular, private, subjective, and quite immune from all ordinary criteria connected with the public objectivity of knowledge, of valid argumentation and the provision of reasons for conclusions.

So ratiocination and ordinary experience, fail altogether to clinch the theist's position. The deductions leak and the inductions don't add up. But then the agnostic observes that the atheist isn't proving anything very positive either; all he *seems* to be doing is spotlighting weaknesses in the theist's claims – weaknesses most theists are aware of, but urge belief anyhow. Hence the agnostic backs off. He remains aloof from the quarrel – sometimes assuming for himself the ostentatious wisdom of the non-combatant referee. But this won't do. Far from his being a referee, the agnostic so described is actually seeking to win 'the game' by playing on both sides at once – a simple breaking of the rules of dialectic. Consider:

The claim 'All A's are B's' ranges over a potential infinitude of A's. Thus no claim of this form can ever be completely established by recourse to a finite number of observed A's being B's. True, 'All bats are viviparous' receives a higher probability every single day; since men began to notice such things there never has been observed an egg laying bat! But the probability of 'All bats are viviparous' being true must *always* be less than one, as a matter of logical principle. For one thing, this statement ranges over *all* past, present and future bats, anywhere and everywhere. It has infinite extensibility both in space and in time. Moreover, since this is a factual, contingent and synthetic assertion, it always remains possible that some as-yet-unobserved egg laying bat is holding, or did hold, or will hold sway over some remote corner of some dark cave in some obscure region of space and time. The hypothetical situation can be described consistently. Therefore it is a *possible* state of affairs. Hence the obverse claim 'All bats are viviparous' can never be exhaustively and completely confirmed: the probability of its being true can never equal *unity*.

Such a claim is easily disconfirmed, however: 'easily' from a logical point of view, that is. The discovery of just one oviparous bat would do it! To find one egg laying bat would be quite enough to disconfirm 'All bats are viviparous'. Hence that statement, although it can never be confirmed, is easily disconfirmed.

Consider now the related but different claim 'There exists an *A* which is *B*'. This can never be disconfirmed! E.g., being told that some particular bat is oviparous cannot be disconfirmed by reference to everything *now* known about bats, as well as to all experiences (past, present and future), concerning all the bats there ever have been, or will be. *The* egg-laying *Fledermaus* may be very remote and hidden from man's curious eyes.

However, although 'There exists an egg laying bat' cannot be disconfirmed in experience (since there could always be a place other than where we have looked, and always a time other than when we have looked, where such a remarkable creature resides), we can easily confirm the claim. Again, this is achieved by discovering just one oviparous bat!

So, 'All *A's* are *B's*' can be disconfirmed, but never completely established. Whereas 'there exists an *A* which is B' can be established, but never completely disconfirmed.

The logical moral for an analysis of 'God exists' should be apparent. Such a claim is, at bottom, of the form 'There exists an *A* which is *B*'. [I.e., there exists a supernatural being who is omniscient, omnipotent, beneficent and who created every 'nameable' in this universe.] Again, such a claim can, in principle, be established in fact, but it can never be completely disconfirmed, logically speaking. Confirming that an *X* exists, therefore, is always a matter for experience. The invulnerability of 'X exists' to disconfirmation, however, is not a matter for experience alone. It is to some extent a conceptual truth, to which further experiential encounters may be irrelevant (as we have seen). Thus we the people could *establish* that flying saucers do exist by having diverse visual experiences of one. But the total lack of such experience within our terrestrial community can never (in logic) establish that there are *no* flying saucers; only, as we noted, *that* is no good reason for believing in flying saucers. The U.S. Air Force has disconfirmed (*pro tem*) the claim that flying saucers exist by exposing as inconclusive or chimerical every account given by people who say they've seen them. *That* is what practical disconfirmation of '*X* exists' is usually like.

'There is a God' has never been factually established – not with anything like the universal agreement which supports claims like 'There is a Coelacanth', 'There is fire', 'There is pain and suffering', 'There is a positron', 'There is death' and 'There is beauty'. Any descriptive account of natural phenomena which seems at first to require God's existence for its explanation, turns out always to be scientifically explicable *via* some alternative account requiring no supernatural reference whatsoever. That is just an historical remark. Most things which once needed God's intervention for man's comprehension of their existence – e.g. lightning and thunder, good fortune, life and death, differences in species, the flight of birds and the disappearance of dinosaurs – all these are now more profitably discussed in terms untinted with the supernatural. Since an appeal to God, and to his divine will, constitutes a terminus to all further inquiry, the alternative appeals have been much more effective. Indeed, the entire history of science can be a narrative dealing with the finding of accounts for phenomena other than any mere appeal to God's existence.

Any person who denied the existence of fire, or of pain, or of life and death, would be thought addled. But such a verdict could never be levelled at a person *just* because he was unconvinced of the existence of a God. This would be logically no different from characterizing a person as addled just because he was unconvinced of the existence of goblins, witches and devils. Therefore the claim 'God exists' concerns a matter of fact, just as with the existence of devils or goblins. These latter have never been factually established. And it is far from clear that 'God exists' has ever been factually established. What *is* clear is that it has not! Because, if it ever *has* been, it would be as irrational and benighted of one to deny the existence of God, as it would be to deny the existence of fire, and of life and death. But this is not so. An atheist may offend in many ways by questioning the existence of God, but he is not offending logic by so doing.

There is no single natural happening, nor any constellation of such happenings, which establishes God's existence – not in the way that witnessing a bat laying an egg would establish, 'There is an oviparous bat'. Again, if the heavens cracked open and the Zeus-like figure referred to before made his presence and nature known to the world, *that* would constitute such a happening. But nothing like that, or even remotely like that, has ever occurred in such a way as to commend all rational men to acquiesce in this belief-commitment.

So the claim 'God exists' although factual, has never been fully and overwhelmingly confirmed for all thinking men.

What about disconfirming 'God exists' then? At this point the agnostic should face the logical music, but he never does. What he does do, and what (as an agnostic) he must do, is something like the following:

The agnostic treats 'God exists' (i.e. there is a Being such that he is omnipotent, omniscient, benevolent, the creator of all things… etc.), as he should – as a factual claim the supporting evidence for which is, at this present time, insufficient for complete and objective verification. However, the agnostic treats the denial of that claim quite differently – asymmetrically, indeed. The agnostic now impales himself upon the logical point we have been at pains to sharpen above. No finite set of experiences which fail to support claims like 'Oviparous bats exist' and 'God exists' can conclusively disconfirm such claims. Perhaps we've not been looking in the right places, at the right things, in the right way, at the right time. We cannot even know what it would be like exhaustively to disconfirm such claims since we cannot possibly have been involved in all the relevant experiences (which are infinite in number). But we *do* know what it would be like conclusively to *establish* that 'God exists'. Theologically-more-subtle-variations on the alarming encounter with the thundering Zeus-like God imagined earlier could conclusively confirm such a claim. For many believers that claim *has* been personally confirmed in just that way.

Now the logical criterion invoked when the agnostic argues that 'God exists' cannot be falsified – this applies to all existence claims. Hence, if looking and not finding does not constitute grounds for denying the existence of God, then looking and not finding does not constitute grounds for denying the existence of goblins, witches, devils, five-headed Welshmen, Unicorns, mermaids, Loch Ness monsters, flying saucers, Hobbits, Santa Claus… etc. *But there are excellent grounds for denying the existence of such entities.* They consist not simply in the failure to find and identify such remarkable creatures. Rather, these grounds consist largely in the fact that there is no good reason whatsoever for supposing that such creatures *do* exist. There is no good reason for supposing that Santa Claus exists, all innocent expectations of our children to the contrary notwithstanding. There is no good reason for supposing that unicorns exist, or witches, or devils, or flying saucers. We don't *need* these entities to explain the things that certifiably do exist – not as we do need force fields, anti-matter and quasars.

When there is no good reason for thinking a claim to be true, *that* in itself is good reason for thinking the claim to be false! We know what it would be like to fish up the Loch Ness monster, or to encounter a five-headed Welshman, or to trap a unicorn. This is the stuff of TV science fiction. It just so happens that there are no such things.

'How can you be so sure' comes the retort to which the response must be: I *am* so sure because

(1) people have looked, and they have not found, and
(2) there is no good reason for supposing that there are still good reasons (circumstantial evidence independent of looking and not finding) for supposing that such things do exist. We infer beyond appearances to the existence of gravity, the positron and life on Mars. But what appearances require us to infer beyond them to monsters and unicorns? Indeed, what possible reason could *you*, dear retorter, have for supposing there to be the slightest chance that such creatures do exist? Science now possesses the best factual grounds for denying precisely this.

Believers will, of course, feel that 'God exists' is far better off than any of these other claims concerning monsters and unicorns. Some will even think it to be a confirmed claim, indeed a *completely* confirmed claim. But, if they do think this, they must also grant that the evidence *could* go in the opposite direction. For if certain factual evidence can confirm a claim, other possible evidence must be such that, had it but obtained, it would have disconfirmed that claim. That is part of the logic of 'evidence', and of 'confirmation'.

It is precisely here that the agnostic slips badly. For, while he grants *that* some possible evidence could yet confirm that God does exist (but that such evidence hasn't turned up as yet), he insists that *no possible evidence* could disconfirm this claim. The agnostic thus shifts his logical ground when he supposes that evidence against the 'God exists' assertion can never be good enough, although evidence for it can be. We would never argue that the claim 'There is life on Mars' is such that certain evidence *could* support the claim, whereas no possible evidence could ever disconfirm it! No, if we land on Mars and find nothing whatever that is indicative of life, it is just *that* 'looking and not finding' which will enable us to deny that there is any good reason for holding to the claim 'There is life on Mars'; this is now very close to the assertion that 'There is no life on Mars', although there are distinctions to be made still. [E.g. if we discover strong concentrations of sulphuric acid vapor in the Martian atmosphere, the positive evidence in support of 'There is no life on Mars' will distinguish that case from finding no such lethal atmospheric component there, but still having no reason to hypothesize life.] In the case of 'God exists', however, the agnostic allows that certain possible (but unavailable), evidence *could* be good enough to support the claim, whereas he then suggests that no possible accumulation of evidence could ever *dis*confirm that claim. This is to insist that the believer need only refer to the faintest experiential 'evidence' to lend color to his contention, whereas the atheist must provide nothing less than a *logical* proof for his denial of the believer's claim. That's not cricket!

Although this is unfair, and a confusion in argument, the agnostic must do something like this in order to remain an agnostic at all. Otherwise he could never achieve his posed posture of 'perfect indecision' concerning whether God exists. Because ordinarily, when the evidence is not good enough for us to conclude that X exists, we leap to the conclusion that X does *not* exist. When the evidence fails to convince us that there is a Loch Ness monster, or a unicorn, or a Santa Claus, we often conclude directly that such beings do *not* exist – especially in those cases where accepted laws or nature must be abrogated in order to sustain the existence claim. The grounds usually offered for saying of something that it does not exist are just that the available evidence does not establish that it does exist, *and* that the supposition requires upsetting other well-established knowledge. [The anatomy of mermaids, centaurs, angels and devils is conceptually revolting!]

Thus does the agnostic don the mantle of rationality and wisdom in the theist vs. atheist dispute. He seeks to appear as the one whose dispassionate objectivity and cool maturity lifts him above the passionate battle of believers vs. non-believers. But he can only maintain such an attitude, and such an altitude, by being to some extent *un*reasonable, i.e. by shifting his ground midway in the argument. *If the agnostic insists that we could never disconfirm God's existence, then he must also grant that we could never confirm the claim either.* (In this posture 'God exists' could never be more than an *hypothesis.) But if he feels that we could confirm the claim, then he must grant that we could disconfirm it too.* (As, indeed, does often happen with hypotheses). Put another way: to play the logician's game when saying 'There are no oviparous bats' cannot be *established*, one must play the same game with 'There *is* an oviparous bat'. Were a bat to lay an egg right before the agnostic's eyes, he should grant that (in strict logic) 'There exists an egg-laying bat' was no more finally confirmed than its denial was disconfirmed. The positive claim would only have been 'probabilified', not established. But this is absurd. To see an egg being laid by a bat *is* to have been fully justified in claiming that there *is* an egg-laying bat. By this same criterion we *do* assert today that There are no oviparous bats'. No one has ever seen a bat lay an egg; we have no good reason for supposing that any bat ever would; there are strong biological (i.e. anatomical and physiological) reasons for supposing that no bat ever *could* lay an egg. We take this claim to be confirmed in precisely the strong sense appropriate within any factual context. Ultimately, this is the kind of context in which the claim 'God exists' must itself be construed.

The agnostic's position is therefore impossible. He begins by assessing 'God exists' as if he were a fact-gatherer. He ends by appraising the claim's denial not as a fact-gatherer, but as a pure logician. But consistency demands that he either be a fact-gatherer with both the claim and its denial, or else play logician with both. If he would do the former, then he must grant that there is factual reason for denying that God exists – namely that the evidence which purports to favor his existence is *just not good enough*. If he would play the latter game, however – if he would make logical mileage out of 'It is not the case that God exists' by arguing that it can *never* be established – then he should treat 'God exists' in precisely the same way. He must say not only that the present evidence is not good enough fully to establish the claim, but that it never could be good enough.

That is the way that propositions work. One cannot confirm (or disconfirm) *P* according to one set of criteria, and then confirm (or disconfirm) not-*P* according to a different set of criteria. But this is precisely what the agnostic chooses to do. He indulges himself in the luxury of Scottish legal practice, as manifested in the possible verdict 'not proven' (i.e. 'the case against the defendant is not well-made'). From this the Scottish magistrate does not infer that the defendant is innocent of that which the prosecution has accused him, but only that the prosecution has not made out his case. In Anglo-American legal practice, however, for the prosecution *not* to have made his case against the defendant is equivalent to the defendant's having proved his innocence. The analogy with the theological issue is striking. The agnostic would like to preserve the verdict of 'not proven' in the theist vs. atheist controversy. But this is tantamount to the claim that the theist has not made his case. And since the onus must be on the theist to do just that (for it is *his* claim which does violence to ordinary canons of evidence and sound reasoning), to have failed in establishing his case is tantamount to proving the position of the atheist! The conclusion would thus seem to go against the claim that God exists.

The moment that the agnostic opts for consistency, he will very likely become an atheist. For, as either a fact-gatherer or as a logician, he will soon discover that there are *no* conclusively good grounds for claiming that God exists. That is, the atheist has never been exposed as deficient either in data or in reason.

The only alternative for the agnostic (if he craves one), is to give up trying to be consistent and reasonable, and simply to assert that *God exists* as a matter of *faith*. This is the strongest standard appeal anyhow. Personal experience, ineffable mystical encounters, heightened emotions and psychological drives may be quite sufficient to make one completely confident that God does exist. But such a person, if *that* is the nature of the backing for his belief, will just have to give up trying to convert others on the ground that his belief is the only rational one to hold. Because, what may be psychologically compelling, or emotionally uplifting, or mystically inspiring, for one person may leave the rest of the world quite unmoved. This is notorious.

Besides, the individual who ultimately grounds his belief in God in personal experiences such as these, will soon have to take off the mantle of rationality which so attracted him (and the agnostic), when he adopted his original position. This the believer may be quite willing to do, given the advantages of his newfound faith. But he should be quite clear about the logical geography concerning where reason ends and faith must begin.

The drift of this entire argument is not all new. Yet it has much to do with this author's present position concerning what he does believe, and what he does not believe. Since I do *believe that* the data and arguments in support of the claim 'God exists' are wholly deficient, I do not *believe in* God. In my view, it is just not reasonable to believe in the existence of a God, be it Zeus, Jehovah, Allah or the Father in heaven. Reflective people throughout the short history of man may have had many kinds of motivations for believing in a God's existence, but these hinge not on the 'having of good reasons' such as are familiar through science, logic, philosophy, or even those of reasonable everyday affairs.

Indeed, one of the genuine anomalies of our time consists in the religious enthusiast's contention that all onus of proof rests on the *non*-believer to make *his* case. This must be the neatest trick of the millennium. But, so it seems to me, the atheist's case is well made simply by the facts. The heavens have never opened. God has never made his existence unequivocally clear to all men. Quite different Gods hold the minds of members of different religions. *Whose* God exists? The Catholics'? The Jews'? The Moslems'? Interminable turmoil over ill-made question! Given all this, the onus really rests on the religious believer to make *his* particular case. The believer in flying saucers does not insist that the *onus* rests on the U.S. Air Force to establish that they *don't* exist. (And certainly the U.S. Air Force's failure to provide a *deductive* demonstration that they don't is not itself a positive reason for believing that they do!) The believer in devils does not urge that the onus rests on the skeptics to establish (deductively) the non-existence of devils. The believer in Santa Claus does not insist that the onus rests on almost every 'grownup' to prove that Santa Claus does not exist. (And, again, the absence of a formal disproof of the existence of devils and elves is not itself a positive reason for believing that devils and elves *do* exist.) Yet the believer in God quite often assumes *his* position to be the rational, defensible and constructive position. Apparently the onus rests always on the atheist to establish that God does *not* exist! Each failure of this latter variety is then somehow construed as an achievement for the theist. This is totally wrong; the religious double-talk of our civilization has blinded us to the simple fact that the Emperor wears no clothes.

A cluster of further trivia often punctuates the believer's panoply. It is sometimes urged that, without a reasonable belief in God, there can be no morality amongst men. This is nothing other than the view that all moral codes must ultimately be anchored in and sustained by a Supreme Being. This is false and naive. There have been excellent moral codes in the history of mankind which in no way depended on supernatural reference for its justification. Ancient Stoicism, Buddhism and Unitarianism are examples of moral codes which owe little to any supernatural framework.

It is occasionally urged that a moral relativism – indeed an *anarchy* of beliefs concerning what is right and what is wrong – will follow the abandonment of ortho- dox belief in God. This again seems just not to be true. Some atheists have given their entire mature lives to philosophical reflection concerning moral, ethical and political issues, as well as to specific social matters concerning what is right and what is wrong. The names of G. E. Moore, George Santayana, Bertrand Russell, not to mention David Hume, may be cited as cases in point. However much in error may be the cumulative issue of all this thought, it cannot be characterized as lacking in standards, or steeped in a spineless relativism, just because it is anti-supernaturalistic in its orientation.

'But how do you bring up your own children? Surely it would be less of a "brain- washing" on your part to educate them within a religion when they are young, and *then* to let them make up their own minds later when they are mature and fully grown?'

Again, this is back-to-front thinking. How can *indoctrinating* a child within a supernatural framework leave him *neutral* for making his own choice later on? That conception is just cracked! In my own family, the children are *informed* about the religions of others, just as they are educated about the customs, practices and behav- iors of others in far away lands and nations. The *fact* of religion has constituted a most important component within the development of Western man; no educated person should be unaware of that. The Bible as literature, the several religious sys- tems delineated as alternative ways of life – these are all discussed in my family with care, sympathy and (it is hoped) complete suspension of judgment or appraisal. Aspects of the history of religions, the symbolic and moral content of religious sto- ries and religious beliefs – as well as the salient differences between, e.g. Roman and Greek Orthodox Catholicism, Protestantism in its hundredfold different forms, Judaism, Mohammedanism – all of these are at least as well known to my children as they are amongst their young friends. Indeed, we sometimes suppose that our neutral objectivity has given our offspring a *better* grasp of these facts than is pos- sessed by their indoctrinated schoolmates. Should my son wish to become a rabbi, and my daughter a nun, when they are mature enough to make such a decision for themselves, my wife and I will be happy to encourage them in the belief-commitment at which they have arrived after serious deliberation and which they will feel is right and proper for them. It is *their* life, not mine. They must live it as they see fit. When they ask me what I believe, I am equally direct (I hope) in making clear my position of non-belief and its 'philosophical' genesis – while also stressing that my views are not shared, or even respected, by the majority of mankind today.

The 'ignoramiboid' identification of atheism with communism constitutes a log- ical howler – a confusion of analytical theology with practical politics – which is unworthy of discussion in the present context. Still, many virtuous, patriotic people seem to argue that since all communists are atheists, therefore all atheists are com- munists! 'Godless Communism' thus becomes a kind of universal battle cry – lev- elled at atheists and Marxists without further discrimination. By an identical form of argument they should be prepared to argue that since all priests are human, there- fore all humans are priests – a transparent fallacy. Equally nettling is the supposed conspiracy, within the popular mind, between atheism and general immorality. It almost appears that the otherwise fairminded faithful suppose that atheists aroused Ghengis Khan, de Sade, Hitler, the Berkeley rebels; and that they eat babies for breakfast every morning! This need hardly be acknowledged a general impression the non-believer should worry about seriously. And yet, many atheists throughout history have been roasted for their unpopular convictions by the virtuous meek that inherited the earth.

The gentle reader of 16 pages ago is probably reading these words (if he has come this far) in a state of hot frustration, righteous impatience and genuine sorrow for, or even anger with, the writer. How can this have happened in so short a time, and in so short a space. My objective was to write on what I don't believe, and this I have done as candidly and in-offensively as possible. Yet some of you readers are certain to have been offended – experience has taught me that. But *why* are you offended? Why should *my* expression of my views and my reasons for holding them, be construed by you as an attack? Undeniably, this is a standard response of believers to an exposition such as the foregoing. For reasons that are not at all clear to me, the fact that I believe what I do believe, for reasons I do hold, seems somehow to constitute a threat to those who believe otherwise. Why? Perhaps it is because we assume that, in this issue, there is but one right answer – all others being hopelessly wrong. That being presupposed, it follows that my articulation of views which constitute the right answer for me indirectly reflect on all orthodox religious positions as being wrong, antediluvian, illogical, unthinking and woefully 'pancreatic'.

Please let me disclaim responsibility for any such reaction on your part. Will it return you to your earlier gentle state if I reiterate once again that these are *my* convictions and my reasons alone? We must all make our own way through this vale of tears, alone and standing on our own two feet. I am prepared to stand on principles such as I have articulated here, and prepared also to pay the loser's forfeit if I've erred. But I grant that others may choose their stand in quite a different way. Since they steer their course by a star different from mine, I wish them 'bon voyage', and even 'Godspeed'. [Dare I hope for the same courtesy and goodwill?]

Part VI
The Theory of Flight

Chapter 22
Introduction

What is presented here is essentially the series of lectures on flight in the history of thought that Norwood Russell Hanson delivered at Northwestern University in 1965. The lectures were taped and, after Hanson's death, were transcribed by Mrs. Margo Dillon.

The amount of editing that was needed varied from lecture to lecture. The first lecture was essentially a reworking of materials that Hanson had presented in 'Actio in Distans' (1964) and in 'Aristotle (and others) on Motion through Air' (1965). For the second and third lectures there were no similar supplementary articles. These lectures were delivered in a very informal style, often relying on the use of model airplanes and diagrams to illustrate points. Since this was not captured on the tape and was not always intelligible in the transcription a certain amount of rewriting was necessary, especially in the third lecture.

To the degree that was reasonably possible any rewriting done was based on a utilization of some of the sources Hanson himself relied on. The references cited in the first lecture gave an initial indication of these sources. Some further guidance came from the fragmentary notes and the few books on aerodynamics which Hanson left behind, and which Mrs. Hanson made available to me. However, the bulk of the research books Hanson had collected had been returned to the Yale library before this editing was begun. The references in lectures II and III, as well as some of those in I, were supplied by the present editor. Though they undoubtedly represent an impoverishment of the scholarship Hanson would have supplied they do attempt to clarify the sources he relied on.

Scholarship, however, was not the primary purpose of these lectures. Hanson had no intention of competing with Sir Charles Gibbs-Smith and other historians of aeronautics. These lectures really represent a fusion of Hanson's varied interests, his zest for flying, his professional concern for the historical evolution of scientific theories, his attempts to relate conceptual revolutions to the problem of scientific explanation. These concerns come to a special focus in the present work. To many

© Springer Nature B.V. 2020
N. R. Hanson, *What I Do Not Believe, and Other Essays*, Synthese Library 38,
https://doi.org/10.1007/978-94-024-1739-5_22

the history of flight seems like a recent and rather bloodless chronicle of drawing boards and equations, of propulsion formulas and construction techniques – all punctuated by episodes of daredevil escapades by barnstormers and test-pilots. Hanson radically disagreed with this evaluation. He intended to show that modern aerodynamical theory is the final outcome of a long complex intellectual evolution interrelating many different aspects of theory and practice. As such it represents, he believed, one of the greatest triumphs in man's attempt to understand, copy, and ultimately control the forces of nature.

These lectures, accordingly, are basically a story with a purpose. Their style is informal, occasionally anecdotal. To the degree possible the printed lectures preserve this informality and even the occasional flamboyance that characterized the original lectures. The lectures as delivered were entitled: 'The Classical Period', 'The Heroic Period', and 'The Modern Period'. Hanson left a paper containing one paragraph summaries of the lectures under these titles. In the margin of this paper he had written in new and more suggestive titles and those are the ones we have used.

The editing of these lectures was begun in San Francisco and completed in Boston. I wish to thank Miss Peggy Bender, Mrs. Sophie Yore, and Mrs. Marie Allen who each typed the final copy of one lecture, and Mrs. Sandra Rasmussen, who assisted me in obtaining needed reference works. I wish especially to thank Mrs. Fay Hanson for her cooperation in making available her late husband's extant notes and books.

Edward MacKinnon, S.J.

Boston College
May 19, 1970

References

Hanson, N.R. 1964. Actio in Distans. *Yale Scientific Magazine* 39 (3): 15–20.
Hanson, Norwood Russell. 1965. Aristotle (and others) on motion through air. *The Review of Metaphysics*. 19 (1): 133–147.

Chapter 23
Lecture One: The Discovery of Air

Einstein told a charming fish-story. It concerned 2-dimensional sardines swimming along straight lines in a plane. They noted an 'attraction', a mystifying deflection toward a certain region in their flat sea. They were drawn toward it, their swim-paths 'bent' in the direction of that peculiar point. Some fish were drawn so close, so sharply deflected from their 'natural' rectilinear paths, that they described circles around the magic spot – *ad indefinitum*. The brainiest of these creatures resisted the usual accounts of this inscrutable attraction; something rational just *had* to structure such a phenomenon, even for sardines. He imagined a hill – a shallow cone – rising up into a '3rd dimension'. Paths near the base of the cone were bent, he argued, by being tilted; paths cutting near the top of the hill, however, became continually more circular, or were at least twisted into tight helices. Such swim-paths were inexplicable, indeed incomprehensible, in 2-dimensional terms – save by reference to mysterious effluvia, evanescent currents, subtle attraction and diaphanous influences. Einstein heroically invites his reader to imagine that we 3-dimensional beings might have *our* paths of travel 'bent' towards large masses in some analogous way. We, and ours, may even end up circling some such large objects – somewhat as our planet encircles the sun. The 'explanation' here may also lie in there being a bump, a cone, or a hill, jutting somehow into a 4th dimension – which is to us what the 3rd dimension was to the 2-dimensional fish. Large masses, then, may generate protuberances, bumps, or ripples, in the space-time continuum within which we live and calculate. They may affect the basic geometry of the vast spaces they inhabit. Perhaps, then, gravitation can be understood without mysterious appeals to miraculous influences – 'attractions', gravitational influences – which operate over great distance (e.g., 93 million miles!), through fields of force themselves as difficult to comprehend as the *action at a distance* they were meant to replace.

Conclusion? Perhaps we hominid '3-D-ers' should take General Relativity seriously, even though it lies far beyond normal, everyday experience – just as 3-dimensional reality lay beyond the '2-D'ers' of Einstein's fish fable. Beyond immediate experience, but not beyond experiences as shaped by brainy sardines, and by brilliant scientists.

© Springer Nature B.V. 2020

N. R. Hanson, *What I Do Not Believe, and Other Essays*, Synthese Library 38,
https://doi.org/10.1007/978-94-024-1739-5_23

We, too, shall begin with a fish story because, in another way, the full experience of *flight* lies beyond the normal experience of most people.[1]

Suppose we lived on the bottom within a world of light, clear liquid. We crawl like crabs on the floor of that liquescent medium. After eons on this rough, weedy bottom, some amongst us will perchance look up into the liquescent, luminescent world above. Some will wish to wend their way aloft. How to lift oneself up into the vast wet space above – to move, to propel oneself – to steer, to climb, to glide and soar? How to gain the 3rd dimension upward and beyond the two which define and control the geometry of the weedy bottom?

Through the random innovations of evolution, one of these floor-crawlers might someday find himself just off the bottom – hovering, floating, paddling, even crudely swimming perhaps. But evolution is fickle, and it takes forever! Imagine instead that we crawlers were given a deadline – by some Kingfish perhaps – a 'crash program' and a final date by which we *had* to have mastered the medium all around us. Bottom-dwellers could not, after that injunction, just wait for Nature to chance-fit some 'baby crawler' with a bubble-sac, or with a fin. We would now be obliged to *think* ourselves off the floor, somewhat as the flat fish had to think his way towards an hypothesis of non-rectilinear paths – as Einstein himself had to think all of us into a kinematical account of non-rectilinear 'natural' paths (contra the spurious dynamical fare we are still served at school).[2] Floor crawlers must now 'puzzle-out' the properties of their liquid world – its resistance to objects moved through it and its buoyancy for such, its internal currents, turbulence-traits and its sometimes smooth, laminar (line-like) flows: they must determine the deflecting effects of forcing certain shapes through the surrounding liquid – and they must learn of pressure changes as one moves off the bottom up through the liquescent medium.[3]

The floor-crawlers, then, to swim, must *think* about (1) their objectives – e.g., whether to flap their way aloft or slither, or soar, (2) the physics of the medium around them – e.g., whether it is always and everywhere the same continuous *plenum*, or whether it may become discontinuous and jagged under certain conditions, and they will have to reflect on (3) the shapes of objects which lift them – e.g., should they be flat vanes, spherical floats, tubular ducts, whirling blades... or what? The challenge, then, is intellectual, philosophical and ultimately scientific for those who won't wait for a caprice of nature to change the world-medium, or to change them.

Some of the incautiously brave will neither wait nor think; scrambling up atop a tall sub-oceanic rock, they will jump – with only crude fins or flaps to support them. They may die for their courage. Later, after the crawlers (as a society) have successfully met this challenge of 'sink or swim', one of them will scribble down 'A History of Swimming', or perhaps the title will be 'The Concept of *Swimming* in the History

---

[1] "That a medium so light and unsubstantial as the air can support a heavy weight is, even today, hardly comprehensible to many..." (Davy 1948, 1).

[2] Cf. Hanson (1964a).

[3] This may have consequences for the dynamics of objects shaped and designed on the bottom. Not all these will deflect uniformly at great distances up. There may even be biological consequences for the crawlers ascending; they may not find the rarer liquids congenial.

of Thought' – to delineate and record the *intellectual* achievements of those who had so quickly (and without special dispensations from Nature) lifted themselves off the submarine bottom by devices and processes the very conceptions of which require the utmost in mental effort.

The analogy with ourselves is exact and complete. Until yesterday *homo sapiens* had always been a floor-crawler, inching his way along the rough, weedy surface of this earth, beneath a light, clear liquescent medium known only to a few winged animals – whose secret of flight was perplexing to philosophers, priests, poets and painters through the ages. The aspiration of human flight, although it dates from our antiquity, is yet quite recent (for what *we* call 'antiquity' is but a few blinks past). This idea of *flight*, the possibility of man moving off this terrestrial platform in emulation of birds and angels, this ranks with, and is attendant to other profoundly perplexing concepts in our intellectual past, such as those of *life* itself, and of *death*, of *mind,* and *spirit*, of the *soul*. All of these have had their full portion of winged representation.

When one reflects the history of *our* 'aerial swimming', a serious activity of barely 100 years' duration, the intellectual nature of man's achievement appears breathtaking. My own father was born before the first powered flight, and yet flies often today at supersonic speeds; the same is true of the parents of many here. Waiting for wings to have sprouted could have been done 'till death', without our lifting one inch off this terrestrial floor. No man really flew until he *thought* very hard about his objectives, the nature of his subject matter and his methods.[4] No man understood flight fully until he reflected, as does a natural philosopher, about the birds and the bees – and their familiar but deeply perplexing movements. The full understanding of insect and bird flight still lies in our future.

The greatest part of this entire achievement consisted just in comprehending the completeness of the analogy we have just developed – that our atmosphere *is* a sea filled with a light, but ponderable liquescent medium. This may be called 'the discovery of the air itself'; for discovering what is subtly everywhere can be more difficult than finding things that are scarce, singular and in hiding.[5] Through this invisible medium birds, bats and bees *swim* by means closely related to the manner in which fish *fly* through their somewhat denser but fundamentally similar atmosphere. (Doubtless the sea around them is not detectable *per se* to the fish within. The brightest Barracuda has still not discovered the sea, as man once discovered air.) The greatest discoveries of aerodynamics, itself a branch of general hydrodynamics, all spill from this single droplet of insight: THE SKY IS AN OCEAN, THE AIR IS A LIQUID. The nature of air *currents*, of turbulence, of aerial-*buoyancy,* of airscrew theory, of circulation, lift and drag, of propulsion climbs and banks and rolls, of stalls

---

[4] Lilienthal flew, and he wrote an important book, *Der Vogelflug als Grundlage der Fliegekunst*, [Bird flight as the basis of aviation] (1889). The Wright brothers flew, and despite all the popular mythology about them being simple bicycle mechanics, they were astute and industrious aerodynamicists [Cf. Baker 1951].

[5] This insight, the detection of air, incidentally is clearly articulated in Aristotle, although it certainly antedates him. Hero and Archimedes were surely cognizant of this datum.

and spins and glides; of compressibility and 'Mach 1', and of 'supersonic booms' –
all of this pours from man's deepest studies of air-as-liquid; air-as-a-physical-
plenum – some people today have hardly learned of this fact. They still think of our
atmosphere as 'nothing at all'. To fly through air at all, however, is to swim – not as
men and dogs swim, but as fish do; it is to move not as soap bubbles fly, but as birds
do. That is The Discovery of The Air Itself – a disclosure made early in the story of
Man, but not for that reason less significant. In a vacuum we could not hear anything,
our baseball pitchers could never throw a curve, our Air Force could only become an
army of fast-taxi-ing rocket sleds – jetless, propellerless and flightless.

To fly well, then, is to swim well in the air – minimizing drag and turbulence (as
lobsters and owls cannot do), while maximizing the available lift and propulsive
power (as crabs and canaries cannot do). Small wonder that penguins, birds which
cannot fly, swim so well – and that flying fish, which swim rather poorly, fly or glide
so spectacularly – albeit nothing like so well as birds. These creatures live at the junction of – the interface between – air and water, the common physics of which man's
mind has at last mastered in ways that even Thales could scarcely have imagined.

In these lectures I invite you to explore with me the intellectual challenge to
ground-crawlers such as we. As with Dante and Virgil let us visit the ideas and theories and experiments which made it possible for us to construct shapes within which
we can now swim up through almost all of our local sea of air – nearly up to *its*
surface! Recently some American and Russian heroes have emulated the Marlin and
the Sailfish, bursting up through the surface of our sea of air into what politicians
call 'space' – a medium as new and different for us as our air is to the deep-sea fish.
Alas, those few who have gone beyond have done so without fully comprehending
the complete story of how we thought and fought our way into and through the 'old'
medium – the air around us. That is true, indeed, of most of us. Let us linger a little
with the liturgy of levitation. Give your thoughts wings, to carry you back to when
there were no thoughts of wings at all.

We shall linger hardly at all with the Neanderthal and Cro-Magnon mythology
concerning human flight. Doubtless the attractions of floating through air 'with the
greatest of ease' dates from the time when man first envied birds – and when groups
of men first battled each other. To survey the enemy from above, to hover and dive
down unsuspected, this would have constituted a great advantage in any fray. This
is the plot of many of our dream-episodes even as we sleep during the night of 1965.
Winged animals supplied the idea; war supplied a need: man's *psyche* provided
imaginative wings. The remains of ancient civilizations, in Mesopotamia, Egypt,
and in the Far East, serve as strong support for this speculative suspicion; monuments, tablets and graffiti memorialize this primitive passion for flight. But there is
little evidence that the dream was ever matched by any successful deed.

A Chinese Emperor, of the third millennium, B.C., by name, Shun, is recorded
as having escaped captivity in 'the work clothes of a bird'.[6] 'Bird-men' are often
encountered in ancient mythology. Daedalus and Icarus 'escaped' from Crete by
flapping aloft on artificial, feathery wings; since 'Daedalus' means 'cunning artifi-

---

[6] In 'Annals of the Bamboo Books' cf. Laufer (1928, 14).

cer', the legend at least suggested that keen wit and industry might be sufficient for the mastery of flight. This attitude ranks with the discovery of air as any essential step in the intellectual mastery of flight. It is already a departure from the primitive investiture of birds, and their wings and feathers, with occult powers permitting motion through empty space.

King Bladud of Britain apparently died in an attempt at flight, in the nineth century B.C.[7] Archytas of Tarentum (fourth century B.C.) may have constructed a wooden pigeon or dove which, like our own modelcraft, flew quite well. A contemporary of Confucius is reported to have done likewise with a model magpie, which remained aloft for three days. Even had these stories some basis in fact, which is very doubtful, their real importance resides in the signal they give of man's interest – 3 millennia ago – in the problem of artificial flight, and in the concept of flying. The interest seems almost innate; the challenge is perennial; the recognition of human flight as a physically-soluble problem dates from remote antiquity.

We should note also that this interest almost certainly derives from the *de facto* existence of birds. Who would have thought of ascension through so diaphanous, invisible and immaterial a medium as the air around us – were it not for the obvious fact that some animals – beasts which are thousands of times heavier than the air they displace – are quite at home in that medium. Who, indeed, would have thought of air as a medium at all were it not for the soaring eagle and the omnipresent fly? Some bold souls would not allow such as these to serve as signals only of the occult and the inexplicable; rather, the flights of birds, bats and bugs signalled the air itself as something subtle but ponderable – perplexing but wholly explicable.

Appeals to bird flight are significant in the history of aeronautics as we shall see; da Vinci, Borelli, Cayley, Marey, Le Bris Mouillard, Lilienthal, Rayleigh and Chanute looked directly to our winged friends during the most critical moments in the development of the idea of flight. Of these we will say much more tomorrow.

Many stalwarts let their imaginations carry them away; they leapt from very high places – clifftops and towers – equipped with perverse wing-like appendages, hoping therewith to achieve flapping or gliding flight. They achieved catastrophe. The 'Saracen of Constantinople' and the British monk, Oliver of Malmesbury were would-be airmen of the eleventh century A.D.: the first with a wide cloak, the second with feathered wings, jumped from tall campaniles and broke their bones. G. B. Danti and J. Damian were like-minded enthusiasts of the sixteenth century. Besnier in the seventeenth, the Marquis de Bacqueville in the eighteenth, and Berblinger in the nineteenth century also had elevating thoughts upon which they acted with energy and resolution. Not surprisingly, they were rarely emulated in such over-confident enterprises. Their cracked bones and bodies seemed an inevitable outcome of ventures so incautiously conceived. But, as announced earlier, our concern is with desk and drawing board, not the pilot's cockpit. It is the *concept* of flight, not the actual achievement thereof, which shall be our central interest. As in

---

[7] This is reported by Geoffrey of Monmouth in *Historia regum Britanniae* [This episode is also described in Laufer (1928, 14). The history of Bladud, who was also father of King Lear, can be found in *ii.10* of Geoffrey of Monmouth's history – *MDL*].

all physical, philosophical and conceptual matters, one can rarely do better than to begin serious study with the great Aristotle (384–323 B.C.).

I

In some celebrated passages, the *Physica* conveys to us that:

> Again, as it is, things thrown continue to move, though that which impelled them is no longer in contact with them, either because of 'mutual replacement' as some say, or because the air which has been thrust forward thrusts them with a movement quicker than the motion by which the object thrown is carried to its proper place. (IV. 8. 215ª14–17)[8]

And again:

> Everything which is moved is necessarily moved by something. If it has not the source of motion within itself, then it is clear it is moved by another… (VII. 1. 241ᵇ24–26)[9]

Finally:

> So it is clear that there is no intermediary between the mover and the moved in the case of local movement. (VII. 2. 244ª24–25)[10]

Although well clear of the archer's bow, the arrow continues to move because the air which is parted at its forward tip circles 'round behind, filling the void left by the advancing tail; it thus impinges upon the after part of the arrowshaft, pressing it forward so as to continue its flight.[11]

When not actually being shoved, large stone blocks halt at once. So it seemed that *any* moving object required the continual and continuous application of a moving force pressing on it. The arrow then, although initially impelled forward by the

---

[8] "Amplius nunc quidem moventur proiecta proiecturo non tangente aut propter repercussionem [ἀντιπερίστασιζ], sicut quidam dicunt, aut ex eo quod pellit pulsus aer velociorem motum illius quod pellitur motu, secundum quod fertur in proprium locum." (1990, 160)) [While Hanson cites Ross's 1936 translation as the source of the English passage, the translation he provides is from Sir Thomas Heath. Hanson's quotation (Heath 1949, 115) omits the italics on 'thrown'. – *MDL*].

[9] "Omne quod movetur necesse est ab aliquo moveri. Si quidem igitur in se ipso non habet principium motus, manifestum est quod ab altero movetur, aliud enim erit movens" (1990, 256).

[10] "Manifestum igitur quod movetur et movens simul, et nullum ipsorum medium est" (1990, 263).

[11] Compare Plato on respiration: "seeing that there is no such thing as a vacuum into which any of those things which are moved can enter, and the breath is carried from us into the external air,… it does not go into a vacant place, but pushes its neighbor out of its place, and that which is thrust out again thrusts out its neighbor; and in this way of necessity everything at last comes round to that place from whence the breath came forth, and fills up the place; and this goes on like the circular motion of a wheel, because there can be no such thing as a vacuum… the hurling of bodies, whether discharged in the air or moving along the ground, are to be explained on a similar principle…" (Timaeus (transl. by Jowett) 1911, vol. II, 570–571). Simplicius elaborates when he notes that in *antiperistasis* "as one body is extruded by another, there is interchange of places, and the extruder takes the place of the extruded, that again extrudes the next, the next the succeeding one (if there are more than one), until the last is in the place of the first extruder" (1895, 1350, lines 31–36) [Hanson's quote comes from Clagett (1959, 508) – *MDL*].

bowstring (this being like ourselves shoving a cornerstone), continues in motion only because the circulated air replaces the string – further impelling the arrow forward in flight just as the bowstring had impelled the arrow forward from the archer's hand.[12]

The deficiencies of this view, as an account of the nature of violent motion, (as against 'natural' motion, e.g., that of the planets), have often been delineated, initially by Aristotle himself. Thus:

> The theory of 'mutual replacement' makes the whole series of things move and cause motion simultaneously, so that they must also all cease to move at the same time; whereas the appearance presented to us is that of some one thing moving continuously. What then keeps it in motion, seeing that it cannot be the same movent [all the time]?[13]

Philoponos is more sweeping in his rejection – attacking both the simple 'antiperistasis' theory of Aristotle's *Physica*, IV, and also the more sophisticated 'air-impetus' theory of VIII, which was meant to substitute for the earlier one: "... suppose... that the air pushed forward by the arrow gets to the rear of the arrow and thus pushes it from behind... [what is it] that causes the air, once it has been pushed forward, to move back... along the sides of the arrow, to turn around once more and push the arrow forward?... how can this air, in so turning about, avoid being scattered into space, but instead impinge precisely on the notched end of the arrow and again push the arrow on and adhere to it?"[14]

However, even allowing that the motion of objects flung through the air may not be fundamentally understood in Aristotle's terms, it remains a fact that the air does circulate around objects moving through it, and that this circulation exerts physical forces on that object.

II

Consider now what an eminent Newtonian urged *vis-à-vis* a view held by some of his contemporaries – a view remarkably like Aristotle's Theory of Peristatic Motion:

> Bodies in going through a fluid communicate their motion (momentum) to the ambient fluid by little and little, and by that communication lose their own motion (momentum) and by losing it are retarded. Therefore the retardation is proportional to the motion (momentum) communicated; and the communicated motion when the velocity of the moving body is

---

[12] Preliminary to criticizing Aristotle's theory, Buridan develops the idea of mutual replacement still further: "... 'antiperistasis' holds that the projectile swiftly leaves the place in which it was and nature, not permitting a vacuum, rapidly sends air in behind to fill up the vacuum. The air moved swiftly in this way, and impinging upon the projectile, impels it along further." From his commentary on Aristotle's *Physics* Book VIII, Q. 12. The quotation is from (Clagett 1959, 532–533).

[13] "Antiperistasis autem simul omnia moveri facit et movere; quare et quiescunt. Nunc autem unum aliquod quod movetur continue a quolibet, non enim ab eodem", *Physica*, VIII. 10. 266 b. 27–267a. 20. (ff. 158v–159r.) [Hanson takes this quotation from Heath (1949, 156). While Heath translates all the lines in Hanson's citation, the text Hanson quotes is only from 267a.17–20. – *MDL*].

[14] This criticism is developed in the late tenth century: cf. Pines (1953).

given, is as the density of the fluid; and therefore the retardation or resistance will be as the same density of the fluid, nor can it be taken away, *unless the fluid coming about to the hinder parts of the body restore the motion lost.* Now this cannot be done unless the impression of the fluid on the hinder parts of the body be equal to the impression of the fore parts of the body on the fluid; that is unless the relative velocity with which the fluid pushes the body behind is equal to the velocity with which the body pushes the fluid; that is, unless the absolute velocity of the recurring fluid be twice as great as the absolute velocity with which the fluid is driven forward by the body, which is impossible. (Cotes [1713] 1960, XXXI)[15]

From Aristotle came the view that the air is, fundamentally, a liquescent medium, a very light, but definitely ponderable fluid. *That* is a most important insight in the History of Aerodynamics. From Newton comes the criticism (the one we note in an Appendix as a difficulty internal to Aristotle's own antiperistasis doctrine of *Physica* IV), that the resisting-assisting effect of this fluid on any body moving through it cannot possibly constitute the total explanation of that body's continued motion.

Nonetheless, *some physical effect the medium certainly does have.* Indeed, it is this effect of the liquescent medium on the body which is alone responsible for the lift, the 'upward swerve', long known to boatsmen and surfboarders, to hydrody-namicists and aerodynamicists. With particular respect to the phenomenon of flight, any proposed explanation of a body's motion in terms of inertial forces alone, would be completely inadequate. Aerodynamics and ballistics are two quite different dis-ciplines – a lesson apparently not yet learned by those writers of histories of flight who devote their final chapters to NASA's Cape Kennedy achievements.[16]

So Aristotle was speaking responsibly about something, but not perhaps about what he thought he was concerned with. Regarding an object's motion through an aerial medium, Aristotle correctly noted the latter's resisting and assisting effect on the former.

Aristotle's doctrine of projectile motion as founded on the assisting action of the air (itself construed as a fluid medium), was opposed in the sixth century by Philoponos as we saw earlier. This Greek grammarian urged that the casting of a projectile *imparts* to it a certain energy, or impetus, which is transported from the throwing agent and somehow 'transfused' into that which is thrown; this view became independently prominent in the fourteenth and fifteenth centuries and is genetically connected perhaps with the more modern theory of inertia. Philoponos felt that it was this energy, or impetus, which maintained the arrow in motion during its trajectory long after having left the bowstring. The air *only* resisted its progress forward. For Aristotle the source of motion was always external to that which is moved; Philoponos felt that it somehow became 'internalized'. Thus, when the bowstring is no longer effective through direct contact, Aristotle saw something else (the circulated air) as continuing to be effective through such direct contact. For, since the arrow continues to move, something must continue to push it. Philoponos

---

[15] Cotes is certainly interpreting Newton's own position as against a view which is at once continuous with Aristotle's *antiperistasis* theory, as well as d'Alembert's paradox – which it antedates by a century [Hanson has added the parentheticals and the italics to this passage. –*MDL*].

[16] Cf., e.g., *The American Heritage History of Flight* (Josephy 1962), especially Chap. 10. See also the review of this work by N. R. Hanson (1964b).

views a body's *impetus*, however, as internal to the body – which thus becomes the source of its own motion. For this he was roundly criticized by St. Thomas Aquinas.[17] Aristotle's *medium* theory and Philoponos' first intimations of an *impetus* theory, remained in opposition to each other until the collective insights of Descartes, Galileo and Newton established the Law of Inertia beyond question.[18] By the late seventeenth century, all serious controversy on the matter was virtually terminated; the air was thenceforth conceived of only as a factor of resistance to the motion of thrown or flown objects. No one continued to argue, *à la* Aristotle, that the aerial flow could actually assist in some manner the forward and upward flight of a projectile, or of a bird. The latter are often described in occult, mysterious terms, due to their capacity to ascend through 'nothing at all' – or rather through something which usually only impedes flight instead of helping it. As we shall see, this Aristotelian lesson had to be relearned once more.

Leonardo Da Vinci was an interesting interim figure, albeit historically ineffectual. Before 1506 he had assumed that some action of the air did assist motion. After this date, however, he thought of air largely as a resisting medium – which resistance he ascribed to its 'compressibility'. It was this compressibility, indeed, which Leonardo identified as the cause of lift in bird flight. Local condensation of the air just beneath the bird's wing is what supports the animal as it glides through the air as on a slightly downward-inclined plane: thus Da Vinci's view – which is still compatible with Aristotle's insight. Some physical property of air supports birds just as some physical property of water supports ships.[19]

Galileo, whose Mechanics crushed the Aristotelian theory of the 'pushing' medium, recognized a proportionality between the air's compressibility and the velocity of objects moving through it. One might hardly have expected anything else: the porpoise would ply through thick oil, or liquid mercury, or a tank of sand, much slower than through water. Galileo's work with pendulum-pairs, involving precise calculations for the vacuum condition, led to this conclusion that projectile paths, and pendulums, could be described in a manner unaltered by air resistance. This was because the projectile's velocity, and air compression itself, were so related as to make the single calculation equally valid for a vacuum – not unlike the constant relationship which obtains between thrust and velocity, irrespective of what is being thrust, and the medium through which it moves.[20]

---

[17] "However, it ought not to be thought that the force of the violent motor impresses in the stone which is moved by violence some force (virtue) by means of which it is moved.... For [if] so, violent motion would arise from an intrinsic source, which is contrary to the nature (*ratio*) of violent motion. It would also follow that a stone would be altered by being violently moved in local motion, which is contrary to sense. Therefore, the violent motor impresses in the stone only motion and only so long as it touches it. But because the air is more susceptible to such an impression...." (Book III, 2, lect. 7, 305 c. 1) [Again, Hanson quotes Clagett's translation (1959, 517). – *MDL*].

[18] Actually, it is unlikely that Philoponos's early version of the impetus theory deeply influenced the later medieval versions of de Marchia, Buridan, Oresme, Albert of Saxony and Marsilius of Inghen. See (Clagett 1959, 514 ff.)

[19] See Giacomelli (1930).

[20] Galileo suspended two identical leaden spheres on cords of equal length. They were made to

III

The second book of Newton's *Principia* explores the resistance, compression, and hydrodynamic forces exerted on moving objects by media such as water, air, oil… etc. There Newton's views, as expressed earlier in this paper, are found. A long history of distinguished hydrodynamicists developed Newton's insights: The Bernoullis, D'Alembert, Robinson, Bossut, Euler, Lagrange, Dubuat, Borda, Avanzili, Robons, Hutton, Vince, Cauchy, Poisson, Bessel, Plana, Reynolds, Stokes, Helmholtz, Kelvin, Rayleigh, Rankine, Wenham, and Langley. They all recognized that our atmospheric air, like water, exerts an appreciable force on objects slightly inclined to 'the wind'. As with water skis, so it is with kites, gliders, flat stones and soaring birds. That is, this long series of investigations consisted not merely in delineating the numerical ratios of resistance to forward motion. It was also concerned with establishing that, at certain angles of incidence, objects actually rise through the aerial medium.[21] Granted, this was almost always by analogy with the forceful resistance offered by air to the arrow at the arrowhead. Birds and kites alike seemed capable of ascending through the air only because of positive pressure being exerted on their undersides – as the fountain's jet of water keeps the ping pong ball aloft by pushing it up from underneath.

Thus it was that the remarkable physical insight, and theoretical tenacity of Lanchester and Prandtl brings aerodynamic theory around full circle to something not wholly unrelated to Aristotle's remarks in *Physica,* IV.

Lanchester perceived[22], as had others, that a large door, pushed open quickly, would be stoutly resisted by the air it moves into (i.e., against). He perceived also that the same door, if dropped horizontally from a high ceiling, would build up a similar resistance in the air beneath it. In short, the air *pressure* beneath the falling

---

oscillate with amplitudes of 10° and 160° respectively; their velocities he expected then to be in the ratio of 1:16. But "we see", he said, "that the two numbers are equal, which is the proof that the two pendulums have been resisted by air proportional to their velocities". [See Galileo ([1638] 1946, 244) for the discussion Hanson here references. Hanson oversimplifies the experimental setup, since Galileo only indicates that one ought to have the two pendulums oscillate through widely different amplitudes and doesn't uniquely specify 10° and 160°. Furthermore, the translation Hanson provides could not be found elsewhere, and is different enough from that of Henry Crew and Alfonso de Salvio that it is worth giving their translation here of the passage in question: "…if two persons start to count the vibrations, the one the large, and the other the small, they will discover that after counting tens and even hundreds they will not differ by a single vibration, not even by a fraction of one." (244) Then, to conclude, "…all motions, fast or slow, are hindered and diminished in the same proportion." (245) –*MDL*]

Galileo doubted that the proportionality between resistance and velocity would hold without limit. He feels that, at very high velocities (e.g., those of musket balls) the path as calculated for a vacuum would differ from that described through air.

[21] The signal discovery of modern aerodynamic theory is that certain objects, shaped in special ways, will lift up through the aerial medium at a zero angle of incidence. This will be discussed in more detail later.

[22] As early as 1891. He reviewed the development of this basic insight in his Wilbur Wright Lecture before the Royal Aeronautical Society in 1926.

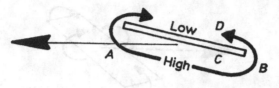

Fig. 23.1

horizontal door will be greater than the pressure immediately above the door; – just as the air pressure immediately ahead of Aristotle's arrow-in-flight will be greater than that just aft its notched stern.[23] It is just as if the door were descending through a room packed full of soap bubbles. They will build up densely beneath, and seem considerably more rarified above. *And* the high pressure air beneath will tend to 'spill around' the edges and corners of the door 'in order to' move into the lower pressure region above the door. This is, in effect, the principle of the parachute, from Leonardo to yesterday.

The same thing would obtain if the door were horizontal half way between the ceiling and the floor (in a large gymnasium, for example), and moving toward one wall with its leading edge slightly lifted. Increased pressure will again build up beneath the wing, just as when the door was imagined to fall from the ceiling. Again, the air will tend to spill upward to the low pressure region above the door (see Fig. 23.1).

This high pressure region is not localized directly beneath the door; it extends in all directions – both forward and behind and at either side of that door. Thus, as the door moves towards the wall the high pressure air immediately ahead and below the door's leading edge (A) will tend to swerve upwards toward the low pressure region just above (Fig. 23.2).

Similarly then for the high pressure region beneath[24] the trailing edge of the door (B); the air there will also tend to move upwards behind the door. It is beneath the aft, lower side of the door (C) that the air would be directed downward. And again, it is above the aft trailing edge of the door (D) that the air will also be moving downward albeit in a somewhat turbulent fashion.[25]

---

[23] On this point, incidentally, Aristotle was wholly correct. Schleiren photographs of arrows in wind-tunnels reveal increased pressure immediately ahead of the forward tip and lower pressures just behind. Aerial circulation of some kind is therefore inevitable.

[24] As Lanchester's reflections develop he finds it advantageous to ignore the air spilling up around the shorter sides of the door – those two edges parallel to the door's forward motion. He thus imagines a door possessed only of a leading and a trailing edge – with no shorter tip sides at all. In short, he speculates on the aerodynamic properties of a door of infinite aspect ratio, i.e., all span, a finite chord (from leading edge to trailing edge), with no tips over which the high pressure air below can spill 'inwards' at right angles to the direction of flight. This transforms his problem into one of two dimensions: "the two-dimensional parachute thus became a two-dimensional glider" (Giacomelli and Pistolesi 1934).

[25] As Lanchester explained it, the air particles will receive an upward acceleration as they approach the aerofoil and will have an upward velocity as they encounter its leading edge. While passing instead under or over the aerofoil, the field of force is in the opposite direction, *viz.* downward, and

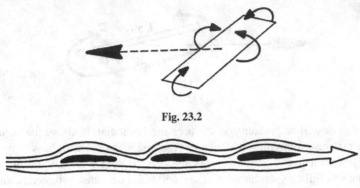

Fig. 23.2

Fig. 23.3

To Lanchester's basic insight concerning aerial pressure-differentials around the door we now add a refinement: the flat door, of our example, will churn up turbulence in the air as its stubby cornered leading edge moves forward at an angle. Imagine what a turbulent, bubbling, frothy wake the door would leave, were it moving through sea water in the same way. The challenge, clearly, will be to keep the air as little turbulent as possible – as 'laminar' and smooth as possible – so as to dissipate a minimal amount of energy in mere turbulence and useless frothing. This achievement would maximize work in the form of increased air pressure below the door and decreased pressure above; thus the provision of a 'lift component' to the door's motion – one which tends to counterbalance its natural 'disposition' to fall to the floor. This diminution of turbulence can be achieved by altering the shape of the door – so that the upper surface curves gently from a rounded edge through a thicker section one third aft, thence tapering to a very sharp trailing edge at the rear. The underside can remain flat, or slightly concave, or slightly convex – just so long as it is cleanly curved (Fig. 23.3).[26]

If the curved door, as seen on edge, were a purely symmetrical (fishlike) shape, with identically contoured surfaces above and below, then (when at zero inclinations to the laminar airstream) particles of air going above will be required to traverse the same length of path to reach the trailing edge as will those which travel below (see Fig. 23.4). But when the underside is flat or only slightly curved in a

thus the upward motion is converted into a downward motion. Then after the passage of the aerofoil, the air is again in an upwardly directed field, and the downward velocity imparted by the aerofoil is absorbed. This is treated in Lanchester (1908, esp. Chaps, iii and iv). Figure 23.8 should clarify the point involved.

[26] These shapes Lanchester experimentally determined to constitute the aerodynamically most efficient means of receiving a current of air in upward motion and imparting to it a downward velocity – the better to achieve a *conservative* aerofoil, the energy of the fluid motion being thus carried along and conserved, just as in wave motion.

Horatio Phillips had already discovered the aerodynamic advantages of the 'dipping edge'; he even patented this innovation, once in 1884 and again in 1891. Lilienthal happened upon the same disclosure in 1889 and 1894 – again purely as a result of trial and error (Lilienthal 1889). The same insight also seems implicit in the work of Mouillard (1881), Wenham (1866) and Cayley (1809–1810). But, as Lanchester himself stresses, it appears that all these were independent intuitions.

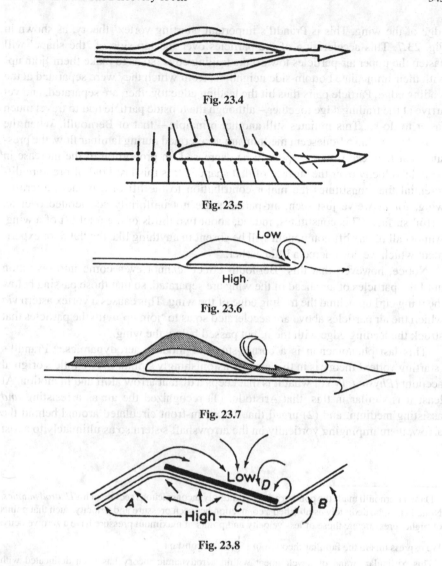

Fig. 23.4

Fig. 23.5

Fig. 23.6

Fig. 23.7

Fig. 23.8

convex or concave manner, the particles below will have a much shorter path to traverse. When one initially accelerates such a shape through the aerial medium the air particles moving over the top of the shape do not reach the trailing edge at the same time as those which traverse the lower side of the 'door' (see Fig. 23.5). Combining this obvious fact with Lanchester's recognition of the high pressure region beneath the shape, it follows that the air particles beneath, those which reach the trailing edge first, will curl up behind the door and circulate back toward the low pressure region immediately above the center of the shape's upper side (see Fig. 23.6). It will, in fact, initiate a vortex the center of which will lie longitudinally along the trailing edge of the 'wing'. Because of this initial vortex the air particles moving more slowly over the upper edge will be accelerated toward the trailing edge of the wing. This is Prandtl's important 'starting vortex' theory, as shown in Fig. 23.7. This acceleration of the particles over the top side of 'the shape' will hasten the upper air-particles toward the trailing edge, so as to make them 'join up' with their immediate bottom-side neighbors, from which they were separated at the leading edge. Particle pairs thus hit the leading edge together, are separated, and yet arrive at the trailing edge together – although the topside particle had to travel much faster to do so. This initiates still another principle – that of Bernoulli. When the 'particles' within a liquescent medium are accelerated during laminar flow, the pressure that they exert on a shape will be decreased, in proportion to the increase in particle velocity over the surface of that shape.[27] It is this last kind of pressure differential that constitutes the major contribution to the lift of a modern aircraft's wing, for as we've just seen, air-particles are not uniformly accelerated over an airfoil surface. This constitutes, indeed, about two thirds of the total lift of a wing. Almost all of *this* lift-component will be absent in anything like the flat door experiment which we considered a little earlier.

Notice, however, that this 'Bernoulli effect' cannot even come into operation until the particles of air ahead of the wing are separated, so that those passing below the wing curl up behind the trailing edge of the wing. This causes a vortex astern *via* which the air particles above are accelerated so as to 'join up' with the particles that struck the leading edge with them, but passed below the wing.

This last phenomenon is a corner-stone of modern aerodynamics.[28] Prandtl's 'starting vortex' theory is in this manner astonishingly similar to Aristotle's original account *(Physica,* IV) of what it is that keeps a distant arrow aloft and in motion. At least it is similar in this, that Aristotle: (1) recognized the air as a resisting *and* assisting medium, and (2) urged that the air-in-front circulated around behind the arrow, there impinging vortically on the arrowshaft's stern so as ultimately to assist

---

[27] Daniel Bernoulli made the point in a disappointingly inconspicuous section of his *Hydrodynamics*. Nonetheless, he definitely established a connection between pressure and velocity, such that points of higher pressure are those of less velocity and points of maximum pressure have a zero velocity. We express this in the familiar theorem, $p + \frac{1}{2}\rho V^2 = $ constant.

[28] This particular strand of development within aerodynamic theory has been delineated with unrelieved simplicity. So much has been deleted, e.g., Lanchester's notion of induced drag, the result of the wingtip vortices which develop in actual wings of finite span, as against the 2-D fictions he speculates about initially (compare Note 24); Kutta's development of Lilienthal's curved laminae, and Joukowski's still more subtle geometric elaborations; Prandtl's investigations of the boundary-layer (of which the 'starting vortex' hypothesis is but one derivative insight). But the oversimplification is justified, I feel, in that the analogies between Prandtl's 'starting vortex' and Aristotle's antiperistasis are striking; in both theories aerial circulation is fundamental. In both, the thus-circulated air has a definite physical effect on the projectile, an effect which prevents the flying object (arrow or airfoil) from collapsing to earth. In both theories the relationship between forward motion and resistance to that motion by the medium has a centrality which makes Aristotle and Prandtl seem more alike than either of them are with the legions of mechanicians who worked during the 2300 years intervening. Some simplification seems in order when such an unsuspected resonance of ideas can be made to stand forth.

it in its motion forward and in its resistance to falling. These two points are analogously embodied in the modern theory of airfoil structure.

So once more we encounter Aristotle succeeding (almost inadvertently) in anticipating fundamental features within the particular examples which he chose to illustrate some general case. The philosopher today who seriously asks 'how can birds, and airplanes, fly' cannot but be struck with the omnipresence of *Physica*, IV in most of the modern aerodynamic answers to that question.

Appendix

Consider further some difficulties internal to Aristotle's theory of motion-through-air, his 'aerodynamics'. Concerning the aerial medium through which the arrow is slicing, there are two possibilities; either it (a) resists the motion of the arrow, or (b) it does not resist that motion. The second hypothesis (b) construes the air ahead as indifferent to the arrow's forward motion. It only circulates behind the shaft, so as to impel it forward. Thus:

> ...the view held that the impelling force of a projectile was associated with forces exerted on the base by the closure of the flow of air around the body. This view of air as an assisting rather than a resisting medium persisted for centuries...[29]

Provisionally accepting this – that the air does not resist but only assists the motion of the arrow – we would be forced to countenance the creation of energy *ex nihilo*. To see this suppose that the arrow's velocity is v, and the parted air in front circulates (without resistance) around behind the arrow, where it also (presumably) has a velocity of v, it would not there convey any 'push' to the aft end of the arrow. The air's velocity against the arrow's stern must then be $v + \wedge v$, in order that the arrow should actually receive some nudge from the air. Otherwise we must simply countenance an arrow moving at velocity v, and the air also moving in behind it with v – in a kinematical succession devoid of dynamical effect, like adjacent twigs floating downstream. This kind of criticism is very much in evidence in Scholastic criticisms of *antiperistatis*.

The Encyclopedia's contributor, however, is historically in error with respect to this interpretation, for Aristotle clearly writes: "Now the medium causes a difference because it impedes the moving thing..."(*Physica* IV. 8. 215ᵃ28)[30] The air in front, then, *does* resist the arrow's forward motion on Aristotle's view. Thus, if the arrow moves with v, by giving the air parted in front a velocity v, it must itself suffer some loss in velocity. Its forward motion (after impact with the air ahead) will be $v - \wedge v$. The air itself, however, will curl in behind the arrow with v. The air aft, then, will have a velocity at any given time greater than the arrow's velocity at that time. It will therefore *push* the arrow forward – as Aristotle requires.

---

[29] *Encyclopedia Britannica*, 14th ed., s.v. "aerodynamics."

[30] The translation is from Heath (1949, 116). – *MDL*

However, one awkward consequence of this presentation is that there will be little difference between the passage of an arrow through air, and its passage through a complete vacuum. If the resistance of the air ahead is exactly compensated for by the air rushing in behind, then the air might as well not have been there in the first place – save for the initial retardation caused by the arrowhead's *first* contact with the air ahead. [This is, in fact, a simplified aerodynamic version of D'Alembert's Paradox which considers a fluid's pressure on the forward contour of a submerged cylinder to be precisely equal to the pressure of the fluid curling back in behind the cylinder, the net result being *no resistance at all!*] If, however, the air impinging on the arrow's stern does *not* exactly make up for the energy lost at the arrowhead, then some discernibly gradual loss of velocity, in the flight of the arrow, should be remarked. Aristotle does not call attention to this. In any event, such loss in velocity as there is would be beyond the capacity of any but the most sensitive instruments to detect.

After Aristotle, contributions to our knowledge of motion through water and air did not cease. But the later contributions are more difficult for the scholar to perceive. Archimedes and Hero of Alexandria accomplished much, although their scholarly pulses were not quickened by problems concerning the dynamics of fluids, as Aristotle's had been. Philoponos, as we saw, opposed the analysis offered in Book IV of Aristotle's *Physica*. His was a primitive impetus theory of motion, a theory which urged that the soaring arrow continues through the air *not* because the air-particles parted in front circulate astern and impinge on the aft members of the arrow, but rather because the action of the bowstring is such as to impart a new property to the departing arrow, an impetus. Thus the arrow resting in the archer's quiver lacks a property which the shot arrow possesses, namely this impetus which is the cause of its continued motion through the resisting air.

On this point Philoponos was criticized by Thomas Aquinas who finds this sixth century theory absurd for two reasons: (1) a simple violent motion is said to make a moving object ontologically different from what it had been at rest – as though a running horse had physically different attributes from what it has when quietly grazing; (2) the theory makes the moving object the intrinsic source of its own motion, which is clearly exposed by Aristotle as an untenable view, since the motion of x is caused always by an external mover of x.

Roger Bacon offered little to this issue of antiperistasis vs. impetus, except to enunciate clearly that, in his opinion, human flight was a real possibility, requiring little more than time, patience, the observation of birds and great skill at fashioning and fabricating mechanical artefacts.

But Robert Grosseteste, John Buridan, Nicholas of Cusa and Nicole Oresme – the great medieval scholars whose criticisms of the Aristotelian philosophy, which had become so dogmatized during the thirteenth and fourteenth centuries, prepared the way for the scientific revolutions of subsequent periods – these men developed the impetus theory into a subtle natural philosophy far beyond anything Philoponos might have hoped for. Indeed, it is doubtful whether Philoponos' work was widely known at this time.

From the medieval impetus explanation of projectile motion through air, it is not a long step to the law of inertia as known to Galileo and Descartes. But it is a step which passes near the person of Leonardo da Vinci, whose contributions to the theory of flight will be considered in the next lecture.

References

Aquinas, Thomas. 1952. *In libros Aristotelis de caelo et mundo exposition, Opera omnia.* Vol. 3. Rome.

Aristotle. 1936. *Aristotle's Physics.* Trans. W. D. Ross. Oxford: Clarendon Press.

———. 1990. *Physica: Translatio Vetus, Aristoteles Latinus,* VII 1. Ed. F. Bossier and J. Brams. Leiden: E.J. Brill.

Baker, Max P. 1951. *The Wright brothers as aeronautical engineers.* Washington, D.C: Smithsonian Institution.

Cayley, George. 1809–1810. On aerial navigation. *Nicholson's Journal of Natural Philosophy,* November 1809: 1–8; February 1810: 1–6; March 1810:1–10.

Clagett, Marshall. 1959. *The science of mechanics in the Middle Ages.* Madison: University of Wisconsin Press.

Cotes, Roger. [1713] 1960. Preface to the second edition of *Sir Isaac Newton's mathematical principles of natural philosophy and his system of the world,* by Isaac Newton. Trans. A. Motte and F. Cajori. Berkeley: University of California Press.

Davy, Maurice John Bernard. 1948. *Interpretive history of flight: A survey of the history and development of aeronautics with particular reference to contemporary influences and conditions.* 2nd ed. London: His Majesty's Stationary Office.

Galilei, Galileo. [1638] 1946. *Dialogues concerning two new sciences.* Trans. H. Crew and A. de Salvio. Chicago: Northwestern University Press.

Geoffrey of Monmouth. [1136] 1983. *History of the kings of Britain.* Trans. L. Thorpe. New York: Penguin Books.

Giacomelli, R. 1930. The aerodynamics of Leonardo da Vinci. *The Journal of the Royal Aeronautical Society* 34 (240): 1016–1038.

Giacomelli, R., and E. Pistolesi. 1934. Historical sketch. In *Aerodynamic theory,* ed. F.W. Durand, vol. I, 305–394. Berlin: Springer.

Hanson, N.R. 1964a. Actio in Distans. *Yale Scientific Magazine* 39 (3): 15–20.

———. 1964b. Book review: The American heritage history of flight by Alvin M. Josephy. *Isis* 55 (2): 230–234.

Heath, Thomas. 1949. *Mathematics in Aristotle.* Oxford: Clarendon Press.

Josephy, Alvin M. 1962. *The American heritage history of flight.* New York: American Heritage Publishing.

Lanchester, F.W. 1908. *Aerodynamics: Constituting the first volume of a complete work on aerial flight.* New York: D. Van Nostrand.

Laufer, Berthold. 1928. *The prehistory of aviation.* Field Museum of Natural History, Publication 253, Anthropological Series, Volume 18(1).

Lilienthal, Otto. 1889. *Der Vogelflug als Grundlage der Fliegekunst: ein Beitrag zur Systematik der Flugtechnik.* Berlin: R. Gaertner.

Mouillard, Louis Pierre. 1881. *L'empire de l'air: essai d'ornithologie appliquée a l'aviation.* Paris: G. Masson.

Pines, S. 1953. Un précurseur Bagdadien de la théorie de l'impetus. *Isis.* 44 (3): 247–251.

Plato. 1911. *The dialogues of Plato.* Trans. B. Jowett. New York: Charles Scribner's.

Simplicius. 1895. *In Aristotelis physicorum libros quattuor posteriores commentaria.* Edition of H. Diels. *Commentaria in Aristotelem graeca.* Vol. 10. Berlin: Reimer.

Wenham, Francis Herbert. 1866. *On aerial locomotion and the laws by which heavy bodies impelled through the air are sustained.* First Annual Report of the Aeronautical Society of Great Britain, 10–46.

Chapter 24
Lecture Two: The Shape of an Idea

Socrates said those who come at the second night are truly lovers of wisdom. It isn't certain that Socrates ever said that, but it is quite clear that it is the sort of thing that he might have said, and if he had I would have quoted him tonight. Last night we considered the discovery of the air, and what this meant for the History of aerodynamic theory and the development of the concept of flight. This evening I want to discuss the shaping of an idea. The idea is this: flight is a subject matter which can be treated objectively and scientifically and consideration of this idea might sooner or later actually lead to the construction of a proper flying machine. Erwin Schrödinger once said "Nature will tell you a direct lie if she possibly can".[1] Nature told such a lie to Daedalus and to Icarus; nature told the same lie to the Chinese Emperor Shun and to the Saracen of Constantinople. And nature also told that lie to Leonardo Da Vinci. The lie was this: all birds flap their wings; therefore flapping wings are somehow essential to flight. Further, any theory of flying through air requires the idea of flapping as a primary premise in the argument. In that collection of notes and random jottings that remain of the literary works of Leonardo, we find something resembling a monograph on the flight of birds, a tract written in 1505 (1893 and 1894–1904).[2] Here, Leonardo was quite sympathetic to the Aristotelian doctrine of antiperistasis which we considered last night. It is the doctrine that a body moving through air is assisted in its forward progress by the circulation of those particles separated by the arrowhead which come around behind and impinge on the aft section of the arrow. But after 1505 Leonardo lost his sympathy for this theory. From that point on he saw air as fundamentally a factor of resistance, something that tended to slow moving objects down, and thought that this was due to a

---

[1] Hanson likely was thinking of Darwin here, who is reported to have said, "Nature will tell you a direct lie if she can." (quoted in Beveridge 1957, 25). –*MDL*

[2] For a detailed discussion of Leonardo's views on flight, the reason for their lack of historical influence, see Hart (1961, chs. ix, x). Plates 11–126 of Hart's book reproduce the sketches and designs of Da Vinci for flying machines.

© Springer Nature B.V. 2020
N. R. Hanson, *What I Do Not Believe, and Other Essays*, Synthese Library 38,
https://doi.org/10.1007/978-94-024-1739-5_24

property of air which Galileo and others later referred to as condensability. Today we would call it compressibility.

Leonardo's question, as we reconstruct it, is: what is it that birds do in flight which utilizes the air's condensibility, or compressibility to sustain them aloft? His answer was that as the bird's wing is brought down swiftly the air beneath the wing compresses, thereby acquiring the sustaining properties of a solid body. Just as, to use a familiar example, the compressed air in a tire pump forces the handle back up if the handle is released suddenly, so Leonardo reasoned, the air compressed just below a wing of a bird forces that wing back up and thereby succeeds in carrying the weight of the bird. For Leonardo, as for all his predecessors and even for armies of his successors, the attention-getting movement of the bird's wing is *the key factor that* makes heavier than air flights possible. This view directs attention to that which we find in other such movements, e.g., a dog when he is swimming, a large oriental fan as it moves briskly before a damp brow, the oar of a boat as it presses the sea water, or a bellows forcing air into cooling embers. All these rapid motions pressing solid surfaces against liquescent media, whose resistance and quasi-solidity is thereby increased. Thus do birds flap their wings down against the air, the resistance and solidity of which sustains the bird.

Leonardo went further. If the air is motionless, then the wing must be driven down against it. But if the air itself is in motion and is driven up against the underside of the wing, the result will be aerial support in exactly the same sense. The air condenses beneath the stationary wing just as it does against a large barn door. The latter is forced shut just as a gliding bird is forced upward, overcoming gravity in the most effortless manner nature ever devised.

Thus, three great insights seem to derive from the reflections of Leonardo Da Vinci. First, because air is to some extent compressible it has an effect on the object moving through it. This, as we now know, is clearly true. But such effects are only manifested at very high subsonic velocities, usually above 650 miles per hour at sea level. Leonardo construed this particular effect as the general cause of all flight at any velocity however slow and sedate. These so-called compressibility effects were a danger to all pilots in the 1940s and early 1950s, something always to be avoided in flight if at all possible. Leonardo Da Vinci saw these as unavoidable if flight was to be achieved at all. As with Aristotle before him, Leonardo's reasoning disclosed the likelihood of certain phenomena long before they were actually experienced. And both of these geniuses then extrapolated, somewhat too hastily perhaps, interpreting certain necessities as general requirements for all motion through air.

The second great insight of Da Vinci was his perception of the principle of aerodynamic relativity, an insight as important for us today as it was for his own reflections. This principle of aerodynamic relativity proclaims that the physical effect of air in relative motion near a wing is the same whether there is a fast moving wing acting on still air or a fast flowing air acting on a motionless wing. This perception is absolutely valid. It is, in fact, built into the very idea of a modern wind tunnel, and it forms a primitive basis for some of the most useful calculations in contemporary aeronautical theory, e.g., the Reynolds number which will be discussed later. As

Leonardo put it, resistance of a moving object against air at rest is equal to the resistance of the air moving against the object at rest.

The third insight of Leonardo was prophetic indeed. Air lift is generated in a wing moving rapidly through dead air, or in a fixed wing being pushed up by fast moving air, Da Vinci stated emphatically that only the latter configuration promised the possibility of human flight. He based this judgment on a comparison of the bird's pectoral musculature with that of man. The former is what articulates and implements the flapping wings of a bird, and in some species actually comprises almost 50% of the beast's total weight. The insignificant pectoral counterpart stretched across man's thorax seemed, to Leonardo, quite unequal to the task of achieving a rapid local condensation of air, as he calls it, sufficient for even the most intrepid Daedalus ever to lift his own weight even one centimeter off the earth. Significantly, we have not to this very day achieved flapping flight in any form not even in the most sophisticated mechanical system. Leonardo's hope was placed on the bat wing configuration as constituting man's best flight into a brisk wind. Nonetheless, Da Vinci tried his hand at the design of flying machines. One of his more famous sketches, complete with oar handles to actuate the feathery wings and a tail to modify flight attitudes with one's legs[3], has so much captured our attention that it is sometimes difficult to forget it in favor of the real discoveries that Leonardo Da Vinci did make relevant to control of motion through the aerial medium.

But these ideas had no historical influence. Da Vinci's work remained completely unknown to others until 1797; indeed this one passage concerning flight lay largely undiscovered until about 1893. So in the actual history of flight theory, as in all other branches of knowledge, Leonardo's cerebrations affected little in the thought of immediately subsequent generations.

Having said this much about Leonardo, now I hope it discloses why in the last lecture, I did not come clean on the distinction between liquids and fluids. Leonardo, of course, was concerned with the air as if it were pretty much a liquid kind of substance. We will consider later a sense in which it can actually be dealt with quite intelligibly as a fluid, though it is quite different from the ordinary sort of liquid. As natural philosophers, Galileo and Huygens were very much interested in the phenomenon of fluid resistance. Through a series of ingenious demonstrations and experiments, Galileo became convinced that the resistance put up by a fluid to a body moving through it (or a body around which the fluid moved) was proportional to the velocity of that movement. Thus the resistance offered to a fish moving at 4 miles per hour would be twice what it would have been at 2 miles per hour. Huygens argued that the correct relationship was between the resistance and the square of the velocity. Thus at 4 miles per hour the resistance encountered by our fish would be 4 times what it would have been at 2 miles per hour and 16 times what it would have been at 1 mile per hour. This is quite close to the eighteenth century estimates of that same relationship.

---

[3] It is not clear which sketch of Da Vinci's captured Hanson's attention. His description seems to fit a composite image of the different sketches reproduced in Josephy (1962, 26–27), a book which Hanson reviewed shortly before giving these lectures. –EM

This protracted development makes us respect even more Leonardo's solitary and independent reflections. For, he seems to have concluded, a resistance which increases as the square of the velocity might well sustain a light bird whose wing beats many times faster than a human eye can follow. How considerable must be the aerial resistance beneath the wing of a humming bird, if this reasoning is correct.

But Giovanni Borelli, in a work posthumously published in 1680, used the laws of the lever to make systematic measurements of the force of contraction of the skeletal muscles. Comparing men and birds on this basis he concluded that men have a very poor power-to-weight ratio, so poor that the flapping flight of birds is not for relative weaklings like man.

Prior to this time discussions of the possibilities of human flight were concerned with fashioning an ornithopter, a device in which men achieve flight by flapping wings attached to their arms or to pedals pushed by their legs. Borelli's conclusion, "... and therefore wing flapping by the contraction of muscles cannot give out enough power to carry up the heavy body of a man" (quoted in Brown 1927, 18–19) was not at all popular. It seemed to outlaw the only conceivable basis for human flight. Yet, Borelli's work must be considered a significant contribution to natural philosophy. As an anatomical analysis it has few equals in the history of science.

Very rare is it indeed, that a professional anatomist can make some sort of a discovery which will have this kind of an impact in the hallowed halls of physics laboratories. The range of responses to Borelli's treatise is itself quite noteworthy. Some construed it as the death-knell to any plans for artificial flight. Others recognized it as putting flapping flight beyond men but not perhaps soaring, or gliding flight. Whatever it was that flapping did – man of course could not do it. But, perhaps, man might get airborne without flapping. Borelli did not speak to this speculative possibility at all. A hard core of enthusiastic aspirant aeronauts dismissed all of Borelli's pronouncements as the pessimism of an old weak cowardly philosopher. Air machines continued to be built; aeronauts continued to be splashed all over the countryside; enthusiasm waned a little bit but not nearly enough. Thus, for example, did Domengo Gonzales according to Bishop Godwin's legend plan to lash 25 geese to a chaise longue. This was the stated attempt – the result isn't broadcast in literature. Cyrano De Bergerac speculated on the possibilities of aerial elevation by hitching swans to a light container and sailing aloft after sunrise. And many others had even more fantastic ideas. The concept of flight; the application of reason to a physical phenomenon; this became lost in a plethora of broken bones and cracked wings.[4]

If Aristotle can usually be relied upon to generate an arresting conception of lasting value, Sir Isaac Newton, who lived from 1642 to 1727, is likely to have produced a technique of perennial worth for analyzing this conception. Though Newton's fluid mechanics did not always reach the analytical and philosophical heights of his achievements in kinetics and dynamics, we must still recognize this great man as the founder of modern fluid mechanics. The signal insight of the

---

[4]The most notable exception to this generalization was the Jesuit mathematician, Francesco Lana, whose *Aerial Ship* [1670] (1910) introduced the idea of using lighter than air evacuated metal spheres to raise a ship.

Principia Mathematica Philosophiae Naturalis is this: that the motions and mutual interactions of all material bodies could be thought of as the phenomenal manifestation of the collective motions of all the micro particles involved. In his work in optics, in ballistics, in celestial mechanics, and in the mechanics that now bears his name, Newton exploited his novel calculational technique with astonishing success. The corpuscularian approach, championed by this man, analytically reduces the most complex intricate, swirling and turbulent happening to micro-prophecies which seem as conceptually straightforward as plotting the future events on a billiard table. And that, indeed, is the essential clue to Newton's contribution to hydrodynamics. For he continually treats the interactions between fluids and solids submerged within them as but the gross observation of effects due to micro-interactions between the particles of which the fluid is composed and the particles of which the solid is constituted.

If two billiard balls collide head on, dead center, they both recoil backwards along a straight line from the point of impact. If the cue ball strikes a stationary eight ball in the same way, the latter will recoil similarly. A flat door held perpendicular to the flow of a stream will also be forced backward along the line of flow – as it absorbs the impact energy of the fluid particle. From a corpuscularian point of view this is very much the same sort of phenomenon as in the case of the two billiard balls. If, on the other hand, the door is turned edge-on to the flow, and if it is very thin and soft at that edge, the resistance it encounters from the stream will be vanishingly small. Now at any angle between zero and 90° however, the door will be forced to move not only aft in the fluid but across the flow lines of that fluid. If we can imagine the fluid entering in a very laminar and straight series of streamlines, then we are going to have the situation wherein this impact between the particles of the fluid and the particles of this plate will be such not only as to drive the plate back, but also to drive it up and across the streamlines, something well-known to some of us who occasionally water-ski. Consider a man wearing water skis – just starting a run – say from a submerged standstill behind a motor boat. As the boat starts it pulls the rope taut, and then lifts his skis, presumably with him on top, to the surface of the water. Now what Newton would have asked is this: what is the fundamental physical connection between all these parameters involved, the skier's velocity, the water's density, the force of friction, and the angle the skis are held to the water's horizontal line. After myriad calculations, Newton's followers concluded that the resisting force acting on such an inclined plane in a fluid is proportional to the area, the square of the velocity, and to $sin^2\alpha$, where the angle of attack, α, is the angle between the skis and the horizontal surface of the water. The first term was Huygens' contribution and represented his correction to Galileo's formula. The sin^2 term is the infamous contribution of the Newtonians, though Newton himself is usually given the blame. I call it infamous because it has been blamed for holding back the practical development of aerodynamics for at least 50 years.[5]

---

[5] This blame might be a bit misleading, for there was always some uncertainty about the validity of the formula in question. Thus, Octave Chanute, writing in 1894, listed six different formulas that had been proposed. The law he attributed to Newton is $P = P' \sin \alpha$ where P' is the pressure against

Before examining the reasons for this allegation we should say something in defense of Newton. First, the obvious point, Newton himself did not formulate this equation. His disciples did that. This might seem to be another proof of the fact that philosophers have more to fear from their disciples than from their enemies. But in this case his disciples were simply using the analytic tools Newton himself had developed and extending the work that he had done.[6] The second point is that in Volume II of his *Principia* Newton was primarily concerned with the properties of an ideal fluid. An ideal fluid is incompressible, as air is not, is irrotational – the main flow does not have any rotations as tributaries – and is non-viscous, or has no internal friction between the molecules that compose it. Such idealizations, which Euler and Bernoulli also used, allow for an elegant mathematical treatment of an otherwise fearfully complex subject matter. But, it also means that the resultant theory of fluids does not quite fit either air or water.

However, it must be admitted that the Newtonian $\sin^2$ law convinced many people in the nineteenth century that flying machines were altogether impossible. If the force is proportional to $\sin^2 \alpha$; then the lift will be proportional to $\sin^2 \alpha \cos \alpha$ while the drag will be proportional to $\sin^3 \alpha$. Consider what this would mean for the design of a Boeing 707. The maximum lift would come when the wings were at an angle of $60°$ to the airstream. This, in turn, would mean a very large drag. When a plane is in steady flight, or in equilibrium, the drag equals the thrust supplied by the motor. The important fact here is the lift to drag ratio. Since this depends on $\cos \alpha/\sin \alpha$ the *smaller* α is the better this ratio is. Hence one has contradictory requirements. The angle should be quite large, to give an adequate lift, and quite small to give a tolerable drag.

Suppose one went to the other extreme and kept α close to zero to maximize the lift to drag ratio. Then to get a Boeing 707 off the ground it would be necessary either to have wings the size of football fields, since force depends on the wing area, or to develop an incredibly powerful motor, which would undoubtedly mean an impossibly heavy motor. If the Newtonian $\sin^2$ law is correct it is hard to see how anyone could design a flying machine. For that matter it is hard to see how anyone could have designed a flying animal, a bird or a bat. People began to think that the flight of birds was something occult, irrational, miraculous. It was rather strange that Newtonianism, looked upon as the essence of rationality, should lead to the idea that flight is something that reason cannot penetrate.

Thanks to the Newtonian $\sin^2$ law, the possibility of human flight was put into the basket labeled 'Implausible', a problem that all innovators have to cope with. Remarkably enough, it was an often critical, but always dedicated, Newtonian who, in the course of developing some of his master's hydrodynamic equations, put the first crack in the Newtonian armor shielding the impossibility of flight. Daniel

---

a surface perpendicular to the wind and P is the pressure against a surface at angle α. The formula that Chanute himself favored, on the grounds that it seemed to fit the experimental data, was:

$$P = \frac{P' 2\sin\alpha}{(1+\sin 2\alpha)} .$$ This is discussed in the Introduction to his (1894).

[6] The basis these Newtonians build on were the developments in Sect. 7 of Book II, esp. Prop. 34.

Bernoulli (1700–1782), a member of one of the most amazing scientific families in history, coined the term 'hydrodynamics' in his treatise, *Hydrodynamics sive de viribus et motibus fluidorum commentarii* (1738). In this treatise he explained the relation between pressure and kinetic energy in fluid flow over a solid object. In modern terminology his formula is

$$H = p + \frac{1}{2}\varrho v^2 \tag{B}$$

In (B) H stands for the total pressure, p for the external pressure exerted on the fluid, ϱ for the fluid's density, and v for the velocity of the fluid. The term $\frac{1}{2}\varrho v^2$ is really the kinetic energy per unit volume. To see the significance of this formula think of water flowing in a pipe of variable width. Since the same amount of water goes through any section, H is constant and the terms on the right in (B) must have a constant total value. If the velocity increases, as it does when the pipe gets narrower, then the pressure must decrease.

Here lies a part of the secret of flight. It suggests how wings that are curved so that the air moves faster across the top than the bottom might provide a lift. It would accomplish this, not by increasing the pressure below the shape, as Newton's analysis required, nor by compressing the air beneath the shape, as Leonardo conjectured, but by *decreasing* the pressure above the wing. Think of a little boy in a soda parlor with a straw in a glass of Coca Cola. Theoretically he could get the coke up the straw by increasing the pressure on the surface of the coke in the glass. This could be called putting pressure on the object. A more satisfactory method is for him to lower the pressure at his end of the straw and then let the normal pressure on the other end force the fluid up into his face. In short, heavier than air flight might not be so much a pushing up from below as a sucking up from above.

This, as it happens, was one of the great insights in man's attempt to understand flight. To anticipate something that will be discussed in more detail later, fully two thirds of the lift of a bird's wing derive from this upper suction rather than from any of the air pressing on the underside of the wing. It is unfortunate that Bernoulli did not stress this more. It was the answer to the $\sin^2$ proof of the impossibility of human flight.

Into this conceptual arena came a man of remarkable insight and intuition, Sir George Cayley (1773–1857). Some eminent historians of aeronautics, notably the Frenchman Charles Dollfus and the Englishman, Charles Gibbs-Smith, have even claimed that he is the true discoverer of the airplane (Gibbs-Smith 1960, 188). This, of course, does not mean that Cayley actually built a functioning airplane, but that he contributed a fundamental understanding of precisely what it is for an airplane to constitute a genuine practical possibility. He was not a mathematician or natural philosopher of the rank of Newton or Bernoulli. His imagination was not the limitless tool that Da Vinci's soaring genius controlled. Yet, this engineer truly laid the foundations for practical flight and, in doing this, formulated a number of aerodynamic principles with a clarity rarely encountered in the history of thought.

Since some good standard works on Cayley are already available I am not going to belabor the genealogical and biographical details of this great individual (see Pritchard 1962 and Gibbs-Smith 1960, 188–196). I am simply going to digress for a moment and list seriatim some of the things for which he was responsible in his aeronautical innovations. He, of course, opted for the desirability of a fixed wing configuration. He recognized clearly that the idea of flapping was just out. Moreover, Cayley was quite convinced that a wing, in order to be maximally effective, had to be cambered. It has to be such that it was swelled up on the top and rather concave on the bottom. He is the individual who first brought forward the notion that wings should not go out straight at 90° from the fuselage body, but ought to be centered up just slightly, ought to have a positive dihedral, to use the technical term. This is very important for later studies in aeronautical design because the main stability of an aircraft derives simply from this innovation. He was the individual who realized that the empennage, the rudder and the elevator which controls the yaw and the pitch of such a machine, should be fixed in the stern of the main plane. This is a simple inference from the fact that what is called the tail on a bird is where it belongs – at the rear. Cayley argues that this is the way it ought to be with aircraft too.

He worked out the principles for a primitive air screw in considerable detail. Then he made a statement summarizing what he thought was the basic problem of flight: "The whole problem is confined within these limits, viz., – To make a surface support a given weight by the application of power to the resistance of air." (quoted in Gibbs-Smith 1960, 149). That was in 1809– and it had everything in it, absolutely everything. This insight grounded his further conjectures about the kind of propulsion required for flight. As an engineer, quite familiar with steam engines, he knew that the power/weight ratios then available were quite inadequate to the task of propelling a flying machine. To get $1\frac{1}{2}$lbs. of horsepower one needed a 300 lb. engine. He had taken his coachman along on some of his glider flights, over distances of 120–200 yards. But he could never hope to bring him, or anyone else, along in a steam powered flying machine with that power/weight ratio.

Now let's reconsider the significance of what we have seen. Newton had considered a circular cylinder submerged in a flowing fluid. He postulated a flow that was perfectly irrotational, incompressible, and inviscid. Granted such an ideal flow, the total unbalanced force acting on the cylinder would be *zero*, something that was later known as d'Alembert's paradox, after the man who made this feature explicit. With this paradox we have some idea of the conflict between theory and practice in this developing field. Euler and Bernoulli had developed an elegant algorithm, a Euclidean theory of fluid flow. But it applied to absolutely nothing. Practical problems with actual fluids required, first of all, the recognition of irreducibly fluid macroproperties. Such things as turbulence, and defective boundaries on flow were problems that were bothersome to plumbers, to planners of bridges, to architects, to people who worked with real fluids. Such fluids are rotational and viscous. One may easily see turbulent vortices at the stern of a driven ship. Some of these fluids, e.g., air, were compressible, at least under some circumstances.

Hydrodynamics of the professorial variety, was a precious mathematical gain for serving the needs of the natural philosophers. But it was virtually useless to engineers, ship builders and architects. They developed quite a different discipline which they began to call 'hydraulics'. They did this in an unplanned way as just something that began to happen. This discipline of hydraulics was really just an *ad hoc* collection of recipes concerning fluid flow, pressure dropping pipe, viscosity, and so forth. The sort of thing that a carpenter, like a not so skillful housewife, might appeal to in the course of routine work. Yet this led to a series of techniques which we now recognize as the beginnings of statistical approximations, summing over the large classes of data, and general descriptions of fluid phenomena. It also left theoretical hydrodynamics to the universities and to the ivory towers.

Concerning the relationship then between theoretical hydrodynamics and the actual problems of flight and the concept of flight, science and theory seemed in the middle of the nineteenth century to offer virtually nothing. Newton had studied resistance and the underside pressure which had little to do with anything identifiable as aerial flight. Bernoulli's theorem was buried in an obscure and esoteric paper within a learned journal. Newtonian hydrodynamics concerned an entity so refined as to resemble nothing on earth, or even above the earth. The concept of flight hit a blank wall of incomprehension along the route designed by the natural philosophers. Physical theory gave way to experience, to trial and error, and to statistical approximations.

Now all of this is true except for one discipline: and that discipline is classical ornithology, the study of birds by bird-watchers, physiological zoologists, anatomists, and tangentially, the study of what birds do when they actually are in the air. As I said before, Leonardo's tract on the flight of birds was only unearthed much later than this period, only in 1893, so it had no effect. Nevertheless, there was, in the early part of the nineteenth century, a considerable rise of interest in what might be called natural history and biology. These new disciplines took over the earlier work of Buffon, the distinguished natural historian, and Goethe, the poet-scientist. With this renewed interest bird flight became once more an object of serious study. Borelli's treatise was recognized as basic.

In all these interests on the part of ornithologists, however, it was flapping which still seemed to be the clue to bird flights. Thus we find even Sir George Cayley trying his hand at building an ornifactor, a machine that flies by flapping wings. If human muscle power is the energy source I call it an ornithopter – but this was what Borelli had excluded. There were many others in this century who speculated about, or even tried to build, ornifactors. The list of names would include Stringfellow and Henson, who built model aeroplanes driven by miniature steam engines: Marriot and Columbine, who joined Stringfellow and Henson in forming a company for the commercial development of flying machines, a company which – need I say it – failed totally. Others who were ornifactor prone included Le Bris and Pilcher, Horatio Phillips, Prof. Samuel Pierpont Langley, and Otto Lilienthal.

Though the resulting ornifactors were not notably successful the discipline of descriptive ornithology grew. The flight style of sparrows was contrasted with that of eagles. Hummingbirds were contrasted *vis-à-vis* their motions with the flight of condors. The swift was contrasted with the flight characteristics of the albatross. Bats and birds were contrasted with each other. Slowly a puzzlement began to arise concerning how soaring birds like vultures, condors, and osprey, and even feather-less bats seem to rise indefinitely on motionless wings. At the same time interests in comparative physiology began to take hold. Many individuals began to contrast the wings of bats, which as you know, are fundamentally modifications of the index finger, with the wings of birds, which are fundamentally a modification of the entire member from the shoulder out.

These developments raise some fundamental questions about the modifications necessary to get the degrees of freedom for the type of motion we will be discussing in a moment. Physiologists studied musculature, that of the hummingbird as con-trasted with what they reconstructed as the musculature of the pterodactyl, and even of man himself. Feathers were studied, leading to a contrast between the primary feathers, which are out at the wing tips, and the secondary feathers, which are in close to the fuselage, the bird's body. All these comparative studies and contrasts were done at a time when there was no high speed photography. In fact there was virtually no photography at all, except for the remarkable multiple exposure photo-graphs Etienne-Jules Marey made of bird flight.

These developments and complications made it very difficult to detect the lie I spoke of earlier. There was some study of the wing section of birds. Phillips, Marey, and Lilienthal did pursue this subject and examine airfoil shapes and aspect ratios. But a very important landmark in this whole story derives from a kind of extra-professional interest in birds and their flight manifested by Lord Rayleigh, the great English physicist.

Rayleigh wrote over five hundred articles for technical journals, articles which won him the Nobel prize for his work on the density of gases and the discovery of argon. But he spent his Saturday afternoons working on articles for semi-popular journals on the lighter aspects of gases.[7] These articles concerned soaring and glid-ing birds, the swerve of tennis balls, and the problem of why a cricket ball bounces the way it does. He speculated on the phenomenon of swans taking off from still water, trying to explain how in the world they ever make it.

Some of you have probably watched this at close range and wondered about it. It raises the question: what does the wing do in its flapping motion; what is its real accomplishment? Rayleigh answered that it seems to provide a shape which sup-plies a lift through the hydrodynamic effect of air swirling upwards across its rela-tive flow lines. To get some idea of what is involved it helps to see some high speed photos of birds in flight.[8] Their wings are not a single piece of flying equipment, but

---

[7] Rayleigh's original publication on the soaring flight of the birds was in a letter to *Nature* (April 1883). In a letter to Rayleigh written in 1895, Lilienthal claimed that this stimulated his own attempts to solve the problem. See Rayleigh (1968, 335–341).

[8] For high speed photography of birds in flight see the outstanding collection of Aymar (1938). The theory of bird flight, together with further photographs, may be found in Storer (1948).

are made of different units. The parts of the bone are labelled: hand, wrist, forearm, elbow; and upper arm – as they are with us. The hand controls the primary feathers, the outer feathers that spread out in flight and function like propellers. The forearm controls the secondary feathers which provide the lifting surface. The shoulder and upper arm controls the wing as a whole.

On the down stroke the wings move downward and forward; on the back stroke they move upwards and backwards. The outer part of the wings, the fast moving primary feathers, provide the pull, while the slower moving secondaries, which are bunched together, provide the lifting surface. If the wing is to provide a lift on both the down and up strokes the slope must change. On the down stroke the leading edge, for the secondary feathers, is lower than the trailing edge; on the back stroke it is higher.

Now try to think of what this means in terms of the parameter we used earlier, α, the angle of attack, or the angle between the airfoil section and the airflow. When you take into account the different factors involved: the forward motion, the changing slope in the up and down motion, and the slight forward thrust on the downward motion and backward thrust on the upward motion, the net result is that the angle α remains practically constant.

It may be a bit difficult for you to visualize this. It is also a bit difficult for the birds to grasp this. My wife and I once spent a whole summer on the cliffs of Cornwall in what was a gull sanctuary. It was the time of the year when the young gulls were leaving the nest and learning to fly. Most of them were pretty slow learners, and didn't quite get this business of keeping their α constant while initiating a downstroke and then bringing the wing back up. It made their flight rather jerky. But with experience they gradually got the hang of it and their flight smoothed out.

This is what Rayleigh argued, that in bird flight the angle of attack remains effectively constant, at least for the main lifting sections of the wing. This means that, aerodynamically at least, there is no real difference between flapping flight and soaring flight. The real secret of flight lies in the physical relation between the aerofoil section that provides the lift and the relative wind. The principle is the same whether one has a flapping wing, a fixed wing, or even the rotating wings which a helicopter uses for the same purposes. All achieve the same result: a lift component normal to the airfoil cordline. As a matter of fact, even the airplane's propeller works that way, as we'll see in the next lecture.

So nature's lie to man dissolved before the recognition of what flapping and soaring had in common. Just as most of the apparent aerodynamic differences between propellers, helicopters rotor blades, axial flow jet engine blades, and the wings of most supersonic airplanes also dissolve when one recognizes what they all have in common. All that matters is the relative flow over the airfoil section, a point that was obscurely groped for by Aristotle, when he tried to deny the possibility of a vacuum; by Leonardo, when he speculated about the air compression underneath a wing; by Newton, when he wrote that the forces acting between a solid body and a fluid are the same whether the body moves with a certain uniform velocity in a still fluid or the fluid moves with the same velocity against the body. It was finally recognized clearly by Rayleigh, who saw this principle of air flow relativity as constituting the only real difference between flapping and soaring flight, the former being the moving of an airfoil through still air, the latter being the moving of air around a still airfoil.

Remember also that each feather in the bird's wing is itself an airfoil. Each one, especially the primaries, acts like an individual wing. It's almost as if the bird had turned into a Venetian blind of wings. In considering this all that we have done so far – and all that Rayleigh did – is to locate the fundamental interaction in aerodynamics as that between airfoil and flow. Delineating the elements which enter into this does not constitute an analysis of the nature of the interaction. This will be the subject of the final lecture. However, what we have seen so far did indicate to many hopeful aeronauts a basic practical truth. Being thrust through the air was the only practical way for man to achieve flight.

The men, the theories, the observations of bird flight, the practical attempts to fly, all that we have considered so far came together in the work of one man, Otto Lilienthal (1848–1896). His work, *Bird Flight as the Basis of Aviation,* which was mentioned earlier, was a treatise in the great tradition of Da Vinci and Borelli. Along with the ornithological studies of professor Marey (1874, 1890), it served as the inspiration for the intrepid American glider pilots, Octave Chanute, James Montgomery, and finally Wilbur and Orville Wright.

Lilienthal gave a camber to his early gliders, exactly as one finds in the wings of a gull, which served as a model. His wings were flexible so that they could be actually walked in flight by pulling them with wires on pulleys, thereby changing the relative camber and controlling the upward swerve. Lilienthal was not going to emulate the monk, Oliver of Malmesbury, who, on crashing from off the top of a high bell tower, bemoaned his failure to have placed feathers on his posterior parts. Lilienthal placed a large, flexible, and controllable empennage well aft on all his gliders, not unlike the tail construction one sees on the swift or the martin. Every new idea was tested, first in a model, then in a kite. After further reflection and research he would finally incorporate the acceptable ideas into his own gliders and then take to the air from the top of a high conical mound. His 2,000 successful flights are a monument to painstaking observation, and his careful records were a guide for future inquiry. These flights were largely unaided by the professorial hydrodynamicists of the time, most of whom were still playing the Euler-Bernoulli game with increasing subtlety and decreasing likelihood of any ultimate application. Otto Lilienthal's death, the result of an unexpected gusty turbulence, which might have overturned even a modern airplane, constitutes one of the really tragic moments in the history of flight. It so moved Wilbur Wright that he went to his bookshelf and reread the whole of *Birdflight as the Basis of Aviation* as well as Marey's treatise, as soon as he heard of the tragedy. The first Wright glider was soon under construction. In a matter of weeks after that event, flight, although still largely beyond human understanding, began to unfold some few of its mysteries to the disciples of Otto Lilienthal.

References

Aymar, Gordon C. 1938. *Bird flight*. Garden City: Garden City Pub. Co.

Bernoulli, Daniel. 1738. *Hydrodynamics sive de viribus et motibus fluidorum commentarii*. Strassburg: Johann Reinhold Dulsecker.

Beveridge, W.I.B. 1957. *The art of scientific investigation*. New York: Norton.

Borelli, Giovanni Alfonso. 1680–1681. De motu animalium … Alphonsi Borelli neapolitani matheseos professoris. *Opus posthumum*. Romae: ex typographia Angeli Bernabo.

Brown, Cecil Leonard Morley. 1927. *The conquest of the air: An historical survey*. London: Oxford University Press.

Chanute, Octave. 1894. *Progress in flying machines*. New York: American Engineer and Railroad Journal.

Lana Terzi Francesco, Thomas O'Brien Hubbard, and John H. Ledeboer. 1670/1910. *The Aerial ship*. London: Printed and Published for the Aëronautical Society of Great Britain [by] King, Sell & Olding, Ltd.

Gibbs-Smith, Charles Harvard. 1960. *The aeroplane: An historical survey of its origins and development*. London: Her Majesty's Stationery Office.

Hart, Ivor Blashka. 1961. *The world of Leonardo da Vinci: Man of science, engineer and dreamer of flight*. New York: Viking Press.

Josephy, Alvin M., ed. 1962. *The American heritage history of flight*. New York: American Heritage Publishing Co.

Leonardo, da Vinci, and Francesco Brioschi. 1894–1904. *Il Codice Atlantico di Leonardo da Vinci, nella Biblioteca Ambrosiana di Milano, riprodotto e publicato della Regia Accademia dei Lincei*. Milan: U. Hoepli.

Leonardo, da Vinci, Theodor Sabachnikoff, Giovanni Piumati, and Charles Ravaisson-Mollien. 1893. *Codice sul volo degli uccelli e varie altre materie*. Paris: Rouveyre.

Marey, Etienne-Jules. 1874. *Animal mechanism: A treatise on terrestrial and aerial locomotion*. New York: D. Appleton and Co.

———. 1890. *Le vol des oiseaux*. Paris: Masson.

Newton, Isaac. 1729/1960. *Isaac Newton's mathematical principles of natural philosophy and his system of the world*. Trans. Andrew Motte and Florian Cajori. Berkeley: University of California Press.

Pritchard, J. Laurence. 1962. *Sir George Cayley, the inventor of the aeroplane*. New York: Horizon.

Rayleigh, John William Strutt. 1883. The soaring of birds. *Nature* 27 (701): 534–535.

Rayleigh, Robert John Strutt. 1968. *Life of John William Strutt, third Baron Rayleigh*. Madison: University of Wisconsin Press.

Storer, John H. 1948. *The flight of birds analyzed through slow-motion photography*. Bloomfield Hills: Cranbrook Institute of Science.

References

Chapter 25
Lecture Three: The Idea of a Shape

As Marcus Aurelius once said: "Those who come on the third night during rain, are truly lovers of wisdom." It isn't absolutely certain that he said that, but there is some scholarly dispute about it. Last night I considered the shaping of an idea. The idea was this: that flight might be thought of as a scientifically understandable, objectively tractable, and practically achievable phenomenon. Through the sorts of work that Newton and the other ideal hydrodynamicists accomplished, they did shape such an idea, the idea of a discipline concerned with the dynamics of gases like air. Tonight we won't talk about the shaping of an idea, but rather the idea of a shape. And the shape we will be discussing tonight by slow degrees will be nothing other than the airfoil shape which, as shapes go, seems to me one of the more influential that we have had in our time. There is evidence in Leonardo's little tract on the flight of birds which indicates that he understood something of the curvature in a bird's wing. It can even be supposed that he was not wholly in the dark concerning the air flow over the top side of such a convex shape, though this is just conjecture. However, you must remember Sir George Cayley, the man who made the remarkable statement that the whole problem in aerodynamics is simply to make a surface support a given weight by the application of power to the resistance of the air. Shortly after he made that statement, he considered what he called the concave wing of a bird. He explained its purpose as follows: "the air being obliged to mount along the convexity of the surface, creates a slight vacuity immediately behind the point of separation."[1] What he means by vacuity behind the point of separation is a topic we will be considering shortly. But Cayley said this in 1809, almost 100 years before the actual advent of powered flight. His passing reference to the slight vacuity above the wing appears to our 20-20 hindsight as the vision supreme in the story of flight. Because this is truly the germ of the full explanation of what all air foils are meant to do. Whether they flop or are rigid or whether they rotate horizontally as helicopter blades do or spin vertically as do orthodox propeller blades.

---

[1] The paper of Cayley, from which this quote is taken, is discussed in detail in Gibbs-Smith (1962, 45–58).

© Springer Nature B.V. 2020
N. R. Hanson, *What I Do Not Believe, and Other Essays*, Synthese Library 38,
https://doi.org/10.1007/978-94-024-1739-5_25

A few years later, William Samuel Henson (1805–1888) applied for a patent on a "Locomotive apparatus… for conveying Letters, Goods, and Passengers from place to place through the Air…." The model he had in mind was correct, in a limited way, but did not use Cayley's key idea in airfoil shape. Thus Henson wrote[2]

> If any merely flat article be thrown or projected edgewise in a slightly inclined position, it will rise in the air till the force exerted is extended. If it possessed in itself a continuous power or force the article would continue to ascend as long as the forward part of the surface was upward in respect to the hinder part.[3]

Now this, of course, applies to water skis, to surf boards, to the gliders water skiers sometimes use. Almost everything that takes place in the Florida water festivals owes something to Mr. Henson. Although Henson is clearly correct in his basic point that the lift that a moving fluid exerts on an inclined plane does exceed the drag, he does not recognize how much more lift would be possible were Cayley's configuration adopted. To anticipate for a moment, it turns out that with certain airfoil designs the amount of lift is about fifty times greater than would be given by a strictly flat plate, a rather remarkable magnification of the original Newtonian analysis.

At the very first meeting of the Aeronautical Society of Great Britain in 1866 Francis Herbert Wenham (1824–1908) spoke of wings and many other things (Gibbs-Smith 1960, 17). Noting that the major portion of a wing's lift is generated along a narrow strip just aft of the leading edge, he argued that increasing the lift really meant increasing the length of the leading edge.

This, in turn, implied longer and thinner wings, what are now called high aspect ratio plan forms, the sort of things one sees in glider wings. Such a design maximizes the area of relatively low pressure near the front of the airfoil and shortens the line along which turbulence and drag usually occur. Wenham proved this in his own relatively crude, but nonetheless quite effective, wind tunnel. He tested two wings having the same area but very different aspect ratios. The high aspect ratio wing, the one with the longer lifting line and the shorter width, lifted itself through the wind tunnel's airflow many times faster and more efficiently than the low aspect ratio wing. Wenham also advocated, in the course of the same lecture, the use of an arched wing. Here he explicitly recognized the value of Cayley's insight.

The further development of this area of inquiry fell to an individual named Horatio Phillips (1845–1926). He may or may not have appreciated the significance of what Cayley called 'the topside vacuity'. But after a great deal of study, or practi-

---

[2] The basic work on Henson is Davy (1931). A brief summary may be found in Gibbs-Smith (1960, 13).

[3] Henson's Patent Specification of 1842, referred to above, is contained in (Davy 1931). The full passage Hanson seems to have quoted from is as follows "If any light and flat or nearly flat article be projected or thrown edgewise in a slightly inclined position, the same will rise on the air till the force exerted is expended, when the article so thrown or projected will descend; and it will readily be conceived, that if the article so projected or thrown possessed in itself a continuous power or force equal to that used in throwing or projecting it, the article would continue to ascend so long as the forward part of the surface was upwards in respect of the hinder part…" (104). –MDL

cal ornithology, and a good bit of experimental research he secured patents in 1884 and again in 1890 for an airfoil section which later came to be known as 'The Phillips Entry'. This is a double surfaced airfoil of differing thickness and shape, or camber. He proved that this type of wing, curved more on the upper surface than on the lower, is subjected to two forces, a positive pressure underneath and a stronger suction from above. He established this by experiments on different types of wing sections and by the use of models.

At this time theoretical hydrodynamics could offer almost nothing in the way of a practical explanation of why a shape like the Phillips Entry lifted so much more efficiently than a flat plate of the same area. Practical aeronauts were completely convinced that a curved upper wing is essential to all flight, whether birds or machines. This is the kind of practical conviction that plumbers, cooks, and musicians usually formulate much earlier than the theoreticians, working in these disciplines, give them reason for supposing to be the case. Thus Otto Lilienthal argued that the arched wing embodies the whole secret of the art of flying, but an adequate concept of flight, a rational analysis of the effect of different shapes and cambers on lift and drag, was still unavailable. The practical convictions lay like a pack of puzzle pieces which had to fit together, whenever someone could find the inner connection. But as man moved towards the end of the nineteenth century the birds that had flown so effortlessly over the heads of Aristotle, Roger Bacon, Isaac Newton and Sir George Cayley still guarded their ancient secret of flight.

Then there came a man named Frederick William Lanchester (1868–1946) who saw, through somewhat darkened lenses, what had remained completely invisible to the greatest natural philosophers of all time before him. He perceived, at least dimly, why arched shapes lift themselves across airflows. And from this remarkable physical insight, as Prandtl referred to it, most of the earlier pieces began to fall together.[4] One could at last explain why the roofs of Dutch barns, which continental farmers often put over large stacks of hay like the tops of quonset huts, why these roofs raise themselves upward when a wind hits them on one side in the teeth of a gale. They do not simply move backwards. They actually rise; something the farmers knew but could not explain. Lanchester also indirectly explained why Rayleigh was quite correct in remarking that flapping flight and soaring flight do not differ in any significant respect. Another phenomenon noted by Rayleigh in his 1877 paper and clarified by Lanchester's analysis is the tendency of spinning cannon balls and tennis balls to swerve from their plotted paths. But first let us focus on some of the problems which antedate Lanchester's remarkable physical insight.

Back to Newton. In 1672 he noted that a cylindrical rod set at right angles to a liquid flow and rotated, swerves in the direction where the flow and the rotation are

---

[4]Lanchester's original paper on the theory of lift was submitted in 1897 before the flight of the Wright brothers and the analyses of Kutta and Joukowski. The Royal Society of London and the Physical Society rejected this paper primarily, it would seem, because Lanchester lacked the training to express his ideas in proper mathematical form. Lanchester was also the inventor of operations research, another idea that was neglected when first introduced. For a brief account by an eyewitness of the relation between Lanchester and Prandtl see (von Kármán and Edson 1967, 60).

the same.[5] To see why this is so think of a cylinder rotating clockwise and an airstream hitting the long side of the cylinder from the left. Above the cylinder the rotating air dragged around by the cylinder is in the same direction as the airflow, leading to an increase in relative air speed. Below they are in opposite directions, leading to a decreased airspeed relative to the cylinder. As we noted earlier, when discussing the Bernoulli equation, increased air speed means decreased pressure and vice versa. As a result, the pressure is decreased above the cylinder and increased below it giving a net lift. This is called the 'Magnus effect' in honor of a German professor who published a study of this problem in 1851. To circumvent this dispute about priority we will call it 'the pitcher's effect', for it is the principle that explains how a good pitcher curves a baseball – or how a hack golfer slices a drive. The further complications that come from boundary layer theory will be considered later.

Now back to Lanchester. He knew the kite effect, or the lifting force of a fluid on an inclined flat plane and he knew the pitcher's effect. He also knew that neither of these alone was sufficient to account for the flight of birds. Where Lanchester went beyond any of his predecessors was in seeing how to combine the two. Prior to this no one had thought that the pitcher's effect was relevant – because nothing in the bird's wing rotates. Lanchester combined these by posing the basic questions in a way that uncovered the fundamental problems concerning the forces that sustain flight. Let's attempt to reconstruct his thought.[6]

He accepted Wenham's idea that the greater part of the lift on a bird's wing came from the forward part of the airfoil section. This lift implies a decrease in pressure and this decrease, in accord with Bernoulli's theorem, indicates that there must be an increase in the velocity of the airflow over the top of the wing. The question Lanchester asked is: how can such a velocity increase result from nothing more than the translation of an airfoil through the air? An airfoil, whether of a bird or a plane, is not a rotating rod. What is there about its sheer shape that can accelerate the air particles it moves under? In posing the question this way Lanchester made the circulation of air around the wing the basic issue. The lift in the pitcher's effect is due to the difference in air circulation above and below the spinning baseball. Thus, through the way he posed the question, Lanchester united for the first time two effects that had always been treated separately, the kite effect and the pitcher's effect.

---

[5] In this statement and in attributing the discovery of the Magnus effect to Newton, Hanson may have read into the documents some conclusions that were only potentially present. Newton treated these problems in Book II of his *Principia*. In Propositions XXXIV and XXXVIII he treats the resistance of cylinders moving linearly in a fluid.

In proposition LI he considers rotation of a cylinder. In Proposition LIII and the following Scholium he applied these principles to bodies moved by a vortex and showed that the Cartesian theory of vortices could not explain the motion of the planets. I could not find any explicit expression of the position Hanson attributes to Newton, though the conclusion could be deduced from the propositions cited. In any case Newton's approach to fluid dynamics was a simple extension of his particle mechanics and is essentially different from modern approaches. – *EM*

[6] This is in Lanchester (1909, ch. iv, under the title, 'Motion in the Periptery'). Lanchester's neologism 'periptery' meaning 'about the wing' never won acceptance.

This is the heart of Lanchester's contribution. To grasp the details we need a double digression. The basic digression is into modern airfoil theory, generally referred to as the Lanchester-Prandtl theory, of airfoils. But before we can talk about this theory we must digress and talk about two other chaps, Wilhelm Kutta and Nicholai Joukowski (sometimes spelled Zhukovski).[7] To understand the theorem that bears their joint name let's return to the simple case considered earlier, a cylinder rotating clockwise with an airflow coming in from the left. We will add a couple of simplifying factors. Firstly, we ignore the ends of the cylinder, so that we are considering essentially a two-dimensional problem. Secondly, we will consider an irrotational fluid, that is a fluid in which there is no internal turbulence. With these simplifications we may express the Kutta-Joukowski theorem in a precise formula. Let ϱ stand for the density of the air, V for its velocity, and Γ for the strength of the circulation. Then the lift is[8]

$$L = \varrho V \Gamma$$

If we really wished to be precise we would write this in vector form to show that the lift is perpendicular to the velocity. But this is not necessary for our limited purposes. The relation of this to the pitcher's effect is clear. But we should also note some significant differences. First, the manner in which the Kutta-Joukowski theorem is formulated makes it independent of the properties of any particular body. It is not necessary to have a rod, an airfoil, a birdwing, or anything in particular in the fluid. All that the formula treats is a closed plane curve, which could be a mathematical fiction.

Secondly, unlike the pitcher effect, the Kutta-Joukowski theorem is restricted to two dimensions. If we had wings of infinite length we could consider the shape while ignoring what happens at the edge. The consequences of this limitation will be considered later. Thirdly, the Kutta-Joukowski theorem is for ideal fluids. It is, in fact, one of the triumphs of the theory of ideal fluids. An ideal fluid, as you may remember, is inviscid and irrotational. If there is no viscosity there is no friction drag. If this is true, and if the circulation is known then one can readily calculate the lift. This calculation can be checked against observations for different shapes.

Here, however, complications ensue. Some airfoil shapes are quite complex. It is difficult to represent them mathematically. Instead of attempting this directly the mathematicians have developed some useful tricks. One of the most useful is finding mathematically simple shapes that are equivalent to complex shapes. This, in essence, is what Joukowski and others did here, deducing the complex flow around airfoils from the simple flow around cylinders. As many of you know, this is a mathematical problem involving functions of a complex variable. I'll try to make the basic idea intelligible using as little mathematics as possible.

---

[7] Among his lecture notes Hanson had a mimeographed copy of a paper by A. T. Grigoryan, 'The Contribution of Russian Scientists to the Development of Aerodynamics' with no source of publication indicated. Many of his ideas on Joukowsky seem to have come from this paper. – EM

[8] This is developed in more detail in Millikan (1941, 28–36).

Consider first the type of Mercator projection you are all familiar with from the maps in your grammar school geography books. This is the map that makes Greenland look as big as South America and the North Pole as wide as the Equator. This map is developed by a projection technique. Get a large piece of paper, roll it into a cylinder and then put it around a globe of the earth so that the paper touches the globe at the Equator. Now imagine radii being extended outward from the surface of the globe until they reach the cylinder. This gives one spot on the cylinder for each spot on the globe. It works well near the equator and less well as you get away from the Equator. You would need a paper of infinite length to get the North Pole in. This projection distorts distances away from the Equator. But it has a distinct advantage. Latitudes and longitudes remain perpendicular so that one can use this map to plot routes, though it cannot be used directly to determine distances.

Now we wish to consider a different type of projection, one that is harder to visualize. Ultimately we need a projection that transforms circles, or the cross-section of a cylinder, into airfoil shapes, or the cross-section of a wing. Just as the Mercator projection sacrifices distances and distorts shapes to preserve directions, so our new projection must be willing to make whatever sacrifices are necessary to preserve fidelity of surface effects. Let's get at this by steps beginning with a projection that transforms a circle into a line. This seems simple, a coin seen sideways looks like a line. But this simple projection does not give the properties needed. To get this we will have to resort to a little mathematics. Let us call the plane with the circle in it the z plane and label its coordinates x for the horizontal direction and iy for the vertical direction. Here i, which is the square root of minus one, is introduced to facilitate the mathematics of the transformation. We will call the surface on which the circle gets projected the w plane and label its coordinates u and iv. Now the projection we want is given by the formula.

$$w = \frac{1}{2}\left(z + \frac{a^2}{z}\right),\tag{25.1}$$

where a is the radius of the circle. One can get the projection of any point by substituting values of x and y in (25.1) and calculating the resultant values of u and v. Since we are working with circles it is more helpful to transform (25.1) into polar coordinates of r, the distance from the center, and θ, the angle of revolution from the positive x axis. By a little manipulation this gives.

$$u = \frac{1}{2}\left(r + \frac{a^2}{r}\right)\cos\theta\tag{25.2}$$

$$v = \frac{1}{2}\left(r - \frac{a^2}{r}\right)\sin\theta.\tag{25.3}$$

From Formula (25.1) it is clear that when z is zero w is infinite. Just as the North Pole gets projected to infinity with the Mercator projection so the center gets projected to infinity here. That is O.K., since we are interested in what happens around the surface. First, let's consider what happens to the surface itself. In (25.2) let r

equal a to get $u = \cos\theta$. So u goes from $+a$ to $-a$ as θ goes from zero to 180°. In formula (25.3) $v = 0$ when $r = a$. This means that a circle of diameter $2a$ gets projected into a line of length $2a$. But when r is not equal to a, v is not always zero. For example, when r equals $2a$, or for a circle twice the radius of our base circle we have

$$u = \frac{5}{4} a \cos\theta, \quad v = \frac{3}{4} a \sin\theta \tag{25.4}$$

This circle gets mapped into an ellipse around the line representing our original circle. Similarly for r equals $\frac{1}{2}a$ we have

$$u = \frac{5}{4} a \cos\theta, \quad v = -\frac{3}{4} a\sin\theta. \tag{25.5}$$

This is the same ellipse as in (25.4) but now one goes around it in a reversed direction.

This may be interesting, but it still does not give us airfoil shapes. Here is where Joukowski introduced a clever trick, one illustrated in Fig. 25.1. Consider the circle with radius $a + e$, which is tangent to our original circle at $x = +a$. The way in which this circle gets projected can be seen in a more or less intuitive way. At the right side the new circle begins by coinciding with the old one and gradually diverges. So its projection must begin like the old one as a straight line and gradually diverge. At the left of the x axis the new circle goes a distance $2e$ beyond the old one. If it were replaced by a concentric circle with radius $r + 2e$ its projection would be an ellipse around the line representing the original circle. Looked at from the left the projection of the displaced circle must begin like an ellipse and gradually taper off till it is like a straight line at the right end. The net result is an airfoil shape. What we have, accordingly, is a way of projecting airfoil shapes onto circles.

The transformation that maps a circle into an airfoil shape also fits the flow lines around both shapes. The resulting mathematical problem can be solved in the case of the cylinder. By using the appropriate transformations we can determine the airflow around an airfoil. This is what Joukowski contributed – though he used geometric methods rather than the analytic methods sketched here.

This approach has some simplifications built in. Viscosity is ignored, which is not too crude an approximation for moderate angles of attack. But the Joukowski profiles are rather unrealistic. A plane with such tad-pole shaped wings would be in trouble before it ever got out of the hangar. Here later modifications introduced by Prandtl, Von Kármán, von Mises and others extended the Joukowski technique to generate airfoils of any thickness and any camber or shape. Finally, the Joukowski technique is essentially two dimensional. This is all right for wings of infinite length, where any cross section is like any other. But it does not fit finite wings by taking into account what happens at the wing tips. We will get to this as soon as we get back to Lanchester.

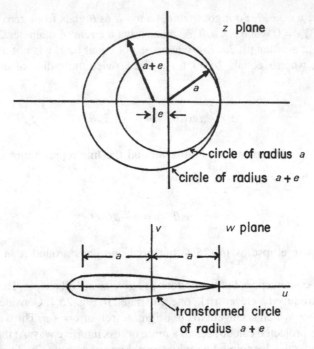

Fig. 25.1 Example of Joukowsky transformation

In spite of these simplifications Joukowski's work was a significant advance. There is one special aspect of it that should be reconsidered. We can study the flow of an ideal liquid around an airfoil by studying the flow of the fluid around a rotating cylinder. But the cylinder itself does not enter into the equations; only the rotary flow around the shape does. A cylinder of air in air or a rod of water in water would do as well. One could discuss the rotary motions of a hypothetical vortex motion inside the flow.

Now one of the insights Lanchester had was to treat the wing itself as a hypothetical airfoil, a special kind of bound vortex. The accompanying illustration, Fig. 25.2, should make this a bit clearer. However, the vortex illustrated is not a hypothetical entity. It is a real thing, a genuine flow around an airfoil section. I regard this as one of the most brilliant insights in the history of aerodynamic theory.

Lanchester treated vortices and their effect on lift in a rather intuitive way. To say this is not to imply that he did not know mathematics. Eventually he acquired a fair degree of competence in mathematics. But he used mathematics to express and develop his remarkable physical insights. Thus, instead of beginning with the standard idealizations of ideal fluids and two dimensional problems he began by considering a real plane traveling through a real medium. This means that the wing span considered is finite and the medium through which it travels has viscosity. Lanchester tried to think through the problem of what happens to the air when such a plane travels through such a medium.

Figure 25.2, a somewhat updated illustration of Lanchester's illustrations, brings out the basic ideas. The airflow around the moving plane forms a complex pattern

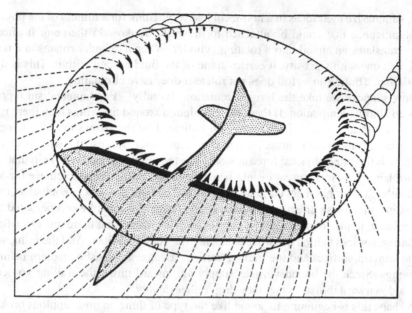

Fig. 25.2

which can be broken down into three different systems of vortices. First, there is the starting vortex. When the plane starts the air going under the wings has a shorter path than the air over the top. Some of the lower air tends to sweep over the trailing edge and meet the topside air at a point (or line) called the stagnation point. The turbulence that forms behind this point tends to create drag and reduce lift. When a well designed aircraft is in steady flight the stagnation point is at the trailing edge so that the resulting turbulence is behind the wings rather than over them. However, this starting vortex returns any time there is a change in velocity or in the angle of lift.

Secondly, there are wingtip vortices. To visualize these as Lanchester saw them, try to think of what happens to the air near the tip of the wings. The pressure is greater underneath than over the wings. It is this pressure difference that supports the plane. Air molecules always tend to go from a high to a low pressure region. This means that the air tends to loop around the wingtips from the bottom to the top. If we were looking at a receding plane from the back and could see this disturbance we would see a counter-clockwise vortex streaming out behind the right wing tip and a clockwise vortex streaming from the left tip. In hydrodynamics there is a theorem, stemming from the work of Helmholtz, concerning constancy of circulation. The same amount of an ideal fluid flows around any closed loop surrounding a vortex. If the radius is smaller the speed is greater. Accordingly, as the plane recedes the wingtip vortices slow down but their size increases. This wake turbulence extends a long way behind the plane and can be quite a hazard for a small plane flying behind and below a large heavy plane.

Finally, there is the so-called horseshoe vortex. Here, as with the other aspects of Lanchester's theory, Prandtl later gave it a more rigorous analytic treatment. We will try to summarize their ideas in a non-technical way. Think for a minute of the physical significance that could be attached to the Kutta-Joukowski theorem. It effectively translates an airfoil into a rotating cylinder. When a cylinder rotates in a real fluid, i.e., one with viscosity, it carries some of the fluid around with it. This is the *circulation*. Though an airfoil does not rotate it does have circulation.

Now suppose we take the term 'circulation' literally – Fundamentalism represents an abiding temptation. If there is a circulation around the wings then there is a vortex, one that would go in the direction indicated by the arrows over the wings. This is called 'the bound vortex'. There is no real vortex because the wing is there, but there is the same physical force at work. The theorem of constancy of circulation applies here. This was a great insight which Lanchester and Prandtl achieved independently. Still air has a net circulation of zero. The bound vortex is equivalent to a vortex in a clockwise direction. To balance the books this must be compensated by a circulation in the counterclockwise direction. This means that there must be a line of trailing vortices behind the wing balancing the bound vortices. Add the influence of the wingtip vortices and the trailing vortices assume a horseshoe pattern behind the wings. Since the bound vortices supply an upward thrust the trailing vortices have a downward thrust. In fact they drop rather rapidly.

Perhaps this is beginning to sound like the type of thing an unscrupulous broker would engage in, taking real money to cover imaginary transactions. How can a real vortex, or line of vortices, behind the wings balance an imaginary vortex, the line of bound vortices within the wing? Lanchester had the physical intuition to see that this had to be. Birds can fly, so can machines. If there is to be lift there must be circulation. If there is circulation there must be the after-effects required by the conservation laws. Such intuitive physical reasoning led to the first adequate explanation of the lifting power of the fixed wing.

This lift is related to the fact that the fluid has viscosity. Lanchester knew this, but it was Ludwig Prandtl (1875–1953) who worked out a way of relating the treatment of real fluids, which have viscosity, to ideal fluids, which have nice mathematical properties. He did this by introducing what he called 'boundary layer theory'. An ideal fluid washes over a smooth surface with no friction and no sticking. But a real fluid does not do this. No matter how low the viscosity the layer of fluid immediately adjacent to the surface sticks to the surface without any slip. This is what Prandtl calls 'the condition of no slip'.

Think of the air above the wings as if it were in layers. The lowest layer, right next to the wing surface, obeys the condition of no slip. Its velocity relative to the wing is zero. Far away from this surface the effective velocity will simply be v, the velocity of the still air relative to the moving wing. In between these two extremes the relative velocity will go from zero to v depending on how much it is slowed down by the dragging action of the plate. Prandtl worked out a mathematical analysis to show the drag that different wing surfaces exert.[9]

---

[9] Millikan (1941, 28–36) provides a treatment of this analysis. It is also discussed in a more technical way in Prandtl (1952, 105–121). These seem to have been the sources Hanson relied on. – *EM*

One could learn the significance of this boundary layer by studying the works of Prandtl or von Kármán. I learned it the hard way – and it almost cost my life. Let me tell you about it. About 3 years ago I found myself somewhat short of money and could not afford the hundred dollars a month needed to keep my F8F Bearcat in a hangar. I decided to keep it out of doors and arrange for some tight fitting covers, the Cono covers that you often see on sports cars, to protect the vital parts of my plane, especially the piping which I did not wish to leave exposed to the wind and the cold. To secure a tight fit for the covers I put some Cono snaps on the first third of the wing without thinking that this might affect the plane's performance in the air.

When I took the plane up the snaps did not seem to make any difference – at first. At about 240 knots (or approximately 265 miles per hour) a strange vibration began, one that I had never felt before. I increased the speed to 310 knots and then almost lost the plane completely. My head was thrown up against the canopy; the control stick began to smash back and forth against my knees; and I really thought I was about to cash in my chips. There is a trick for such situations, one familiar to most old fighter pilots. I simply shut everything off. I thought that when the power was gone whatever was causing the trouble would stop. It did. Everything stopped except my heartbeat; that was racing. As I slowed to 300 knots I hit the same vibrations and dropped my dive breaks. When I came down to 240 knots I hit it again.

I finally got the plane back down and, thanks to a few liquid libations, my courage back up. In a more relaxed frame of mind I began to think of my problem in the light of J. S. Mill's canon of concomitant variations. There was something happening to the plane now that had never happened before. There was also something added to the plane which had never been on before. The two must be related. After a few more libations to re-enforce this reasoning I removed the clips and took the plane back up. It behaved perfectly.

When I returned home I went back to Prandtl looking for answers. He had them. The snaps I had affixed to the wings extended into the boundary layer. What I had done in effect was to spread the boundary layer and introduce turbulence in the interface. Prandtl's analysis showed that the thickness of the fluid layer affected by viscosity is inversely proportional to the square root of the speed. This is the boundary layer we spoke of earlier. At the right speed – or in my case at the wrong speed- the boundary layer was lowered until the clips poked through it. The resulting turbulence caused the vibrations I felt.

Modern aerodynamicists have developed a variety of tricks and techniques to control spreading of the boundary layer. The goal is to keep the air flow laminar over the wings and to keep the stagnation point at the trailing edge of the wings. This is done by varying the angle of attack, by having variable flaps at the back edge of the wing, or fixed slots at the front edge which serve to shape the airflow. Recently there have been more sophisticated developments such as boundary layer suction control. This involves a series of special perforations on the aft edge of the wing and a suction pump which sucks the turbulence through the perforations and leaves the air flow over the wings laminar. I know someone who developed a similar but simpler system for his glider. To illustrate its effect he switches on the pump in the right wing only. Then, without touching anything else, the machine begins to do aileron rolls. This, of course, is due to the fact that the drag on the right wing has been decreased and its lift increased.

When one gets to supersonic flight the problems are even more complex. I believe that I was one of the first in this country to hit the sound barrier. I did it in early May of 1942 in an F4U Corsair, which was completely demolished by this encounter.[10] However, I do not wish to go into the technical details here but simply to indicate the relation of these complexities to the conceptual problems we have been considering. Modern aerodynamics is a composite study drawing on many other technical disciplines. It builds on hydrostatics and hydrodynamics, on gas theory and kinetic theory. It uses particle mechanics, statistical mechanics, thermodynamics, electronics, solid state physics, and meteorology. It is so complex that no one man, not even a von Kármán, can master all the technical details involved.

These complexities and the even greater complexities that will surely follow in the future are the outcome of the conceptual evolution we have endeavored to sketch. The sciences seem to have a life process that is all their own. There is a long obscure period of incubation, a groping attempt to separate the real from the pseudo-problems, to ask the right questions and determine what sorts of evidence should or could count in favor of proposed answers. Finally, after many false starts, after failure and frustration, the key concepts are clarified and their interrelation seen. Then the subsequent growth can be explosive.

So it was with the conceptual foundations of aerodynamics, the problem that has concerned us. Man first had to discover the air and get some understanding of its nature as a medium that both supports and resists motion. Bird flight suggested the possibility of human flight – and inevitably led to a futile concentration on flapping wings as the basis of flight. The failure of all ornithopters seemed to prove the impossibility of human flight. So too did the consequences of Newton's $\sin^2$ law. But the birds still flew and man's imagination still soared and so the search went on. Theories had to be advanced on fluid flow and pressure differentials, on rotation and turbulence, in shapes and speeds. New mathematical techniques had to be developed to fit these conceptual advances. On a more practical level man had to learn the properties and weaknesses of various materials, work with gliders, observe air currents, and study bird flight. Above all else he had to experiment and learn from his failures, including a few fatal failures. Eventually, inexorably these parts began to cohere into a workable whole, and heavier-than-air machines flew. In this complex and protracted development we have one of the finest chapters in the story of how man has studied nature, succeeded in copying it, and finally through reason bent it to his will.

---

[10] Hanson's listing of May 1942 is incorrect. The year should be 1944, and it is likely August, not May, of 1944, since then the incident would match up with Hanson's military personnel record. One might wonder how Hanson could have claimed to have hit the sound barrier in a WWII-era piston engine aircraft since the top speed of such planes is far below the sound barrier. However, as is made very clear in these lectures, the velocity of the air over the wings is much higher than the plane's absolute velocity. Thus, sound barrier effects could be experienced at the highest velocities that WWII-era piston engine aircraft could reach. See (Lund 2010, ch. 1) for more detail about Hanson's flying career. – MDL

References

Davy, Maurice John Bernard. 1931. *Henson and Stringfellow: Their work in aeronautics.* London: His Majesty's Stationery Office.

Gibbs-Smith, Charles Harvard. 1960. *The aeroplane: An historical survey of its origins and development.* London: Her Majesty's Stationery Office.

———. 1962. *Sir George Cayley's aeronautics: 1796–1855.* London: Her Majesty's Stationery Office.

Lanchester, F.W. 1909. *Aerodynamics: Constituting the first volume of a complete work on aerial flight.* New York: D. Van Nostrand.

Lund, Matthew D. 2010. *N.R. Hanson: Observation, discovery, and scientific change.* Amherst: Humanity Books.

Millikan, Clark Blanchard. 1941. *Aerodynamics of the airplane.* New York: Wiley.

Prandtl, Ludwig. 1952. *Essentials of fluid dynamics.* New York: Hafner Pub. Co.

Rayleigh, John William Strutt. 1877. On the irregular flight of a tennis-ball. *Messenger of Mathematics* 7: 14–16.

von Kármán, Theodore, and Lee Edson. 1967. *The wind and beyond: Theodore von Kármán, pioneer in aviation and pathfinder in space.* Boston: Little, Brown.

Printed in the United States
by Baker & Taylor Publisher Services

Printed in the United States
by Baker & Taylor Publisher Services